00-WSM-6b

Moving Toward Sustainable Water Resources Management: A Framework and Guidelines for Implementation

Technical Report

by:

Kent W. Thornton, Ph.D.
Christina Laurin
Donald Catanzaro, Ph.D.
FTN Associates, Ltd.

James Shortle, Ph.D.
Ann Fisher, Ph.D.
Pennsylvania State University

Jose Sobrinho, Ph.D.
Kristyn Stevens
Perot Systems Government Services

2006

WERF

The Water Environment Research Foundation, a not-for-profit organization, funds and manages water quality research for its subscribers through a diverse public-private partnership between municipal utilities, corporations, academia, industry, and the federal government. WERF subscribers include municipal and regional water and wastewater utilities, industrial corporations, environmental engineering firms, and others that share a commitment to cost-effective water quality solutions. WERF is dedicated to advancing science and technology addressing water quality issues as they impact water resources, the atmosphere, the lands, and quality of life.

For more information, contact:
Water Environment Research Foundation
635 Slaters Lane, Suite 300
Alexandria, VA 22314-1177
Tel: (703) 684-2470
Fax: (703) 299-0742
www.werf.org
werf@werf.org

This report was co-published by the following organization. For non-subscriber sales information, contact:

IWA Publishing
Alliance House, 12 Caxton Street
London SW1H 0QS, United Kingdom
Tel: +44 (0) 20 7654 5500
Fax: +44 (0) 20 7654 5555
www.iwapublishing.com
publications@iwap.co.uk

Library of Congress Catalog card Number: 2006921383
IWAP ISBN: 978-1-84339-756-4

FTN Associates, Ltd.

The research on which this report is based was funded, in part, by the United States Environmental Protection Agency through Cooperative Agreement No. CR-827345-01 with the Water Environment Research Foundation (WERF). Unless an EPA logo appears on the cover, this report is a publication of WERF, not EPA. Funds awarded under the Cooperative Agreement cited above were not used for editorial services, reproduction, printing, or distribution.

ACKNOWLEDGMENTS

Report Preparation

Principal Investigator:
Kent W. Thornton, Ph.D.
FTN Associates, Ltd.

Project Team:
Donald Cantanzaro, Ph.D.
FTN Associates, Ltd.

Ann Fisher, Ph.D.
Pennsylvania State University

Christina Laurin
FTN Associates, Ltd.

James Shortle, Ph.D.
Pennsylvania State University

Jose Sobrinho, Ph.D.
Perot Systems Government Services

Kristyn Stevens
Perot Systems Government Services

Project Subcommittee

Robert Harriss, Chair
National Center for Atmospheric Research

Robin L. Autenrieth
Texas A&M University

Laura Gabanski
United States Environmental Protection Agency

James William "Willy" Owens
The Proctor and Gamble Company

Peter Ruffier
Eugene/Springfield Water Pollution Control

Water Environment Research Foundation Staff

Director of Research: Daniel Woltering, Ph.D.
Program Director: Margaret Stewart

ABSTRACT AND BENEFITS

Abstract:

Integrated water resource management has been discussed since at least the Civil War; yet there is still no integrated framework for managing water in a sustainable manner. Recognizing this need, the Water Environment Research Foundation (WERF) funded a research project to develop an integrated, conceptual framework for sustainable water resources management. Through WERF funding, this framework was developed over the past 4 years. Development of the framework was guided by the United Nations Agenda 21, Global Water Partnership, the Enlibra Principles, and Panarchy Theory.

The conceptual framework for sustainable water resources management considers water as a renewable, but finite resource with global and regional constraints. It integrates ecological, economic, and social considerations through institutional and legal/regulatory constructs to move toward sustainable water resources. Implementation of the framework is guided by a process flow-chart that considers both crisis management and proactive management activities. We believe that sustainability is as much an outcome as a goal. If water resources are viewed within a total systems context and monitored, assessed, and adaptively managed *through time*, sustainable water resources are the outcome.

Benefits:

- Demonstrates how ecological, economic, and social elements can be integrated to adaptively manage water resources sustainably.
- Provides guidance to help subscribers move toward sustainable water resources management, from concept to practice.
- Provides insight into how policy, law, research, and social mindsets influence sustainable water resources in the United States.
- Provides criteria for determining if water resources management is sustainable.
- Provides indicators for monitoring progress toward achieving sustainable water resources.

Keywords: Sustainability, water allocation, water supply, water resources planning, water policy, environmental indicators.

TABLE OF CONTENTS

LIST OF TABLES

LIST OF FIGURES

LIST OF ACRONYMS

α	reorganizing phase of Panarchy adaptive cycle
Ω	release phase of Panarchy adaptive cycle
7Q10	the lowest 7-day average flow that occurs (on average) every 10 years
ADEQ	Arizona Department of Environmental Quality
ADWR	Arizona Department of Water Resources
AES	Agricultural Experiment Station
AMO	Atlantic Multidecadal Oscillation
ANC	acid neutralizing capacity
ARASA	Associacion Regional Ambiental de Sonora y Arizona
ARS	United States Department of Agricultural Research Service
ASCE	American Society of Civil Engineers
AWWA	American Water Works Association
BLM	Bureau of Land Management
CCCER	Cochise College Center for Economic Research
CDC	Center for Disease Control
CGSDI	Consultative Group on Sustainable Development Indicators
CWC	Clear Water Conservancy
DEA	data envelopment analysis
DEM	digital elevation model
DOE	U.S. Department of Energy
DOT	U.S. Department of Transportation
ENSO	El Niño Southern Oscillation
ESI	Environmental Sustainability Index
FWS	Fish and Wildlife Service
GEMI	Global Environmental Management Initiative
IBI	Index of Biotic Integrity (Steedman, 1988)
IWRM	integrated water resources management
K	conservation phase of Panarchy adaptive cycle
NAWQA	National Water Quality Assessment Program
NGO	nongovernmental organization
NMFS	National Oceanic and Atmospheric Administration National Marine Fisheries Service
NOAA	National Oceanic and Atmospheric Administration
NRC	National Research Council
OECD	Organization for Economic Cooperation and Development
PAGE	The Pilot Analysis of Global Ecosystems
PDEP	Pennsylvania Department of Environmental Protection
PDO	Pacific Decadal Oscillation
PFBC	Pennsylvania Fish and Boat Commission
P-PET	Precipitation minus potential evapotranspiration
PSU	Pennsylvania State University
QOL	quality of life
r	exploitation phase of Panarchy adaptive cycle
SCBWA	State College Borough Water Authority
SCERP	Southwest Center for Environmental Research and Policy

SCWC	Spring Creek Watershed Commission
SETAC	Society of Environmental Toxicology and Chemistry
SGCD	Subcommittee on Global Change Research
SPRNCA	San Pedro Riparian National Conservation Area
SWCBD	Southwest Center for Biological Diversity
SWRM	sustainable water resources management
TMDL	total maximum daily load
TNC	The National Conservancy
UNESCO	United Nations Educational, Scientific, and Cultural Organization
USACE	U.S. Association of Civil Engineers
USDA	United States Department of Agriculture
USDA NASS	U.S. Department of Agriculture National Agricultural Statistics Service
U.S. EPA	U.S. Environmental Protection Agency
USGS	U.S. Geological Survey
USPP	Upper San Pedro Partnership
VEA	value efficiency analysis
VOC	volatile organic carbons
WRC	U.S. Water Resources Council
WTA	willingness to accept compensation
WTP	willingness to pay
WWPRAC	Western Water Policy Review Advisory Commission

EXECUTIVE SUMMARY

ES.1 Sustainable Water

Water, Air, Earth, and Fire! We have known for almost 2,500 years (Empedocles ~ 450 B.C.) that water is an essential element of life. Hydrology shows us that water cycles between the atmosphere, surface water, and groundwater. Engineering shows us how to move water, treat water, and store water for various uses and how to discharge water following use. And ecology demonstrates that aquatic ecosystems provide both goods and services and that there are minimum water requirements to maintain these ecosystems

But are these all the pieces we need to solve the puzzle of sustainable water resources management? We think of hydrology, engineering, and ecology as major pieces of the puzzle, but knowledge of other disciplines are needed to solve the puzzle, and gain a more complete picture, of sustainable water resources management (Figure ES-1).

Because water is essential for survival, it touches every sector of our lives—ecological, economic, and social/cultural. Water resources management in the 21st century must address each of these sectors as well as the links and trade-offs among ecological needs and services, human welfare, and desired human uses of water.

What is needed is a framework that integrates the many competing facets of water so that water can be managed as a sustainable resource. Unfortunately, most water management efforts are incomplete because they do not couple a firm conceptual framework with the processes necessary to integrate multisector influences and their cumulative effects on water resources.

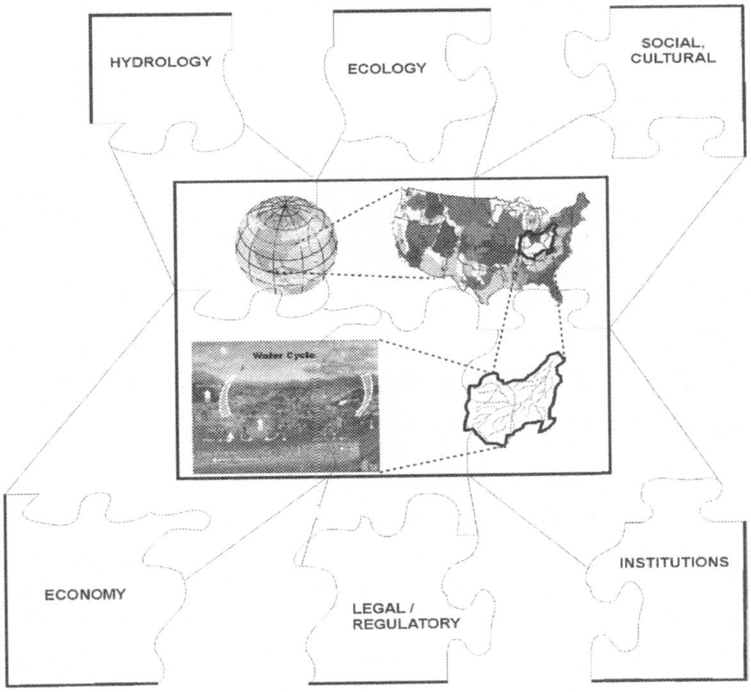

Figure ES-1. Elements of Sustainable Water Resources.

Recognizing this need, the Water Environment Research Foundation (WERF) funded a research project to develop a conceptual framework for integrated sustainable water resources management (SWRM). Much of the information needed to move toward SWRM is currently available but it is widely dispersed throughout the literature. Through the conceptual framework and guidelines developed as part of this project, these pieces of the puzzle can be identified and placed in perspective within the SWRM picture (Figure ES-1).

Sustainability is not an easy term to define. An early definition of sustainability was "the ability to meet current needs without compromising the ability of future generations to meet their own needs" (World Commission on Environment and Development, 1987). As this project progressed and our thinking evolved, we came to realize that sustainability is an outcome of using a total systems approach to monitor, assess, and adaptively manage water resources through time, as much as it is a goal.

ES.2 Purpose of this Report

This report presents a framework, a set of guidelines, and supporting information for developing an implementation plan for SWRM.

ES.3 Project Objectives

The WERF project had two objectives:

1. develop a comprehensive, integrated framework for developing and implementing SWRM plans from an interdisciplinary perspective; and
2. demonstrate how this framework can lead to sustainable water resources.

This study integrated existing information from a range of disciplines to form a framework for SWRM that is applicable from local to regional scales. It builds on the foundation of emerging concepts, theories, principles, and frameworks for managing water quantity and quality in the United States and globally.

ES.4 Elements of Sustainable Water Resources

To move toward SWRM we must identify not only the elements that should be considered as part of any management plan, but also their interactions. To develop sustainable water resources, we must know not only which pieces of the puzzle are important, but also how those pieces fit together to form a complete picture (Figure ES-1).

Hydrology

The global hydrologic cycle is the largest movement of any substance on earth. Water exerts a major influence on earth's climate, which itself determines the distribution of water around the globe. Studies on global climate change have significantly increased our understanding of the global hydrologic cycle. Understanding the global hydrologic cycle is important because

- It is only at this scale that the hydrologic cycle represents a "closed" system.
- The global interactions among ocean, atmosphere, and land mass control regional and local hydrology.
- It is critical to consider the long time scales associated with global hydrologic phenomena in planning and managing regional and local water resources.

SWRM requires regional management units that are defined by the hydrologic cycle. The hydrologic cycle represents the interactions among surface water, groundwater, and atmospheric vapor. All three components must be considered if sustainable water is to be realized. Regional hydrologic landscapes represent management units that are large enough to include regional groundwater aquifers, surface water basins, and large-scale patterns of precipitation and evapotranspiration. Regional scale management units are needed for SWRM.

The amounts of available and useable water in a system are important considerations in SWRM. The amount of available water is a function of climate and hydrology. Not all water present in a system is necessarily suitable for all uses, because of water quality or accessibility considerations. To get a true picture of water resources, both availability and usability must be considered.

Understanding water use—and overuse—by region is another of the pieces of the puzzle needed to develop a picture of sustainable water resources. About 80% of all the freshwater withdrawals in the United States are from surface water sources. In the eastern United States, the greatest withdrawals are for thermoelectric power cooling water; in the western United States, the greatest withdrawals are for irrigation. A major difference between these two uses is that although 98% of the thermoelectric withdrawals are returned to receiving waters and are available for other uses, over 60% of irrigation and livestock withdrawals are consumed and are not available for other uses. Total withdrawals of water are greatest for California, the south Atlantic-Gulf of Mexico, and the mid-Atlantic regions. Although groundwater use information is not readily available at national and regional scales, groundwater depletion is a serious issue in the High Plains—from Nebraska to Texas—and in parts of Arizona and California. Water conservation, reuse, and recycling are feasible approaches for increasing available surface water and groundwater supplies.

Ecology

Ecological systems are not only affected by the hydrologic cycle, but they also modify the hydrologic cycle. Clearly aquatic ecosystems (e.g., streams, lakes, wetlands, estuaries) depend on water, but nearly every aquatic ecosystem also depends on its watershed and groundwater aquifer. Terrestrial systems modify both the quantity and quality of water received by aquatic ecosystems through surface runoff, infiltration of precipitation into groundwater, evapotranspiration back into the atmosphere, and groundwater discharge to maintain base flow in streams.

Land use management is one of the dominant factors affecting both water quantity and quality. Cumulative effects of land use change on water resources include changes in available water quantities, shifts in frequency and timing of runoff and infiltration, and shifts in the spatial distribution of available water affecting both terrestrial and aquatic ecosystems. Land management programs, such as hubs and corridors, working lands, and aquifer protection programs, all affect water resources and their sustainability. Land use patterns on the landscape can impede or contribute to sustainable water, so SWRM also must include integrated watershed management.

Aquatic ecosystems provide a number of important goods and services that contribute to the economy and well-being of human society. Determining the water needs of ecosystems is critical for developing sustainable water resource allocation. Decisions about maintaining aquatic ecosystems and the services they provide need to be consciously made, and based on understanding all of the trade-offs involved (i.e., environmental, economic, societal, cultural, local, regional, national, and global).

Society

Because water is a basic element for human survival, it also is an integral part of the social and cultural fabric of society. International declarations, covenants, and agreements on human rights have stated that water is a basic human right. Water as a basic human right has not been adopted in the United States. There is recognition in the United States, however, that sustainable water resources will contribute to satisfying basic human/societal physical, physiological, and psychological needs.

Mental models represent our perception of truth and reality. These mental models are formed through experience, education, and interactions with others and the environment. These models form the basis of our opinions and beliefs. Our different opinions, in many cases, reflect the differences in what we perceive to be true, even if there is not supporting information or data. Communities and cultures also have collective mental models referred to as social mindscapes. It is rare that any individual or community identifies and documents their mental models, but these models can have a profound affect on how decisions are made. For example, there are mental models that indicate water comes from an infinite source because every time the tap is opened, water comes out. Another mental model is based on the belief that using recycled wastewater will make everyone sick. With these two mental models or community mindscapes, proposed tax increases or bond issues to raise money for new water sources or to invest in water recycling programs are unlikely to pass. Documenting these community mindscapes and individual mental models needs to be part of the process of moving toward sustainable water resources.

Social marketing is an approach that has been used by the human health community to change mental models and subsequent human behavior. For example, the social marketing campaigns against smoking had a significant effect on how society now views and reacts to smoking. Another example is the national campaign led by Mothers Against Drunk Driving for the use of designated drivers. Similar social marketing can be used to change fallacious mental models or community mindscapes about water and water management.

Economy

Historically, water has been underpriced, with government investments and subsidies being used to develop supplies (e.g., build reservoirs). Economic approaches are available for estimating both the market and nonmarket value of water. There are, however, challenges in valuing social and ecological benefits: the social issues of fairness and intergenerational equity, and the ecological issues of potential irreversibility and cumulative effects. Economic incentives can help manage the demand for water and manage water quality. They can be used to ameliorate allocation issues between areas where water is highly valued but in short supply (such as urban areas), and areas where water is the economic value of water is lower (such as irrigation water in agricultural areas). Economic incentives can also be used to allocate more equitably costs for maintaining water quality (i.e., water treatment). Economic incentives for management using equimarginal principles, transactions costs, and other approaches are needed for sustainable management of water resources, integrating conservation, supply, and demand management practices.

Legal, Policy, and Regulatory Constraints

There are currently 22 federal executive agencies in 10 cabinet departments, including the Executive Office of the President, that are actively involved in developing policies and programs to manage and protect both water quantity and quality in about 25 separate water programs. There are also 12 standing committees in Congress that have direct jurisdiction over various components of federal water resources policy and legislation, in addition to the Budget Committees, Appropriation Committees, Ways and Means and Finance Committees, House Government Reform and Oversight and Senate Government Affairs Committees, which also have ongoing interests in water reflecting the competing interests of their constituents.

Existing legal doctrines are not necessarily compatible with sustainable water resources. For example, one of the purposes of the Prior Appropriation Doctrine is to encourage predictability and certainty in water use to support economic investment in consumptive water uses. The doctrine does not reward efficiency, and can exact a legal forfeiture of water rights if the appropriated water is not used within a prescribed period. Many of the existing laws and regulations were needed historically to protect and manage water. Societal needs, norms, and mores have changed over the years, however, resulting in some of these existing laws taking on "blue law" status. Developing sustainable approaches to water resources might require changes in existing policies or laws before SWRM practices can be implemented.

In contrast to laws, water policies are intended to maximize the benefits that can be derived from water and its rational management. Effective policy formulation is fully consistent and compatible with alternative futures and adaptive management approaches. Rather than consider the universe of policy alternatives, a few alternatives are selected for comparison. Generally, decision-makers base policy implementation decisions on which policy has the greatest marginal benefits. In many cases, it is possible to get different groups to agree on a needed policy change, even if they do not agree on why the policy change is needed.

Institutions

According to the London School of Economics, civil societies include organizations such as registered charities, development nongovernmental organizations (NGOs), community groups, women's organizations, faith-based organizations, professional associations, trade unions, self-help groups, social movements, business associations, coalitions, and advocacy groups.

Civil society organizations, institutions, and their affiliates are critical to moving toward SWRM. These groups, with their partnerships, communication networks, education and outreach programs, and government liaisons, exemplify the way things will be accomplished in the 21st century. Studies show that local, state, and federal government agencies are not trusted by the public (WERF, 2004). It will be through civil societies that new regulations, technologies, and practices will be implemented. Peter Drucker (1999) has stated that the more economy, money, and information become global, the more community, will matter. And only civil society organizations, and institutions perform in the community, exploit its opportunities, mobilize its local resources, and solve its problems.

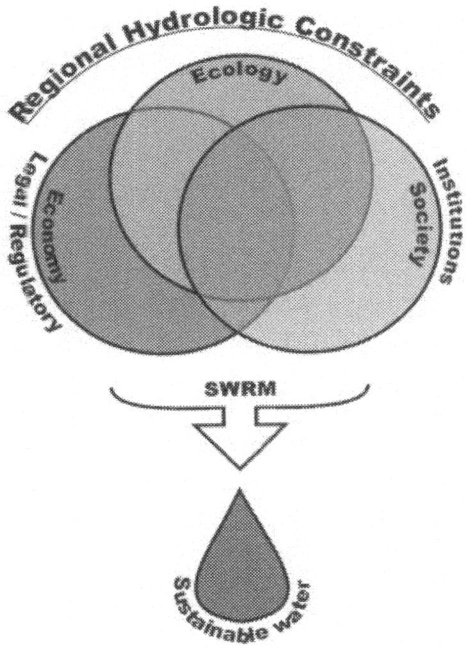

Figure ES-2. Sustainable Water Resources Management Framework.

ES.5 Implementation

Implementing SWRM requires planning and an adaptive management approach that uses monitoring and assessment. Planning is necessary to establish direction, to make sure we know where we want to go. Monitoring and assessment are necessary to make sure we are going where we want to go. It is important that appropriate indicators be selected for monitoring and assessment. The indicators being monitored need to be able to tell us where we are, the direction we are going, and how rapidly we are moving in that direction. They must be properly assessed, and the results communicated to stakeholders. The results of the assessment provide insight into the success of management activities that have been implemented, and suggest opportunities for adjusting them to achieve SWRM, which is the foundation of adaptive management.

ES.6 Developing an Integrated Framework

An integrated SWRM framework can operate at multiple scales (e.g., community to region) and focuses on the interactions and interrelationships among environmental, economic, and social/cultural sectors that operate within civil society, legal, regulatory, and policy constructs (Figure ES-2). Because water is essential for life, there are no policies, practices, or procedures in any sector that do not directly or indirectly affect water. The corollary is that every segment of society, economy, and environment benefits from SWRM.

Putting the Framework into Action: Steps in the Process

How do we use this framework to move toward SWRM? Unfortunately, most communities, agencies, and organizations do not consider moving toward sustainable water resources until they are in a crisis. The need for a water crisis to initiate change in water management practices was confirmed at a WERF-sponsored stakeholder workshop held at a September 2004 reference meeting. When asked what would be required to implement a framework for SWRM, stakeholders answered that the following three things would be required (in priority order):

1. a crisis or defining event,
2. leadership, and
3. money.

Therefore, the first part of the process flow chart is applicable to communities or watershed associations that are facing a water crisis (Figure ES-3). In a water crisis, management options are restricted to those that can be implemented quickly. These management options are typically developed and implemented by government entities. The report provides text and questions that illustrate the information needed to address a water crisis (water quantity and/or quality) at the watershed scale and eventually move toward SWRM. This report addresses the watershed scale because watershed management is actively encouraged by many agencies, at this scale. However, the questions and guidance presented can—and must—be applied within communities, as well as at basin and regional scales.

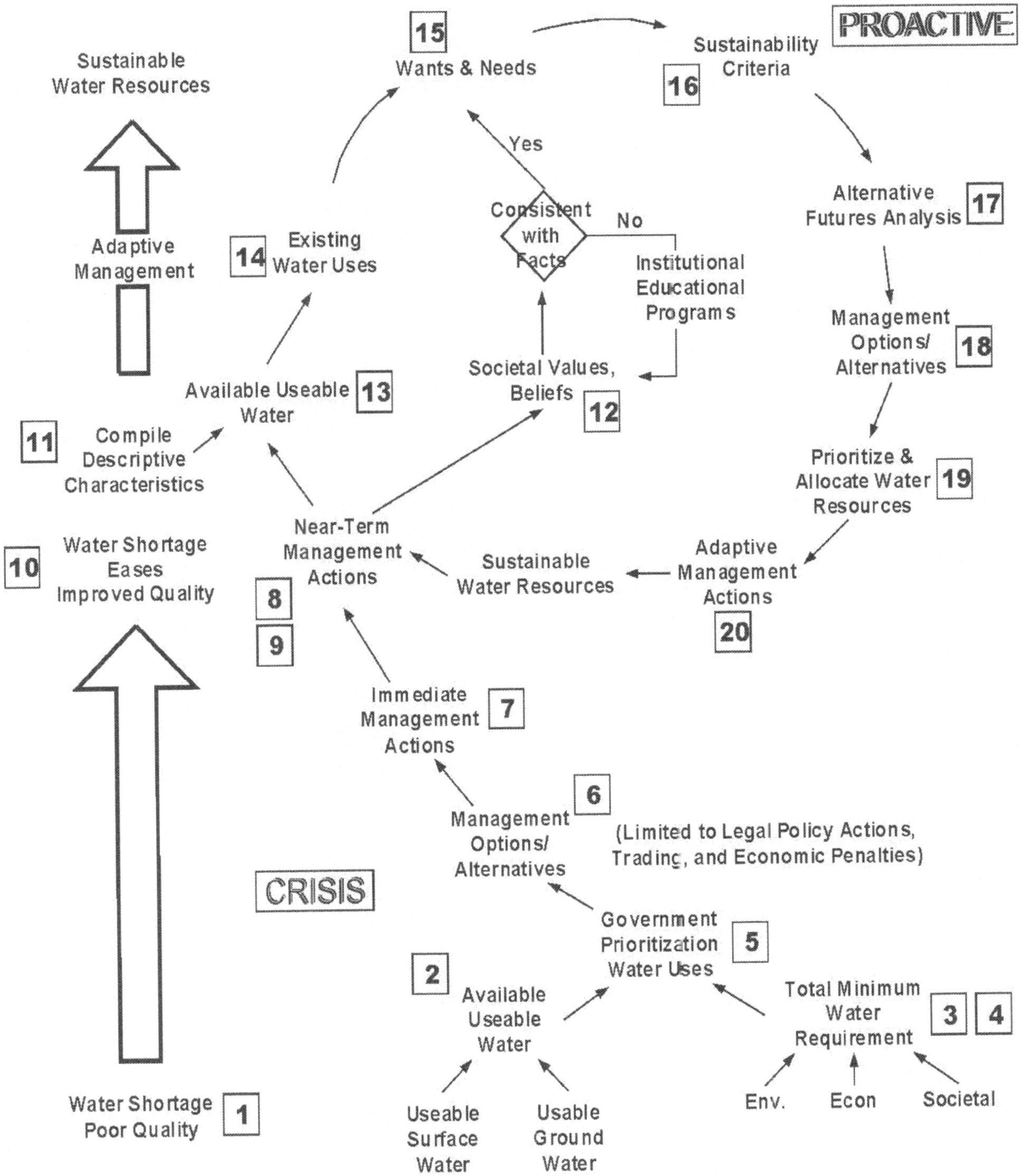

Figure E-3. Sustainable Water Resources Management Process Flow Diagram with Two Paths—a Crisis Chart and a Proactive Chart.

Move to the Proactive Chart

As the water crisis is resolved and longer-term options and alternatives can be assessed, the water resources management process should move from the Crisis Chart to the Proactive Chart so that future crises can be averted (Figure ES-3).

The time to begin implementing SWRM practices is before water needs or water quality issues reach crisis proportions. Management options increase when you are not in a crisis, and stakeholders can actively participate in development of the management plan.

The key is to start the process! There are always water issues in a community, even if there is no water crisis. Start with a water issue of concern in your community, watershed, basin, or region. Water can be related directly or indirectly to almost every management practice in the community or watershed. Starting with community or watershed issues is preferred because you not only know the actors, you are one of them.

Steps in the implementation process need to be considered at each scale, from the local community to the region. Each of the scales, from local community to region, however, is interactive, so it does not matter where along the scale continuum you begin. Cumulative effects of decisions made by local communities contribute to regional scale effects, and regional constraints affect what can be attained in local communities. In addition to the regional constraints, it is useful to determine the hydrologic effects on your region from large-scale weather cycles. These large-scale cycles, such as El Niño and La Nina, influence periods of flooding and drought at regional, basin, and watershed scales.

ES.7 Start Now

SWRM might appear to be overwhelming, but all the information in this report was readily available in the literature. What is new is the synthesis and integration, the framework and the guidelines. The key to SWRM is to start. Pick an issue, form a team, start the process. Nearly all the information you need is available. Work through the guidelines. Move toward a sustainable outcome.

CHAPTER 1.0

INTRODUCTION

Water is unique on Earth:

- It is a prerequisite for life—every living plant and animal (including humans) needs water for survival.
- It has no substitute for most of its uses—the goods and services provided by other natural resources have substitutes (e.g., energy can be produced from coal, oil, or wind) (Postel, 2000a).
- It is a fundamental force in the earth's climate regime—the physical properties of water are responsible for the characteristics of our climate (Webster, 1994).
- It is a renewable, but finite resource—earth's hydrologic cycle is closed, so the global amount of water today is essentially equivalent to what it was millions of years ago.

Over the past several decades, there has been a growing recognition of the water crisis that is developing throughout the world, including the United States [Gleick, 1996, 1998a; National Research Council (NRC), 1992, 2000a, 2001; Postel, 1992, 2000a; United Nations, 1992]. Water quantity and quality are major issues in a world with a population of about six billion people that is expected to grow to 8.9 billion by 2050 (Postel, 2000b). Globally, about 30,000 people a day, almost 11 million per year, die because of water scarcity or waterborne diseases (Hamlin, 2001).

Water resources are expected to be, and in some cases already are, the growth-limiting factor for development at national and international levels [Gleick, 1998b, 2000; Postel, 2000a, b; United Nations Educational, Scientific, and Cultural Organization (UNESCO), 2000]. In this context, water is subject to multiple—and often conflicting—quantity and quality demands. Sustained use of this shared resource requires that the needs of all stakeholders, including industries, municipalities, and recreational users, are balanced by natural limits on recharge, ecological needs, and other natural processes.

When this project was initiated, we used the American Society for Civil Engineers (ASCE; 1998) and UNESCO (1999) definition of SWRM, "Sustainable water resources management is the design and management of water resources systems so that they fully contribute to the objectives of society, now and in the future, while maintaining the ecological, environmental, and hydrological integrity." Sustainability, as initially conceived by the Bruntland Commission, was, "The ability to meet current needs without compromising the ability of future generations to meet their own needs" (World Commission on Environment and Development, 1987). As this project progressed and our thinking evolved, our concept of sustainability changed. Although sustainability can be established as a goal, we now believe that sustainability is the outcome of using a total systems approach to monitor, assess, and adaptively manage water resources *through time.* This report addresses the elements in a total systems approach so the reader will understand why we consider sustainability an outcome, as much as a goal.

Water resources management historically focused on the extreme events of floods and droughts caused by natural processes external to the water resource. Planning strategies were derived from statistical estimates of precipitation and flood frequencies and recurrence intervals, assuming that human activity had little direct or indirect influence on the hydrologic cycle. Only local watershed—or, at most, basin—precipitation and discharge statistics were considered. It is now apparent that the entire global hydrologic cycle must be considered in water resources planning and management.

Over the last two decades, it has become increasingly apparent that there are connections between population and land use change and the hydrologic cycle at scales ranging from local to global scales (Eagleson, 1986; NRC, 1991, 2000a; Postel, 2000a; Vörösmarty & Sahagian 2000). Population growth and land use change, such as deforestation, have contributed to severe flooding and massive slope failures in Venezuela and Africa, and in the United States, particularly in southern California. The 1993 Mississippi River flood was attributed, in part, to the loss of wetlands throughout the upper Mississippi River valley. Similarly, the magnitude of the 1999 flooding in the Neuse River basin, North Carolina, was attributed, in part, to watershed urbanization and forest fragmentation. These two factors increased runoff volume and reduced time to peak flows.

Population growth and land use change have aggravated water shortages during periods of drought throughout the United States. In 2001, farmers in the west challenged federal marshals to gain use of water for irrigation that was needed to support endangered fish species. In the same year, Anheuser-Busch Corporation experienced increased prices and reduced availability of barley and aluminum because of the reduced irrigation water and hydropower generation caused by drought in the western states. Eastern reservoirs that had firm yields before 1970 are experiencing greater project demands because of urban sprawl and increased population growth. Reservoirs in the Mid-Atlantic region were at record low levels in the late 1990s because of drought and the demand from increased population density in the Baltimore–Washington, D.C., corridor and sprawl into surrounding counties. Groundwater depletion and subsidence have reached critical levels throughout the United States (Galloway et al., 1999). For example, some Houston, Texas, streets have subsided by more than 6 feet because of groundwater depletion.

Water resources management in the 21st century must address the linkages and trade-offs among ecological needs and services, human welfare, and desired human uses of water (e.g., irrigation, water supply, transportation, recreation) (NRC, 2000a). Although the concept of integrated water resources management is not new, having been discussed since the Civil War (Reuss, 1993), it has generally not been applied at the basin or regional scale. Several major issues must be considered to formulate and implement SWRM plans at these larger scales:

- historical laws and regulations that constrain water use,
- water protection strategies constrained by stakeholder perceptions of environmental need and integrity, and
- long time frames needed for water resources management strategies that are outside the typical time frame of concern for most Americans.

In addition

- Water management strategies must treat water as a finite, albeit renewable, resource that is subject to multiple demands and uses.
- Although water conservation and reuse have seldom been a major focus of water management, they must become a priority for resource planners in the future.
- Water resources management strategies must be developed at large scales (e.g., regional, global), implemented at local scales, and managed at multiple scales along this continuum.
- An interdisciplinary perspective and approach is required to develop effective SWRM plans and strategies.

What is needed is a framework that integrates the many competing facets of water use so that water can be managed as a sustainable resource. Unfortunately, most water management efforts lack a firm conceptual framework or decision-making process with which to integrate the multiple influences on water resources, and analyze their cumulative effects. Recognizing this need, WERF funded a research project to develop an integrated conceptual framework for SWRM. This study integrates existing information from a wide range of disciplines to form a framework for sustainable water resource management that is applicable nationally from the regional to the local scale. It builds on the foundation of the emerging concepts, theories, and frameworks for managing water quantity and quality globally and in the United States.

> This literature review was not intended to be exhaustive, but rather to demonstrate the wealth of information available to move forward toward sustainable water resources. We can initiate the process of sustainable water resources management now with the framework, guidelines, and information presented in this report.

CHAPTER 2.0

WATER RESOURCES MANAGEMENT

2.1 International Water Management Organizations, Associations, and Institutions

As our understanding of sustainability and its dependence on the global economy and environment has increased, so has the number of international organizations, associations, and institutions concerned with sustainable water resources (Anderson, 2001). The awareness of sustainability and water resources issues at the global scale has grown in proportion to the increase in population and our understanding of global connections and interactions. The 1992 Rio Conference resulted in the development of the United Nations Commission on Sustainable Development and Agenda 21, which is a plan for attaining sustainable development on a global scale (United Nations, 1992). Chapter 18 of Agenda 21 specifically refers to sustainable water resources. A list of 44 of international water management entities, with their Web site addresses and brief summaries of their roles or functions, is shown in Appendix A.

These international associations and international studies have helped reinforce the importance of considering social and ecological factors, as well as economic and engineering factors, in water resources management. Direct and indirect effects of developing sustainable water resources need to be objectively evaluated, factored into decisions, and then implemented accordingly through formal agreements. Dams and reservoirs were the solution to many water resources problems worldwide from the 1920s through the 1970s. In the United States, the peak period of dam construction was during the 1960s when there was an average of one dam closure per day. As the social and environmental effects of large dams became better understood, alternative solutions to water resources problems were sought. Today, dam removal is being considered for some river basins because of the ecological effects of fragmenting rivers. Although dams have provided benefits, they have also contributed to problems.

The World Commission on Dams (2000) was commissioned to review the effectiveness of large dam development and assess alternatives for water resources and energy development, including the development of guidelines for assessing and appraising large dams. The Commission reached three major conclusions from its study:

1. Dams have made an important and significant contribution to human development, and the benefits derived from them have been considerable.
2. In too many cases, an unacceptable and often unnecessary price has been paid to secure those benefits, especially in social and environmental terms.
3. The key decisions are not about dams as such, but about options for water and energy development. There is a need to rethink the management of freshwater resources.

International water law plays a role in sustainable development. The agreements codified in international law are sustained only through the efforts of the individual nations involved in the agreement. Even though there is no international water law court to regulate international boundary water issues, international institutions can be established to monitor and resolve differences. The effectiveness, longevity, and functioning of these international organizations can be illustrated by the Boundary Water Treaty of 1909 between the United States and Canada and the establishment of the International Joint Commission (IJC).

The IJC was established 93 years ago to provide oversight and help prevent and resolve disagreements between Canada and the United States about the use of their shared water resources (Galloway & Clamen, 2001). In 1909, Great Britain and the United States established a treaty to resolve issues along the 5000-mile U.S.–Canadian border. The IJC has the responsibility for approving and overseeing the operation of dams and other structures, and joint water use of waters along the U.S.–Canadian border to protect all interests from the effects of the projects or project operation (Galloway & Clamen, 2001). Six commissioners, three from each nation, who do not receive instructions from their governments, are charged with arriving at the best agreements for both countries. The treaty that established the IJC provides for questions or differences between the two countries to be submitted to the IJC for investigation and reporting on the "reference" issues. The IJC can appoint a task force or investigative board to provide advice on these reference issues as well as continue surveillance to monitor progress on resolving the issues between the two nations. On more than 120 occasions, the IJC has addressed reference issues related to transboundary waters. The IJC offers one example of how issues among nations can be addressed without litigation. Global management decisions to clear tropical rainforests, develop large dams, and similar actions require similar international agreements and institutions if global sustainability is to be achieved.

2.2 Water Resources Management in the United States

A similar list of organizations, associations, and institutions addressing water resources management could be provided for the United States as was provided internationally (Appendix A). There are multiple federal agencies responsible for water resources management (see Chapter 11.0). The issue in the United States is that water resources management is hindered because there is no agency or organization that is coordinating the activities of all these organizations, associations, and institutions (Jackson et al., 2001). From the 1960s through 1983, the U.S. Water Resources Council (WRC) coordinated activities among the major federal planning, engineering, and construction agencies (e.g., U.S. Bureau of Reclamation, U.S. Army Corps of Engineers, Tennessee Valley Authority) and developed principles and guidelines for water and related land resources planning and design.

Because of the decline in dam construction, the WRC was terminated in 1983. Since then, there has been no coordination or oversight of federal agencies' water resources projects. This situation has led to a recommendation for the creation of a National Water Research Board to oversee a national water research agenda for increasing our knowledge of the factors and interactions needed for sustainable water resources (NRC, 2000b). Although this board would oversee research, the NRC (2000b) did not recommend an organization to oversee the development and management of sustainable water resources. Jackson et al. (2001), however, did recommend not only a national panel to coordinate federal activities related to freshwater, but also the development of a coordinated national plan for managing freshwater.

Numerous regional commissions, (e.g., Ohio River Valley Water Sanitation Commission, Delaware River Basin Commission, Susquehanna River Basin Commission, Interstate Commission on the Potomac River Basin, Lower Mississippi Valley River Division, Great Lakes Commission, Northwest Power Planning Council, and Colorado River Compact) have been established to coordinate activities and resolve interstate issues in river basins related to water quality, flooding, conservation and development of surface and groundwater supply, hydroelectric power potential, navigation and similar activities. Fisheries protection, soil conservation, and similar environmental concerns have been added to the charge of these commissions in recent years, but, in general, the emphasis is on ensuring human water uses are protected in the basin. These commissions generally interact with state and federal agencies and have less involvement with local communities and stakeholders.

Recent emphasis on community-based programs and watershed management [United States Environmental Protection Agency (U.S. EPA), 1995, 1996, 1997a,b, 1999, 2001; NRC, 1999] has contributed to the formation of watershed associations. These watershed associations represent a grassroots movement that includes local communities and stakeholders, and NGOs, with some state involvement. Kenney (1999) assessed watershed-level organizations interested in water and land management in western states and noted an emerging era of community-based watershed management. Kenney (1999) reported six findings from his review:

1. Managing water and land resources at large scales is not new, but it is now being widely implemented. River basin management is still largely rhetoric, but hydrologic management at the watershed scale through watershed associations is proliferating throughout the West. Most of these associations are less than a decade old. These watershed initiatives are based on three fundamental elements: (a) the need for both ecological and community sustainability; (b) the democratic process, stressing decentralization, collaboration, inclusion, and participatory government; and (c) pragmatic approaches to management and dealing with bureaucratic red tape.
2. Watershed initiatives have a great variety of structures and functions, but two factors have spurred the formation of watershed associations. The first factor is studies or conferences that articulate problems in the watershed. The second factor is the fear of possible government intervention (usually federal). Five elements that contribute to the success of these associations are strong leadership, member participation, focus, adequate resources, and a process for working through the issues
3. A lack of formal authority for most watershed initiatives did not hinder the functioning of the associations, and in many instances, a reliance on "moral authority" was generally seen as an asset.
4. Watershed initiatives were not closely linked to management programs at the larger river basin scale. These initiatives were nested within major river basins, with little coordination among watershed associations. In part, this was because there were no mandated coordinating bodies for many river basins. With the exception of the specific river basin commissions listed above, there were no incentives or resources for coordination at the river basin scale.
5. The watershed movement, although successful, has received serious criticism on legitimate issues of its representation, decision-making process, and motives. Critics have raised issues about the formation of watershed associations to selectively exclude dissenting voices from the collaborative effort, to circumvent environmental laws or regulations, or to promote the economic interests of selected members of the association.

6. Performance of many watershed initiatives and associations is sufficiently positive to warrant greater consideration and support from all levels of government and the private sector (Kenney, 1999).

The Western Water Policy Review Advisory Commission (WWPRAC, 1998) recommended the formation of regional basin councils to integrate local watershed groups and associations with federal, tribal, local, and state government agencies to govern water resources based on hydrologic systems, including both groundwater aquifers and surface water watersheds. The WWPRAC specifically did not recommend a national oversight board but did suggest using the existing U.S. Geological Survey (USGS) National Water Quality Monitoring Council to provide a continual supply of water data needed for adaptive management. The WWPRAC (1998) cited the emergence of the Northwest Power Planning Council on the Columbia River, and the Bay Delta Accord and the CALFED (http://calwater.ca.gov) process on the Sacramento and San Joaquin Rivers in California as examples of regional councils that are including local stakeholders and water management agencies. The U.S. EPA Chesapeake Bay Program (http://www.chesapeakebay.net) is another regional program that has integrated agencies from five states, nonprofit organizations, and local stakeholders in the management of the Chesapeake Bay watershed. There is still a need for better coordination and cooperation at both the national level and the state level. The need for coordination at the state level is imperative, because most water resources management decisions are made at the state level.

At the regional, state, and local scales, watersheds and watershed associations are recognized as important in managing for sustainability, including sustainable water resources. Although the need for larger-scale management is recognized, coordination among these various entities is currently lacking. An organizing framework is needed that incorporates the multiple elements required for sustainable water resources.

2.3 Organizing Management Frameworks

There currently is no single framework that integrates the socioeconomic, legal and regulatory, hydrologic, ecological, and engineering elements needed for sustainable water resources. However, integrating elements from the four frameworks described below might provide a foundation for developing a sustainable water resources framework:

1. Ecological risk assessment framework (U.S. EPA, 1998)
2. Framework for assessing and reporting on ecological condition (Young & Sanzone, 2002)
3. Economic and Environmental Principles and Guidelines for Water and Related Land Resources Implementation Studies (WRC, 1983)
4. Integrated water resources management (IWRM; Agarwal et al., 2000)

2.3.1 Ecological Risk Assessment Framework

The ecological risk assessment framework provides a flexible process for organizing and analyzing data, information, assumptions, and uncertainties in assessing the likelihood of adverse ecological effects. It provides a critical element for environmental decision-making by giving managers an approach for assessing environmental risk and subsequently incorporating social, legal, political, and economic factors in selecting a course of action. The ecological risk assessment framework has three phases: problem formulation, analysis, and risk characterization (Figure 2-1).

The problem formulation phase includes a preliminary assessment of existing information to determine what is available and what is needed to evaluate potential effects and risks. It includes the selection of assessment endpoints and the development of a conceptual model, or models, linking the assessment endpoints with the proposed management actions in the management regime, and with the pathways and processes that connect sources to effects in the scientific regime.

The analysis phase is the arena in which most scientists and engineers feel comfortable. This phase includes the use of empirical and dynamic models, data analysis procedures, and other analytical techniques to evaluate the effects of various management actions on the assessment endpoints. For chemical stressors, the analysis includes both exposure and effects (dose-response) pathways. The ecological risk assessment paradigm was originally developed to consider chemical effects, but it has been expanded to include potential effects from other stressors such as hydrologic variation, habitat alteration, and invasive species.

The risk characterization phase includes two components: risk estimation and risk description. Risk estimation is the process of integrating potential effects with various management actions and assessing the associated uncertainties. Risk description uses a line of evidence or weight of evidence approach to corroborate or refute the risks estimated for the assessment endpoints, and interpret the significance of the adverse effects on the assessment endpoints from various management actions. Risk description is the area in which other information (e.g., socioeconomic, regulatory, etc.) can be integrated with the ecological effects for decision-makers. The final risk description can be qualitative or quantitative, but for most ecological effects, the risk description will be qualitative.

Ecological risk assessment provides a useful umbrella under which ecological information can be integrated to assess the likelihood of adverse effects, or risk, from proposed management actions. Three other frameworks that are uniquely suited to guiding the acquisition and assessment of water resources information are the U.S. EPA Science Advisory Board (Young & Sanzone, 2002), the U.S. Water Resources Council (WRC, 1983), and IWRM (Agarwal et al., 2000) frameworks.

2.3.2 Framework for Assessing and Reporting on Ecological Condition

A framework for assessing reporting on ecological condition has been developed by the U.S. EPA Science Advisory Board (Young & Sanzone, 2002). This framework is intended to serve as an organizational tool to help decision-makers decide what ecological attributes are important to measure, and how to aggregate these measurements into an understandable picture of ecological condition (Young & Sanzone, 2002). Once the goals and objectives of a study are defined (which is the purpose of the problem formulation phase of ecological risk assessment), there are six essential ecological attributes, each with component categories and subcategories,

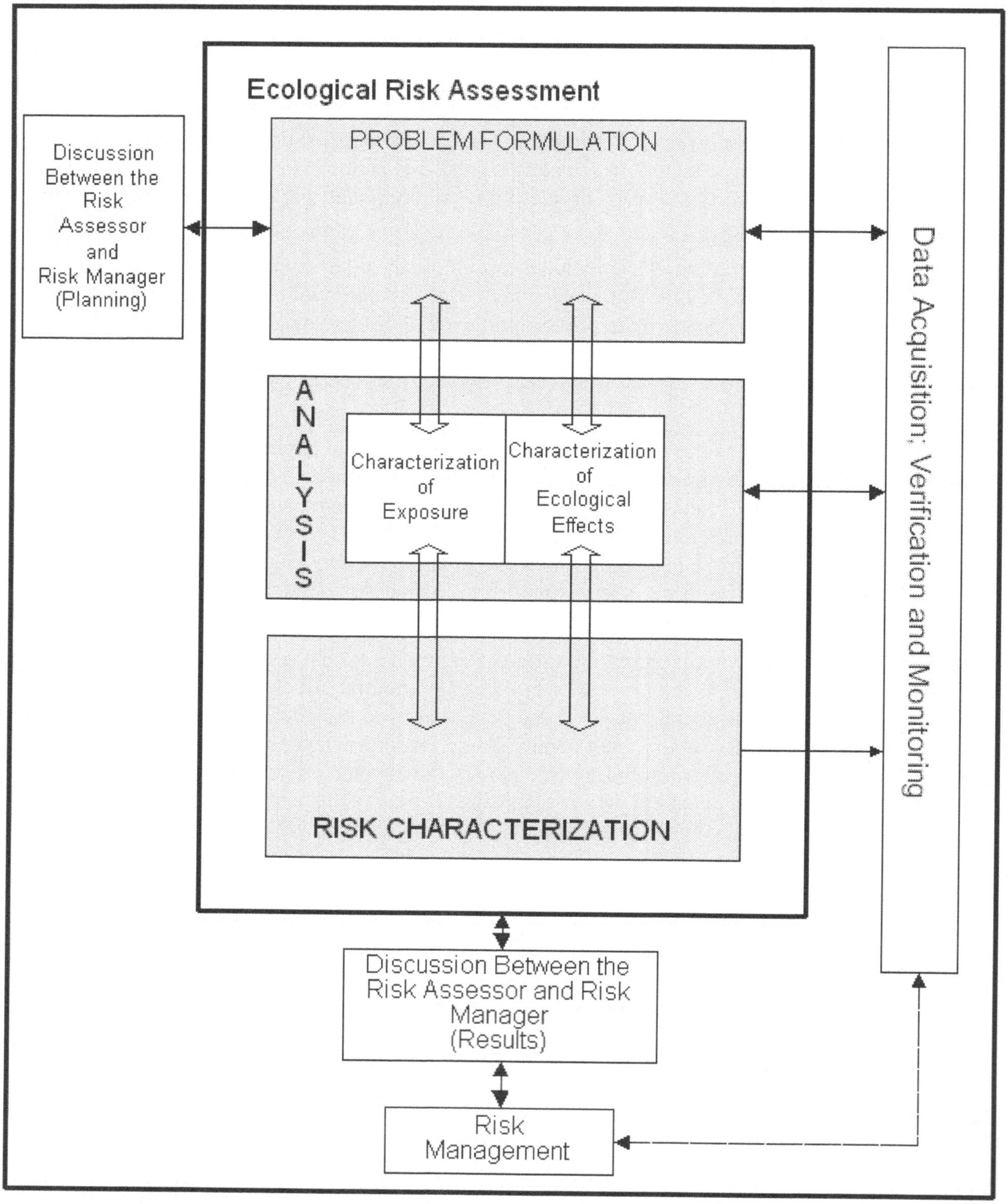

Figure 2-1. Framework for Ecological Risk Assessment (U.S. EPA, 1998).

which can be used as a checklist to help ensure appropriate indicators (information) are collected to monitor and sustainably manage ecosystems:

1. Landscape condition
2. Biotic condition
3. Chemical and physical characteristics
4. Ecological processes
5. Hydrology and geomorphology
6. Natural disturbance regimes

The first three attributes relate primarily to ecological structure and pattern, and the last three relate to ecological processes and function. Table 2-1 lists examples of component categories and subcategories for these essential ecological attributes. The Science Advisory Board's essential ecological attributes incorporate the larger-scale (i.e., landscape) perspective needed to understand the observed biotic, physical, and chemical responses in ecosystems. These observed responses result from underlying processes, and contribute to larger-scale ecological processes. With U.S. EPA's focus on watershed management, hydrologic, and geomorphic indicators are critical for characterizing ecological condition. Finally, natural disturbances, as well as anthropogenic stressors, shape the responses and condition of ecological systems. Understanding both the contributions, and interactions, of natural and anthropogenic factors is important in describing and diagnosing the sources and stressors contributing to ecological condition. This framework is not intended to be prescriptive, but rather descriptive of the kinds of information that are needed to be able to assess ecological condition.

Table 2-1. Essential Ecological Attributes and Reporting Categories (Young & Sanzone, 2002).

Essential Ecological Attribute	Reporting Categories and Subcategories
Landscape Condition	Extent of Ecological System/Habitat Types Landscape Composition Landscape Pattern and Structure
Biotic Condition	Ecosystems and Communities - Community Extent - Community Composition - Trophic Structure - Community Dynamics - Physical Structure Species and Populations - Population Size - Genetic Diversity - Population Structure - Population Dynamics - Habitat Suitability Organism Conditions - Physiological Status - Symptoms of Disease or Trauma - Signs of Disease
Chemical and Physical Characteristics (water, air, soil, and sediment)	Nutrient Concentrations - Nitrogen - Phosphorus - Other Nutrients Trace Inorganic and Organic Chemicals - Metals - Other Trace Elements - Organic Compounds

Essential Ecological Attribute	Reporting Categories and Subcategories
	Other Chemical Parameters - pH - Dissolved Oxygen - Salinity - Organic Matter - Other Physical Parameters
Ecological Processes	Energy Flow - Primary Production - Net Ecosystem Production - Growth Efficiency Material Flow - Organic Carbon Cycling - Nitrogen and Phosphorus Cycling - Other Nutrient Cycling
Hydrology/Geomorphology	Surface and Groundwater Flows - Pattern of Source Flows - Hydrodynamics - Pattern of Groundwater flows - Salinity Patterns - Water Storage Dynamic Structural Characteristics - Channel/Shoreline Morphology and Complexity - Extent/Distribution of Connected Floodplain - Aquatic Physical Habitat Complexity Sediment and Material Transport - Sediment Supply/Movement - Particle Size Distribution Patterns - Other Material Flux
Natural Disturbance Regimes	Frequency Intensity Extent Duration

2.3.3 Economic and Environmental Principles and Guidelines for Water and Related Land Resources Implementation Studies

The Economic and Environmental Principles and Guidelines for Water and Related Land Resources Implementation Studies were developed over a decade by the Water Resources Council (WRC, 1983), a multiple federal agency committee (see Section 2.2). The WRC was intended to coordinate activities of the three major federal water resources construction agencies—Bureau of Reclamation, U.S. Army Corps of Engineers, and Tennessee Valley Authority. The Principles and Guidelines established standards and procedures to be used by federal agencies in formulating and evaluating alternative plans for water and related land resources studies. These Principles and Guidelines arose, in part, from the passage of the National Environmental Policy Act of 1969. The Principles and Guidelines recognized there are two important elements in any federal project: contribution to national economic development, and protection of the nation's environment. The two primary sets of procedures in the Principles and Guidelines were related to national economic development, and environmental quality. Activities addressed by national economic development and environmental quality are listed in Table 2-2.

Table 2-2. Activities Addressed by National Economic Development and Environmental Quality Procedures.

National Economic Development	Environmental Quality
♦ Municipal and Industrial Water Supply	♦ Interdisciplinary
♦ Agriculture	♦ Public Involvement
♦ Flood Damage	♦ Process
♦ Power (hydroelectric)	- Define Resources
♦ Transportation	- Development Evaluation Framework
- Inland	- Inventory Resource
- Deep Draft	- Forecast
♦ Recreation	- Assess Effects
♦ Commercial Fishing	- Appraise
♦ Other Direct Benefits	- Judge Net Environmental Quality
♦ Unemployed/Underemployed Labor Resources	

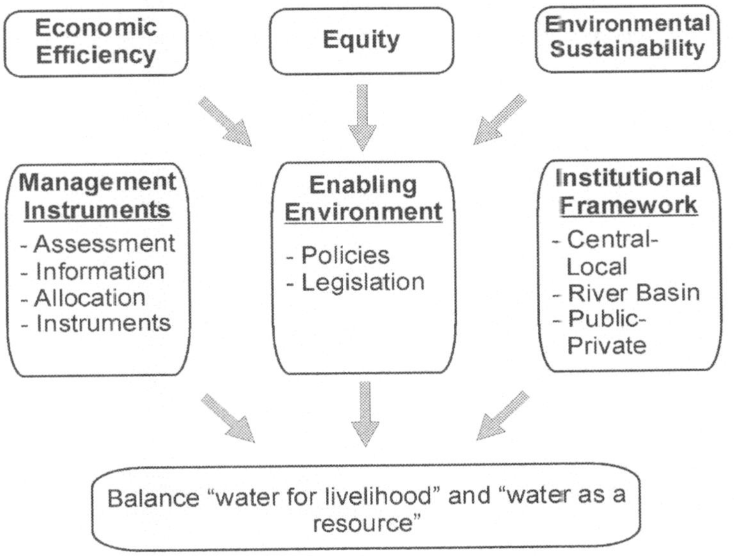

Figure 2-2. IWRM Framework. [Reprinted from Jonch-Clausen (2004).Copyright 2004, Global Water Partnership, Stockholm, Sweden.]

The Principles and Guidelines were developed during an era of active dam construction and reservoir development, which is no longer occurring. The overarching purposes of these guidelines, however, are still relevant—contributing to *sustainable* economic development and *sustaining* ecological systems. Without sustainable ecological systems and the goods and services they provide, sustainable economic development is not possible.

2.3.4 Integrated Water Resources Management

The Global Water Partnership, established in 1996, is an international network open to all organizations involved in water resources management. The purpose of the Global Water Partnership is to foster IWRM, which aims to ensure the coordinated development and management of water, land, and related resources by maximizing economic and social welfare without compromising the sustainability of vital environmental systems (Agarwal et al., 2000).

The Global Water Partnership governance includes a Technical Advisory Committee, which consists of a group of 12 internationally recognized professionals and scientists skilled in different aspects of water management. This Technical Advisory Committee provides technical support and advice to the other governance arms and to the Global Water Partnership. The Global Water Partnership Secretariat is located in Stockholm, Sweden.

Integration of water resources management is necessary for

- the natural system, with its critical importance for resource availability and quality, and
- the human system, which fundamentally determines the resource use, and waste production and pollution of the resource; and which must also set the development priorities (Agarwal et al., 2000).

The framework proposed for IWRM includes consideration of economic efficiency in water use, social equity for water allocation, and environmental and ecological sustainability (Figure 2-2). The pillars of IWRM are the management instruments (e.g., monitoring, assessment, allocation), the enabling environment (e.g., policies, legislation), and the institutional framework (e.g., river basin, public–private organizations) applied to water management (Jonch-Clausen, 2004).

CHAPTER 3.0

INTEGRATIVE PRINCIPLES

One important conclusion from current scientific thinking is that socioeconomic and ecological systems act as strongly coupled, integrated systems (Levinson & Thornton, 2003). However, many of our management practices have considered human and ecological systems as separate entities and attempted to manage them independently (Folke et al., 2002). In addition, there has been an implicit assumption that responses of ecosystems to human activities are linear, predictable, and can be controlled. Evidence, however, suggests that socioeconomic and ecological systems respond in nonlinear ways and exhibit thresholds in their dynamics, and that "command and control" management practices are generally ineffective (Folke et al., 2002). There are two important questions in this discussion:

1. How can we move toward management approaches that consider dynamic, evolving, integrated socioeconomic-ecological systems that are sustainable?
2. Is this even theoretically possible?

This water resources management research project has emphasized the synthesis and integration of approaches, practices, methods, and ideas from other disciplines to move toward SWRM. The following sections describe and document principles and approaches that are being pursued by various organizations throughout the world for integrated water resources management, environmental management, and sustainable development. We have used these principles to develop the framework for SWRM.

3.1 Dublin Principles

The Dublin Principles were formulated through an international consultative process that culminated in the International Conference on Water and the Environment in Dublin in 1992 (Solanes & Gonzalez-Villarreal, 1999). These principles contributed to Agenda 21, adopted by the United Nations Conference on Environment and Development in Rio de Janeiro in 1992, and were reinforced at the United Nations Commission on Sustainable Development in 1998 and 2002.

The four Dublin Principles are

1. Freshwater is a finite and vulnerable resource, essential to sustain life, development, and the environment.
2. Water development and management should be based on a participatory approach, involving users, planners, and policymakers at all levels.
3. Women play a central part in the provision, management, and safe-guarding of water.
4. Water has an economic value in all its competing uses and should be recognized as an economic good (Solanes & Gonzalez-Villarreal, 1999).

Each of these principles is expanded in Global Water Partnership Technical Advisory Committee Background Paper No. 3 (Solanes & Gonzalez-Villarreal, 1999). For example, the first principle of water as a finite and vulnerable resource recognizes that a holistic approach must be used for water management and that water yield has natural limits. Human activities clearly affect water availability over time and space, including upstream–downstream water user relationships. The interaction of humans and natural systems also requires a holistic institutional approach for management.

The second principle is the use of participatory approaches where participation represents real, not token, efforts (Solanes & Gonzalez-Villarreal, 1999). The desired goal is consensus, fully recognizing consensus might not be achievable. However, mechanisms and capacity should be developed to move toward consensus on water management.

The third principle is that women should be involved in the decision-making process (Solanes & Gonzalez-Villarreal, 1999). In many developing countries, women are the primary water users and play a pivotal role in the collection and safe-guarding of water for domestic, and, in many cases, agricultural use.

The fourth principle is that water is an economic good that has value (Solanes & Gonzalez-Villarreal, 1999). The value of water and charges for water use are distinct and different. Understanding that the economic value of water can differ in alternative uses is important for the rational allocation of water as a scarce resource. Water pricing is an economic instrument that can be used to affect behavior towards conservation and the efficient use of water.

3.2 Enlibra Principles

Enlibra, created from two Latin words, means balance and stewardship (Western Governors Association, 2002). The Enlibra Principles were originally developed by Governors Mike Leavitt, Utah (R), and John Kitzhaber, Oregon (D), based on their respective experiences with the Grand Canyon Visibility Transport Commission and the Oregon Plan for Salmon and Watershed Restoration. The success of these two efforts led the governors to articulate, and propose to the 1997 Western Governors Association, a new approach to environmental management based on a set of shared principles. The Western Governors Association agreed to the policy resolution, *Principles for Environmental Management in the West*, in February 1998. These principles were discussed at an Environmental Summit on the West in December 1998, which was attended by more than 400 participants with diverse interests. An Enlibra Advisory and Steering Committee revised the Enlibra Principles based on discussion and recommendations from the Environmental Summit. The Western Governors Association, consisting of 21 members, adopted these revised principles in June 1999. The eight Enlibra Principles are listed and briefly described below. This text was taken directly from the U.S. EPA Web site, www.epa.gov/adminweb/leavitt/enlibra.htm.

3.2.1 Solutions Transcend Political Boundaries—Use Appropriate Geographic Boundaries to Resolve Problems

Many environmental challenges fall within natural geographic boundaries, but most cross political and agency boundaries. Focusing on the natural boundaries of the problem helps identify the appropriate science, possible markets, cross-border issues, and the full range of affected interests and governments that should participate and facilitate solutions. Voluntary interstate strategies, as well as other partnerships, are important tools.

3.2.2 Science for Facts, Process for Priorities—Separate Subjective Choices from Objective Data Gathering

Environmental science is complex and uncertainties often exist. Competing interests usually point to data supporting their view and ignore or attack conflicting or insufficient information. This results in polarized positions, interferes with reconciling problems, and may leave stakeholders in denial. Public confidence declines and the stridency of debate increases. A much better approach is to reach agreement on the underlying facts as well as the range of uncertainty surrounding the issue before framing the choices. A public, balanced, and inclusive collaborative process should be used, with respected scientists in a variety of fields and peer-reviewed science from multiple disciplines. If agreement on scientific facts cannot be reached, decision-makers must evaluate the differing information and make the difficult policy choices.

3.2.3 Collaboration, Not Polarization—Use Collaborative Processes to Break Down Barriers and Find Solutions

Environmental issues tend to be highly polarizing, leading to destructive battles that do not further environmental goals. Goals are best accomplished through balanced, open, and inclusive approaches at the ground level, where stakeholders work together to formulate critical issue statements and develop solutions. Collaborative approaches often result in greater satisfaction with outcomes, broader public support at less cost, and better long-term stakeholder involvement. Public and private interests may need to provide resources to ensure these local collaborative processes are transparent, have broad participation, and are supported with good technical information.

3.2.4 Recognition of Benefits and Costs—Make Sure All Decisions Affecting Infrastructure, Development, and Environment are Fully Informed

Environmental decisions should be guided by an assessment of the true costs and true benefits of different options, including life-cycle costs. These assessments can illustrate advantages of various methods of achieving common public goals. However, not all benefits and costs are measured in dollars. Noneconomic factors, such as equity within and across generations, should also be fully considered in every assessment. Options should consider all social, legal, economic, and political factors, while ensuring that neither quantitative nor qualitative factors dominate.

3.2.5 Markets Before Mandates—Pursue Economic Incentives Whenever Appropriate

Although most individuals, businesses, and institutions want to achieve desired environmental outcomes at the lowest cost to society, many environmental programs include mandates enforced through the threat of legal action, fines, and other penalties. Although the result may be environmental protection, this approach is not always efficient or cost effective. By contrast, market-based approaches and economic incentives often result in greater efficiency at less cost, and may lead to less resistance and more rapid compliance. These win–win approaches reward environmental performance, promote economic health, encourage innovation, and increase trust among all stakeholders.

3.2.6 National Standards, Neighborhood Solutions—Assign Responsibilities at the Right Level

In many instances, it is entirely appropriate for the federal government to establish national environmental standards. However, states, tribes, and local governments should have the flexibility to develop their own plans to achieve the national standards, and to provide accountability. Plans that consider local ecological, economic, social, and political factors often enjoy more public support and involvement and therefore can reach national standards more efficiently and effectively. Governments at all levels should reward innovation and support empowerment for any entity that can meet or exceed standards and goals through local or regional plans.

3.2.7 Reward Results, Not Programs—Move to a Performance-based, Instead of Process-based, System

A clean and safe environment will best be achieved when government actions are focused on outcomes, not programs and processes, and when innovative approaches to achieving desired outcomes are rewarded. Federal, state, and local policies should encourage innovative thinking in the development of strategies to achieve desired outcomes. Solving problems, rather than just complying with programs, should be rewarded.

3.2.8 Change a Heart, Change a Nation—Environmental Education and Understanding are Crucial

Governments at all levels can develop policies, programs, and procedures to protect the environment, but the success of these policies ultimately depends on the daily choices of citizens. Beginning with the nation's youth, people need to understand their relationship with the environment. They need to understand the importance of sustaining and enhancing the natural world for themselves and future generations. If we are able to make environmental progress, it will be because citizens understand that a healthy environment is critical to the social and economic health of the nation. Government has a role to play in educating people about stewardship of natural resources.

3.3 Panarchy Theory

Panarchy Theory was developed as part of the Resilience Project—a collaboration of an international group of ecologists, economists, social scientists, and mathematicians funded by the MacArthur Foundation—to explain how social, economic, and environmental systems are linked and respond to change. Three criteria were established for this integrative theory:

1. It must be dynamic and prescriptive, not static and descriptive. Monitoring the past and the present is static when it is disconnected from policies and actions to evaluate different futures.
2. It must embrace uncertainty and unpredictability. Expect surprises; surprises and structural change are inevitable in socioeconomic and ecological systems.
3. It must be as simple as possible, but no simpler. Even though all these systems are dynamic and complex, there might be a handful of processes that control system dynamics (Holling, 2001).

In part, the complexity of natural, social, economic, and institutional systems has overwhelmed attempts to develop an integrative theory of sustainability in the past. However, complex systems theory and thinking is providing a bridge for integrating ecological, social, and economic sciences, and understanding nonlinear relationships and interactions, and discontinuities and uncertainty from multiple, synergistic stressors (Berkes & Folke, 1998; Gunderson & Holling, 2002; Kasperson et al., 1995; McIntosh et al., 2000).

Complex systems theory holds that the complexity of living systems of people and nature emerges from a small number of controlling processes, rather than the large number of interacting factors (Holling, 2001). Further, complex systems theory states that complex systems are self-organized, with a small set of critical processes creating and maintaining this self-organization. Self-organization is when the macroscopic system properties and patterns that emerge from the interactions among system components feedback to influence the subsequent development of those interactions (Folke et al., 2002). Self-organization and its inherent properties can create systems that are far from equilibrium. These complex, adaptive systems can be characterized by multiple possible outcomes from management actions (Levin, 1999). The term coined by the Resilience Alliance for the evolving nature of complex, adaptive systems was Panarchy (Gunderson & Holling, 2002).

All systems—ecological, social, economic, institutional—are exposed to a changing environment (e.g., climate, societal norms, markets, etc.). Historically, these systems were assumed to respond gradually to gradual change, so linear approximations of these changes were considered reasonable. However, sudden, drastic changes can occur in systems responses that alter system state. Extreme environmental events, such as hurricanes or large-scale forest fires, can result in these sudden drastic changes; however, recent studies also indicate that the slow loss of resilience in systems (e.g., rangelands, coral reefs, forest, lakes, and oceans) can also trigger drastic changes in system state (Scheffer et al., 2001).

Three properties are important in shaping the adaptive cycle and the future state of systems:

1. inherent potential for change (i.e., total capacity to adapt to change),
2. internal connectedness and controllability, and
3. resilience (i.e., adaptive capacity of the system to withstand change) (Holling, 2001).

These three properties are shown on the axes in Figure 3-1. There are four phases that are associated with adaptive cycles—exploitation (r), conservation (K), release (Ω), and reorganization (α). The r and K phases are similar to the traditional concepts of ecological succession where exploitation is the rapid colonization of recently disturbed areas and conservation represents niche diversity and slow accumulation and storage of energy and mass (Holling & Gunderson, 2002). In economics, this might be equivalent to an entrepreneurial market in the r phase and an established business/service market for the K phase. Two other phases are now also recognized in ecology—the Ω or "creative destruction" (Schumpeter, 1950) phase and the α phase. During the Ω phase in an ecological system, the accumulated biomass and nutrients, which are tightly held through interconnections among system components, becomes increasingly fragile or vulnerable and can be suddenly released by agents such as drought, forest fires, insect infestations, or intense grazing pressure (Holling & Gunderson, 2002). During the α phase in an ecological system, processes minimize the loss of nutrients and other essential constituents so that they become available for the exploitation phase. During reorganization, there are transient appearances and expansions of species that form new associations and begin to capture opportunities for exploitation (Holling & Gunderson, 2002).

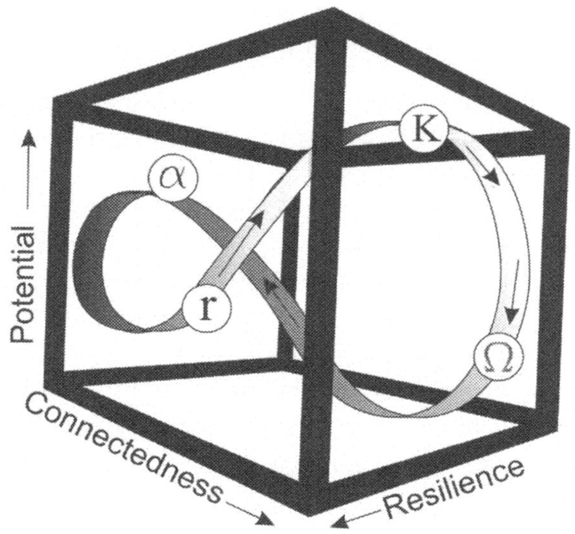

Figure 3-1. Adaptive Cycle of Growth, Accumulation, Restructuring, and Renewal (after Gunderson & Holling, 2002). [Reprinted with permission from Gunderson & Holling (2002). Copyright 2002, Island Press, Washington, D.C.]

As shown in Figure 3-1, potential for change (potential) increases incrementally in conjunction with increased efficiency and connectedness, in addition to increased rigidity, in the r-K phase. As potential increases, slow changes gradually expose increasing vulnerability to such external (environmental) factors as fire, insect outbreak, competitors, or opposition groups (human-ecological systems are linked). These external factors can trigger the release of accumulated potential (i.e., creative destruction) in the K-Ω phase (Holling, 2001). As the system moves to a phase of high potential and high uncertainty, pulses or surges of innovation occur and novel combinations form because connectedness, or control, is low. This is the Ω-α phase. Finally, these innovations are tested, with some failing and some surviving and forming the nucleus for succeeding growth in the α-r phase. Each of these phases is described in detail in Gunderson and Holling (2002).

Hierarchy is used here in the sense of Simon (1974) where semi-autonomous levels in the hierarchy are formed through the interactions of a set of variables and processes with similar speeds and geometric/spatial attributes. In addition, each level of the hierarchy contributes a small quantity of mass or energy or information to the next larger and slower level above (Figure 3-2). Although there are constraints from the level above, hierarchy does not imply authoritative control from above (Holling, 2001). Panarchy provides a way of illustrating the coupling of variables and processes at different scales, such as the movement of water from the global to continental to regional to basin to watershed to catchment scales.

Adaptive management is an integral part of Panarchy. In Panarchy, policies are treated as hypotheses and management practices are treated as experiments. Monitoring provides the information needed to test these hypotheses and refine or develop new experiments to sustain the system in its desired state while accommodating uncertainty and surprises.

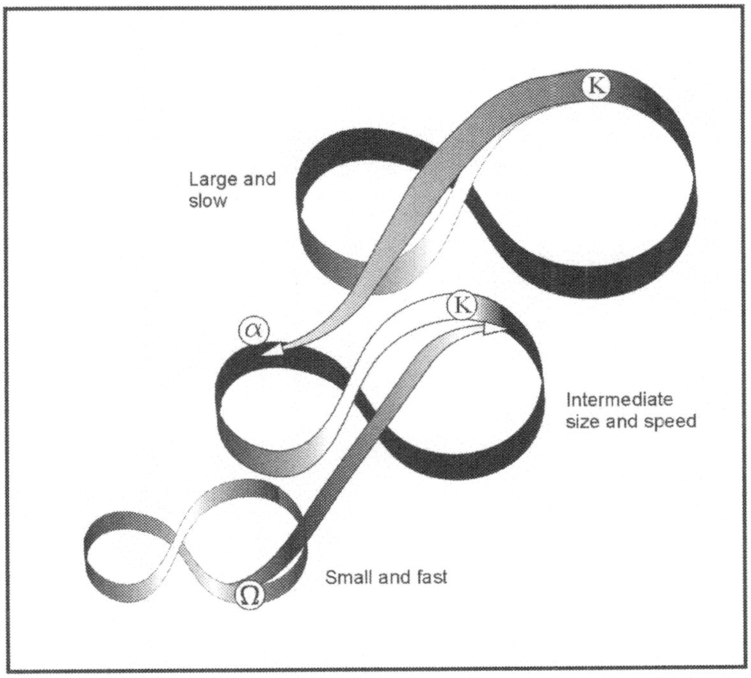

Figure 3-2. Panarchy Hierarchy, Where Larger, Slower Cycles and Processes Constrain Smaller, Faster Cycles, but Also Contribute to Responses. [Reprinted with permission from Gunderson & Holling (2002). Copyright 2002, Island Press, Washington, D.C.]

3.4 Summary

The principles and theories discussed in this chapter provide valuable guidance and insight into how to integrate social, economic, and ecological considerations for SWRM.

CHAPTER 4.0

SUSTAINABLE WATER RESOURCES MANAGEMENT FRAMEWORK

4.1 Introduction

Sustainable water resources result from using a total systems approach for monitoring, assessing, and adaptively managing water through time. Using a total systems approach ensures that humans are considered part of, not apart from, the environment. A framework for SWRM that uses a systems approach thus integrates environmental, economic, and social considerations through institutional and legal/regulatory constructs to provide sustainable water (Figure 4-1). The greatest risk for sustainable water resources is to assume that environmental, economic, and social issues can be resolved independently. It is when the management of environmental, economic, and social sectors not only begin to overlap, but truly intersect and interact, that the probability of sustainable water resources increases. Complete intersection and interaction of these three management sectors would ensure water is a sustainable resource.

It should be noted that moving toward sustainable water resources moves the United States closer to sustainable development and a sustainable society. Sustainable benefits increase dramatically as the intersection of the elements in the Venn diagram (Figure 4-1) increases and nears unity among the three sectors.

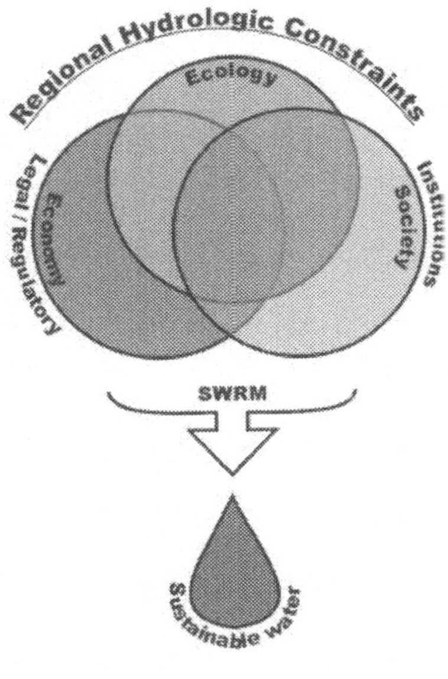

Figure 4-1. Framework for SWRM, Illustrating the Necessity of Integrating Ecology, Economy, and Society within Hydrologic, Institutional, and Legal/Regulatory Constraints and Constructs.

4.2 Scales of Concern

The hydrologic cycle is dynamic, affected by factors operating at a range of spatial and temporal scales (i.e., local to global; hours to centuries). Therefore, it is critical to consider time and space at multiple scales in developing a framework for managing water as a sustainable resource. Information is summarized and synthesized by focusing on the interactions occurring at various scales, from global to local scales:

- Global scale – Global climate patterns and interactions constrain water amount and movement at other scales, both spatially and over time.
- Regional basin scale – Factors to consider at this scale include finite water quantity, hydrologic landscape, basin patterns of water use and demand, water distribution, regional hydrology, regional economic and social constraints, and water law.
- Watershed scale – Historical and current land use, land use change (e.g., urban sprawl, forest fragmentation, land conversion, wetland loss), cumulative effects of multiple local scale demands, landscape patterns (e.g., patches, corridors, connectivity), water transfer, ecological goods and services, social welfare, revenue distributions and costs, and similar factors are important in watersheds.
- Local scale – Local considerations include traditional water management concerns such as flood control, water use demands, recreation, fish and wildlife needs, minimum low flow, water charges and costs, revenue structure, local dischargers, ecological goods and services, and conservation practices.

Panarchy theory indicates that larger scales constrain the possible responses at lower or smaller scales (Gunderson & Holling, 2002). At the same time, some processes that are important at local or smaller scales are considered noise in regional responses or have minimal effect at regional scales (Gunderson & Holling, 2002). Developing sustainable water resources will depend, in part, on distinguishing important signals from noise, and constraints from opportunities.

4.3 Sustainable Water Resources Elements

The definition of sustainability in general, and sustainable water resources in particular, is a much-debated topic (European Commission, 2000; Goodland, 1995; Loucks et al., 2000; Lubchenco et al., 1991; Serageldin, 1995; United Nations, 1992). Gleick (1998a), for example, identified seven sustainability criteria for water planning:

1. A basic water requirement will be guaranteed to all humans to maintain human health.
2. A basic water requirement will be guaranteed to restore and maintain the health of ecosystems.
3. Water quality will be maintained to meet certain minimum standards. These standards will vary depending on location and how the water is to be used.
4. Human actions will not impair the long-term renewability of freshwater stocks and flows.
5. Data on water resources availability, use, and quality will be collected and made accessible to all parties.
6. Institutional mechanisms will be set up to prevent and resolve conflicts over water.
7. Water planning and decision-making will be democratic, ensuring representation of all affected parties and fostering direct participation of affected interests.

There is currently a movement to develop sustainability science as a discipline to meet fundamental human needs while preserving the natural life support systems of earth (Kates et al., 2001). A common theme occurring in nearly all discussions of sustainability is echoed in the following Holling quote:

> "Sustainable development and management of global and regional resources is not an ecological problem, nor an economic one, nor a social one. It is a combination of all three. And yet actions to integrate all three typically have short-changed one or more" (Holling, 2000, p. 1).

To move toward sustainable water resources management we must identify the elements of each of these three areas that should be considered as part of any management approach, as well as their interactions. To develop sustainable water resources requires that we not only know the pieces of the puzzle, but also how those pieces fit together to form a complete picture (Figure 4-2).

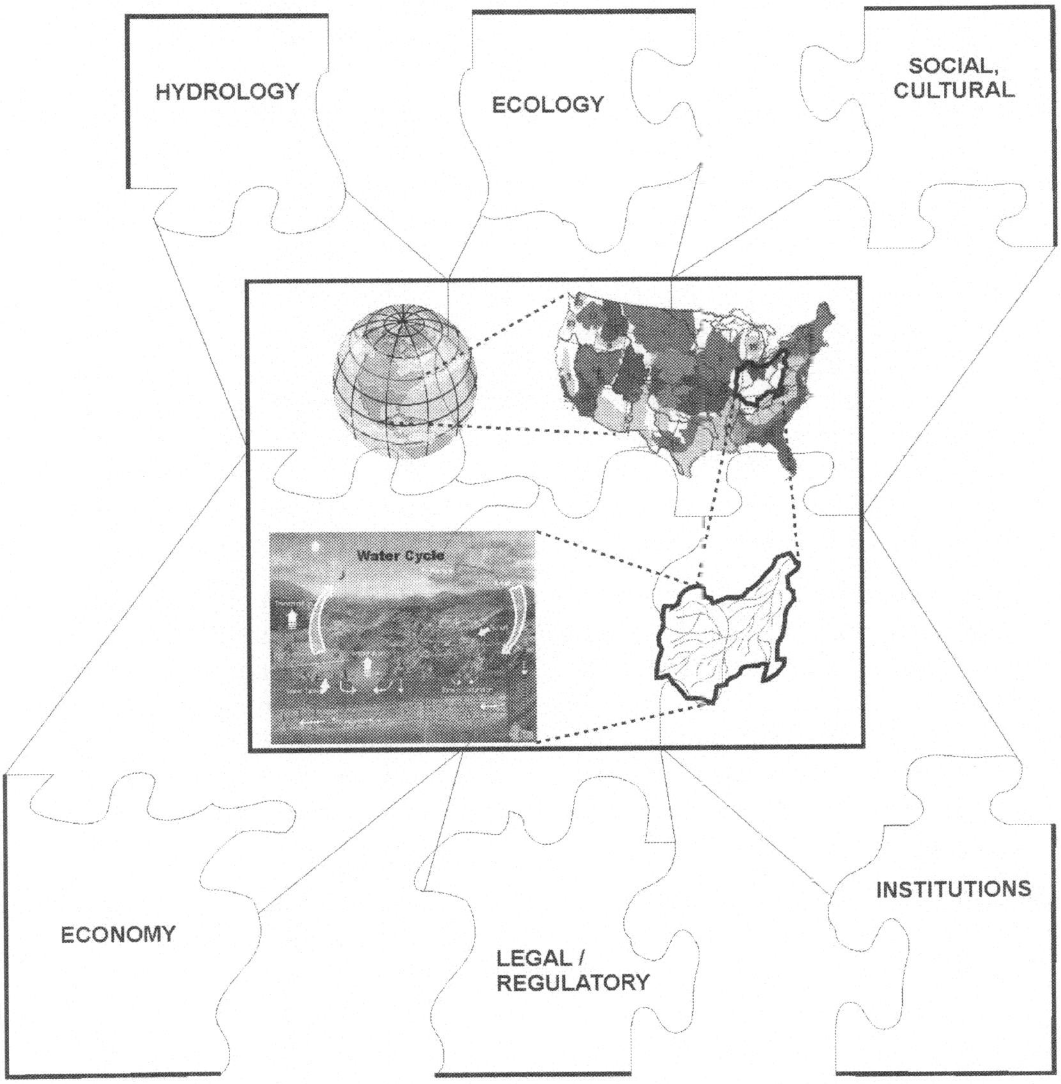

Figure 4-2. Elements of the SWRM Framework and How the Pieces of the Puzzle Might be Organized to Complete the Picture.

Some of the major elements we think must be considered for sustainable water resources are

- Hydrologic cycle – Does the global hydrologic cycle have any relation to local conditions? Are there hydrologic regions that are logical management units? Does distribution of water change over time? How much water is there? How is it distributed geographically? How is water availability likely to change over time? How is this water being used, geographically and by sector? Can water be reused or recycled? How is this likely to change over time?
- Ecological requirements, goods, and services – How much water is needed to support both terrestrial and aquatic ecosystems, by geographic area and ecosystem type? How, and how much, do ecosystems contribute to the hydrologic cycle? What are the potential consequences if ecosystem goods and services are not sustained?
- Social/Cultural – What role does social equity and justice have in sustainable water resources? How can social capital be developed that will contribute to sustainable water resources? Does water contribute to the development of social capital in communities or watersheds? What mental models do various population sectors and segments have about water quantity, quality, and infrastructure?
- Institutions – How can civil society contribute to sustainable water resources? Which institutions have water as an integral part of their vision and mission? How can institutions help move society move toward conservation, reuse, and recycling of water? How can institutions help integrate ecological, social, and economic factors into decision-making?
- Legal and regulatory requirements – What legal, policy, and regulatory constraints and mandates are there for water management in the United States, by geographic area and sector? Do these requirements conflict with one another?
- Economics – What economic gaps, costs, benefits, and opportunities exist in current water resources management? How can ecological systems be valued so their inherent worth is considered in allocating water among various uses? What economic incentives can be used to move toward sustainable water resources? What tools are available for estimating the economic benefits of ecological services in sustaining water?

It is also imperative that new terms for elements in the conventional planning process are not simply substituted under the rubric of "sustainability." Loucks et al. (2000) indicated that some authors of "sustainable management plans" have incorporated many of the criteria and elements that are currently considered in routine water resources planning. Missing from these plans, however, are the contentious parts of the planning process that deal with identifying and distributing costs and benefits; conflicts among local, state, and federal decision criteria; and tradeoffs among conflicting objectives (i.e., the normative evaluation planning components) (Loucks et al., 2000).

The elements of the SWRM framework are associated with the integrative principles used in developing the framework and discussed in detail in the following chapters.

CHAPTER 5.0

ELEMENTS OF SWRM: HYDROLOGY

Dublin Principle 1 – Freshwater is a finite and vulnerable resource
Enlibra Principle 1 – Solutions transcend political boundaries
Panarchy Principle – While there are constraints from above, it is not complete control
Panarchy Principle – The complexity of systems emerges from a small number of controlling processes

Sustainable water resources begins with the hydrologic cycle. Water is a renewable resource, but it is also a finite resource. In addition, as is illustrated in this chapter, readily available freshwater constitutes only a fraction of earth's water. A number of federal agencies collect hydrologic data and conduct hydrologic studies. Hydrology Web (http://etd.pnl.gov:2080/hydroweb.html) is a Web site with a comprehensive list of links related to hydrology and related subjects. Appendix B is a list of some of the links available at Hydrology Web for government sites, data, publications, and bibliographies (accessed 2006).

Catchment and watershed hydrology has received much attention over the years, and is not discussed in depth here. This chapter, instead, focuses on global and regional hydrology and how water movement at these scales affects catchment and watershed hydrology. A global perspective is needed if sustainable water resources are to be realized.

5.1 Global and Regional Constraints

The hydrologic cycle is generally pictured at a watershed scale—precipitation, evaporation, interception, infiltration, runoff, and groundwater recharge (see Figure 4-1, lower panel). These hydrologic processes are important in managing water resources, but these watershed-scale processes are embedded within the global hydrologic cycle. If we are to develop sustainable water resources, then it is critical that we understand this global cycle, its spatial and temporal extent, and its control and constraints on regional and local conditions. Understanding the global hydrologic cycle is critical for developing sustainable water resources because

- it is only at this scale that the hydrologic cycle represents a "closed" system;
- the global interactions among ocean, atmosphere, and land mass control regional and local hydrologic systems; and
- accounting for the long time scales associated with global hydrologic phenomena and processes is critical in planning and managing regional and local water resources.

The global hydrologic cycle follows the largest movement of any substance on earth (Chahine, 1992). Water exerts a major influence on the earth's climate, which itself determines the distribution of water around the globe (Webster, 1994). The hydrologic cycle influences climate through the exchanges of moisture and thermal energy between the atmosphere and earth's surface. These exchanges fundamentally affect the dynamics and thermodynamics of climate systems (Webster, 1994). Water contributes to both heating and cooling of the environment. In fact, 50% of the earth's surface cooling occurs through evaporation (Chahine, 1992).

The interactions among the atmosphere, ocean, and continents control the global distribution of water, resulting in tropical rain forests in some areas and deserts in others. What happens in one area on earth can affect regional and local hydrology in entirely different areas. Eagleson (1986) describes some of these long distance connections, called teleconnections, in his discussion of the influence that destruction of tropical forests in Malaysia can have on regional hydrology in the southwest United States. Understanding these interactions and teleconnections is important for sustainable water resources.

In the past 10 to 15 years, climate change research has increased awareness and study of the global hydrologic cycle. Some of the national and international programs that have been implemented to study climate change and the hydrologic cycle include the International Geosphere-Biosphere Programme, the Global Energy and Water Cycle Experiment, the International Satellite Land Surface Climatological Project, the Tropical Rainfall Measuring Mission, the World Ocean Circulation Experiment, the Hydrologic Atmospheric Pilot Experiment, and the Earth Observing System (EOS), which is a centerpiece of NASA's Mission to Earth and provides information for many of these other programs. These, and similar programs, will increase not only our understanding of climate change, but also of global hydrology, over the next decade.

The hydrologic cycle can be discussed through the interactions of water in the atmosphere (terrestrial and marine), ocean, and land surface (Figure 5-1). The oceans contain about 97% of the world's water; only 3% of the earth's water is freshwater. Almost 99% of this freshwater is stored in ice caps and underground; less than 1% as surface water, soil moisture or atmospheric vapor. Over the open ocean, evaporation exceeds precipitation, while precipitation exceeds evaporation over the continents (Figure 5-1). About 35% of rainfall over land comes from marine evaporation, with the water vapor driven by winds over the continents, and 65% comes from evaporation over the land mass (Chahine, 1992). The excess precipitation over the land surface contributes to runoff, which enters the ocean, completing the water mass budget.

Storage in the atmosphere, ocean, and land reservoirs is one measure of water quantity, but as important, if not more so, is flux or turnover. The atmosphere recycles its entire water content about 33 times per year, for a mean residence time in the global atmosphere of about 10 days. The mean residence time for the oceans is about 3,000 years, but varies considerably by depth in the ocean (Chahine, 1992). The residence time in the upper, mixed layer of the ocean is on the order of days to weeks, with residence times in the deep ocean layers on the order of centuries to millennia. The residence time of surface water on the land masses is on the order of weeks to months, with residence times for ice and snow and groundwater on the order of decades to centuries.

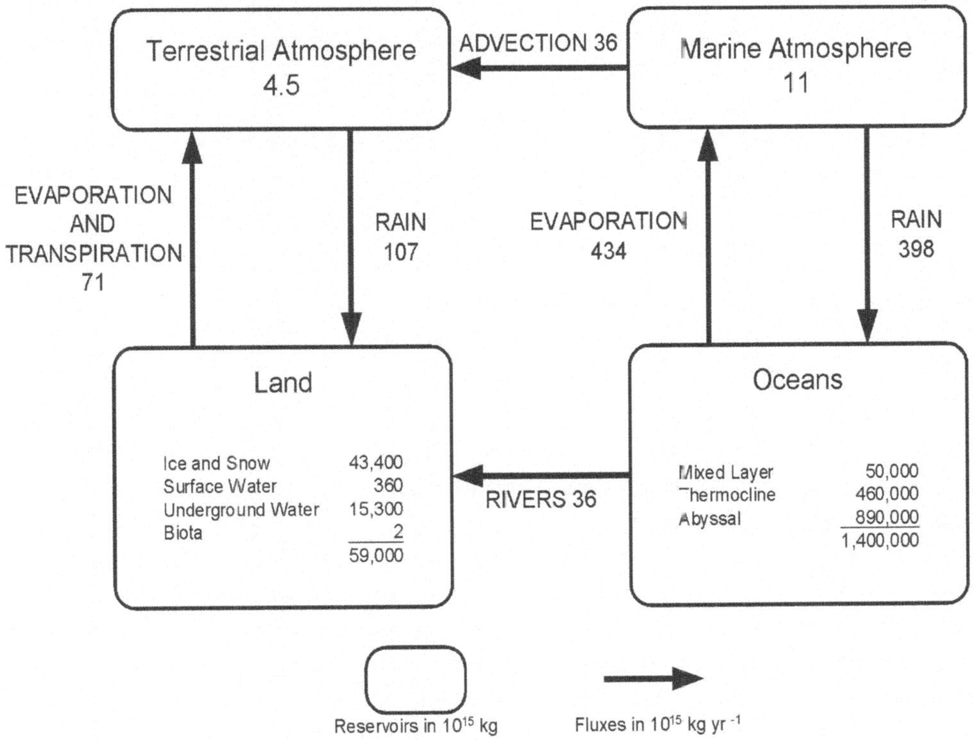

Figure 5-1. Estimates of Reservoirs and Fluxes in the Global Water Cycle. (Reprinted from Webster, 1994. Copyright 2004, American Geophysical Union, Washington, D.C.)

There are two residence time domains, fast and slow (Chahine, 1992). The fast domain consists of the upper ocean, atmosphere, and land surface. The fast domain influences the magnitude and regional patterns of climate change. The slow domain consists of the ice caps, land glaciers, and deep ocean layers. These elements modulate the transient responses of climate change, introducing inertia into the system.

Chahine (1992) discusses the interactions of clouds, radiation, and atmospheric vapor; precipitation and temperature interactions; ocean fluxes; and land surface hydrologic processes that influence the global hydrologic cycle. This discussion is summarized and supplemented below.

5.1.1 Interactions of Clouds, Radiation, and Atmospheric Vapor

Clouds and water vapor interact with solar radiation to influence the thermal structure of the atmosphere and the energy partitioning between the earth's surface and the atmosphere. Water vapor affects solar radiation and global temperatures in the same way as greenhouse gases, and to a greater degree, exerting nearly double the effect of carbon dioxide, methane, and other greenhouse gases on warming of the atmosphere (Manabe & Wetherald, 1967, Raval & Ramanathan, 1989). The annual global mean net effect of clouds is a slight cooling effect on the climate system (Miller & Yates, 2004). Clouds control climate by altering the earth's radiation budget, reflecting incoming short-wave radiation back into space, and cooling the atmosphere.

However, there are significant inconsistencies in cloud effect. Some inconsistencies appear to be based on different sources and sinks of energy. For example, although clouds can reflect short-wave radiation coming from space, cooling the atmosphere, they can also trap long-wave energy being radiated back into space, warming the atmosphere. Cloud height and albedo are also important. Increased cloud height can lead to an increase in surface temperatures at all latitudes (Chahine, 1992). However, increased water vapor content in these clouds can increase their albedo, causing them to reflect more incoming radiation, thereby cooling the surface. Unfortunately, cloud formation and their effects on climate and the hydrologic cycle represent major areas of uncertainty. Cloud formation and effects are the nemesis of many global climate change models. The International Satellite Cloud Climatology Project should provide greater insight into cloud formation and effects on the global hydrologic cycle and climate system.

5.1.2 Precipitation and Temperature Interactions

Annual global precipitation averages from 37–44 in/yr (95–113 cm/yr), but is highly variable spatially, with about two-thirds of the global precipitation occurring between latitudes of 30EN and 30ES, the equatorial zone. This variability influences floods, droughts, and vegetation patterns; and exerts a controlling effect on large-scale atmospheric and oceanic circulation. Spatial variability in the release of latent heat of condensation is the main driver of dynamics among the atmosphere, ocean, and land (Theon & Fugono, 1998). The annual variability in tropical rainfall is strongly correlated with the annual variability in sea surface temperature (Webster, 1994). There is a strong coupling between the atmosphere and the ocean through sea surface temperatures, rainfall, and wind stress. Even though the amplitude of the annual cycle in sea surface temperatures is small, the thermal coupling is strongly magnified by precipitation and the latent heat released by condensation. This latent heat release influences the surface wind field, which induces further changes in sea surface temperature in an interactive feedback loop (Webster, 1994).

It is the interaction of, and periodic changes in, sea surface temperature, rainfall, and wind fields that produce the El Niño/Southern Oscillation (ENSO) and the Pacific Decadal Oscillation (PDO) climatic periodicities. Harrison and Larkin (1998) discuss the interactions among sea surface temperature, rainfall, and wind anomalies associated with ENSO that occur irregularly every few years (i.e., 2–7 years) and contribute to significant flooding and drought in various areas of the world (Figure 5-2). For example, Pizarro and Lall (2002) have observed an increased probability of winter floods in California and Oregon; spring floods in Colorado, southern Idaho, and northeastern Utah; and summer floods in New Mexico and southern Colorado during El Niño years. In contrast, the probability of flooding in Washington, northern Idaho, Montana, and Wyoming is reduced during El Niño periods. Similar observations have been made by Cayan et al. (1999) and Livezey et al. (1997).

Swetnam and Betancourt (1990) correlated fire scars and tree growth chronologies (1700 to 1905) with ENSO in the southwest United States and determined that increased precipitation during El Niño periods was negatively correlated with fire scars (i.e., reduced fire occurrence during El Niño). This dendrochronological record of increased precipitation in the southwest during El Niño periods is similar to that observed by Pizaro and Lall (2002) based on their evaluation of over 130 gaging stations with 70 years of recorded data. Nigam et al. (1999) extracted strong linkages between the PDO, with warmer or positive sea surface temperature anomalies in the North Pacific Ocean, and the severe drought and decreased streamflow in the northeastern United States during the 1960s.

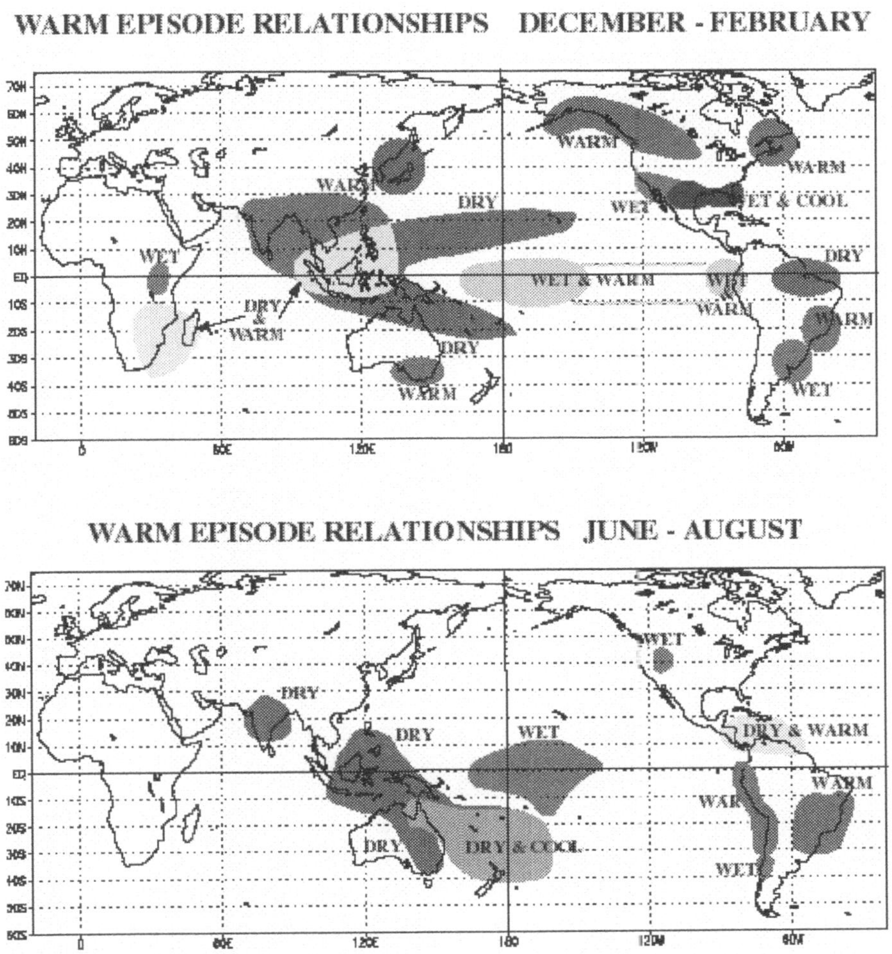

Figure 5-2. Expected Seasonal Effects of El Niño (Warm Episodes) Across the Globe During December – February (top) and Expected Seasonal Effects of La Niña (Cold Episodes) during the Same Time Period (from Climate Diagnostics Center, NOAA, http://www.cpc.ncep.noaa.gov/products/analysis_monitoring/impacts/warm.gif).

Recently, Enfield et al. (2001) investigated the relationship between annual regional U.S. rainfall patterns and the Atlantic Multidecadal Oscillation (AMO). The AMO is a cycle of North Atlantic Ocean sea surface temperature variances with a periodicity of 65–80 years. Enfield et al. (2001) evaluated rainfall variability in the United States for the last century, supplementing the rainfall patterns with streamflows from two representative hydrologic basins. During warm phases of the AMO, the United States received lower than normal precipitation, particularly over the eastern Mississippi River Basin. Other areas, however, such as Florida and parts of the southeast United States, received greater than average precipitation amounts. Enfield et al. (2001) suggested that increasing AMO temperatures over the past several years indicated lower than normal precipitation patterns for much of the United States.

Teleconnections between precipitation, sea surface temperature, and wind affect regional hydrologic patterns. Understanding these teleconnections and incorporating their spatial and temporal variability and periodicity into water resources planning and management is needed for sustainability.

5.1.3 Ocean Fluxes

The oceans contribute to global hydrologic cycles in two ways: (1) heat storage and water vapor evaporation and (2) ocean transport of heat from the subtropics to the poles at a rate equal to the rate heat is transported by the atmosphere (Rasmusson et al., 1993). Oceans move both horizontally and vertically on a global basis because of the influence of winds and density differences.

Stommel (1958) first established the existence of a coherent deep ocean current pattern that linked the ocean basins. Gordon (1986) and Broeker (1991) developed a three-dimensional pattern that completed the circuit and linked deep ocean exchanges with surface currents. Broeker's "conveyor belt" (Figure 5-3) shows how heat is transported from deep layers and mixed with surface layers, influencing sea surface temperatures and subsequent water vapor exchange with the atmosphere (Webster, 1994). In the thermohaline circulation of the northern Atlantic Ocean, for example, warm surface water flows northward, becoming denser through increased salinity and cooling. It then sinks and flows south. Increased precipitation in the equatorial zone reduces the density of the upper mixed layer, which represses mixing and results in increased surface temperatures. Oceanic interactions, such as through sea surface temperatures, are important processes in the global hydrologic cycle.

Unfortunately, the interaction between precipitation and the oceans is poorly understood because of the absence of rainfall measurements over the open ocean. Satellite systems should contribute to improving our understanding of precipitation patterns and ocean interactions as they provide increased and improved remote sensing information.

5.1.4 Land Surface Hydrologic Processes

Evaporation and precipitation over land surfaces are also major components of the hydrologic cycle. The energy needed to convert soil moisture to water vapor and to transport water from the soil through plants comes from energy absorbed from solar radiation. This energy is affected by surface albedo (i.e., the fraction of incident radiation that is reflected by a surface) and other factors. Surface albedo is, in turn, determined by vegetation and soil coverage conditions. Changes in vegetation and soil moisture alter the partition between evaporation and runoff.

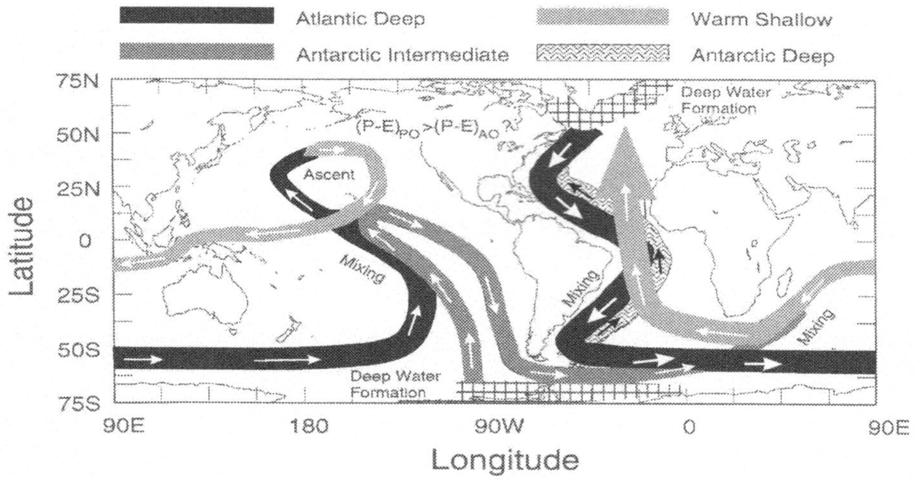

Figure 5-3. Schematic Illustrating the Ocean Current Heat Exchange "Conveyor Belt" Developed by Gordon (1986) and Broeker (1991). (Reprinted from Webster, 1994. Copyright 2004, American Geophysical Union, Washington, D.C.)

Global climate patterns are not related to just atmospheric and oceanic process interactions (e.g., precipitation and sea surface temperatures). Because the heat capacity of the soil is small compared with that of the upper layers of the ocean, the amplitude of the annual cycle of land surface temperatures is much greater than occurs in sea surface temperatures. The difference in temperature between continents and oceans drives the Asian–Australian monsoon system, which is the dominant monsoon circulation on earth (Rasmusson et al., 1993). During the winter phase of the monsoon, there is a flow of dry, cool air from the colder continents over the ocean so precipitation over land is light. During summer, the flow pattern is reversed, with atmospheric moisture from the tropical ocean moving over the warmer land mass, where the upward movement of this moisture-laden, warmer air produces heavy rains during the monsoon season (Theon & Fugono, 1998). Although the monsoon area is primarily equatorial, these circulation patterns can extend as far north as the southwestern United States.

Similar interactions between land surfaces and oceans also affect the East and Gulf U.S. coasts. Multidecadal patterns of summer rainfall in the western Sahel region of West Africa have been linked with the incidence of intense hurricanes making landfall along the United States' East and Gulf coasts and the Caribbean basin (Gray, 1990; Gray et al., 1999, 2000). Gray (1990) analyzed a 50-year record of summer rainfall in the Western Sahel region of West Africa and found a strong correlation with patterns of hurricane intensity in the Atlantic Ocean along the U.S. coast. This pattern had a periodicity of 18–20 years, corresponding with the rainfall periodicity in the western Sahel. More intense hurricanes making landfall along the U.S. coast were correlated with periods of plentiful rainfall over the semiarid region of the Sahel. During drought or low rainfall periods, the intensity of hurricanes was lower. Multidecadal oscillations of precipitation over land surfaces can exert similar influences on the global hydrologic cycle as the influences from precipitation-oceanic interactions.

Land surfaces also modulate the annual distribution of evaporation. For example, in the extratropical regions, seasonal fluctuations in temperature and moisture result in soils storing water received during the winter periods and releasing it to the atmosphere during the summer months. Evaporation releases both water vapor and latent heat to the atmosphere. Current estimates are that evapotranspiration fluxes are about twice runoff fluxes from land surfaces on a global basis (Chahine, 1992). Changes in land use result in changes in evapotranspiration or runoff. These changes can be illustrated by considering the simulated effects of deforestation in the Amazon basin (Table 5-1). Although there are differences in the estimated magnitude of change, all three modeling studies indicate reductions in precipitation, evaporation, and atmospheric moisture convergence. These results suggest that the changes in land cover introduce nonlinear feedbacks that amplify the linear response of the land-atmosphere system to deforestation (Eltahir & Bras, 1996). Reduced evaporation reduces both atmospheric humidity and moist static energy. Moist static energy is important in the dynamics of local convective storms. The reduction in humidity and moist static energy is likely to cause a further reduction in local rainfall, and, because local rainfall triggers a feedback process that modifies atmospheric circulation, result in further reductions in atmospheric moisture and regional rainfall (Eltahir & Bras, 1994).

There are also differences in precipitation recycling ratios among terrestrial systems (Eltahir & Bras, 1996). Precipitation recycling is the ratio of local evaporation to local precipitation over the land surface. It is an approach for estimating the contribution of

Table 5-1. Predictions of the Impact on the Water Cycle Caused by Large-Scale Deforestation in the Amazon Region (from Eltahir & Bras, 1996). (Copyright 1996, American Geophysical Union, Washington, D.C.)

Study	Change, mm/day*		
	Precipitation	Evaporation	Atmospheric Moisture Convergence
Nobre et al. (1991)	-1.8	-1.4	-0.4
Dickinson & Kennedy (1992)	-1.4	-0.7	-0.7
Lean & Rowntree (1993)	-0.8	-0.6	-0.2

*Values are averages over the area of the Amazon basin.

advectively derived water vapor from outside the region to regional precipitation. Estimates of precipitation recycling ratios for the Amazon basin, for example, range from 25–35% (Brubaker et al., 1993; Eltahir & Bras, 1994), which means that from 25–35% of the precipitation that falls over the Amazon basin is derived from evaporation within the basin. For the Mississippi River basin, the estimated precipitation recycling ratio ranges from 10–24% (Benton et al., 1950; Brubaker et al., 1993). These precipitation recycling ratios indicate the importance of the global hydrologic cycle in regional hydrology.

5.1.5 Global Hydrologic Cycle

Sustainable water resources cannot be achieved without considering the hydrologic processes, and feedbacks, operating at a global scale. The atmospheric-ocean-land surface interactions control the global hydrologic cycle, with time scales ranging from 2–7 years for ENSO to decade or century time scales for other oscillations. These longer time scales are not only beyond recent memory, but in some cases, beyond generational memory. Incorporating these oscillations into the water resources planning and managing process, however, is critical if sustainable water resources are to be achieved. With some understanding of the global hydrologic cycle, it might be informative to determine if there are regions that respond similarly to global constraints, and might serve as reasonable management units.

5.2 Regional Hydrologic Management Units

If SWRM strategies are to be developed, a management unit must be defined that integrates surface water, groundwater, and atmospheric vapor. It must be at a large enough scale that it responds to changing global processes, but at a small enough scale that water surpluses and shortages among watersheds can be assessed. It must be large enough to alias the daily noise at the local scale, but small enough to accommodate adaptive management approaches that can respond to global scale changes.

5.2.1 A Preliminary Evaluation

A preliminary approach for identifying regional water resources management units was to look for concordance of surface water regions with the regional groundwater aquifers. We started with the USGS major surface water regions (Seaber et al., 1987) in the United States and overlaid, at the same scale, the USGS major regional aquifers (Sun et al., 1997) (Figure 5-4). In the western and southeastern United States, the major drainage basins generally incorporate the regional aquifers. However, in the Great Plains and Central Region of the United States, the High Plains (Map Code 2) and Central Midwest (Map Code 10) aquifers extend into both the Missouri and Arkansas-White-Red River basins. In the Northeast, the Northeast Glacial Aquifer (Map Code 13) also extends into the New England, Great Lakes, and Mid-Atlantic basins.

Although there was not perfect concordance, the concept of regional hydrology-based management units appears to have merit. USGS has characterized regional hydrologic landscapes (as summarized below).

5.2.2 Regional Hydrologic Landscapes

Regional hydrologic management units must incorporate a complete hydrologic system consisting of surface water-groundwater flow and atmospheric water interactions. Winter (2001) recently defined these units as hydrologic landscapes. Hydrologic landscapes are defined based on land-surface form, geology, and climate (Winter, 2001). Surface water movement is controlled by land-surface slope and permeability; groundwater flow is a function of gravitational gradients and the hydraulic characteristics of the geologic framework; and atmospheric-water exchange is determined primarily by climate (Winter, 2001). The same physical and climate characteristics control the movement of water over the surface and through the subsurface regardless of the geographic location of the landscapes. For example, if a landscape has gentle slopes and low-permeability soils, then surface runoff will be slow and recharge to groundwater will be limited. In contrast, if the soils are permeable in a region of gentle slopes, then surface runoff may be limited but groundwater recharge will be high. Therefore, it should be possible to use these physical and climate characteristics to determine regional hydrologic landscapes.

The USGS initiated the National Water Quality Assessment (NAWQA) program in 1991 to

- assess the status and trends in the quality of the nation's groundwater and surface water resources; and
- link status and trends with an understanding of the natural and human factors that affect water quality (Gilliom et al.,1995).

The original NAWQA implementation plan was based on assessing water quality in 60 study areas throughout the United States. In planning for the second decade of the NAWQA program, one constraint was budget; the current funding outlook for 2001–2011 was estimated to support only about 40 of the original 60 study areas. The wide and representative sample of the hydrologic settings in the nation needed to be retained, but the number of study areas had to be reduced. The challenge, then, was to carefully select study areas so they represented the range of hydrologic conditions in the country.

For the purpose of selecting areas to be studied in the second decade of NAWQA, a map of hydrologic setting regions based on Winter's hydrologic landscapes concept (Winter, 2001) was produced. Winter's hydrologic landscapes concept reflected fundamental hydrologic processes that are expected to affect water quality and other hydrologic characteristics. A hydrologic landscape regions map was generated using readily available spatial data layers, geographical information systems (GIS) tools, and statistical analyses to produce a consistent and reproducible hydrologic characterization of the nation. Wolock et al. (unpublished data) described how these regional hydrologic landscapes were derived. This section draws from Wolock et al. in explaining this derivation. An important caveat is that the hydrologic landscape regions presented below were derived specifically to select large study areas for a national water quality assessment program. The hydrologic validity of these regional landscapes has not been verified, nor would similar groups or clusters necessarily be generated if different variables were considered in the analyses.

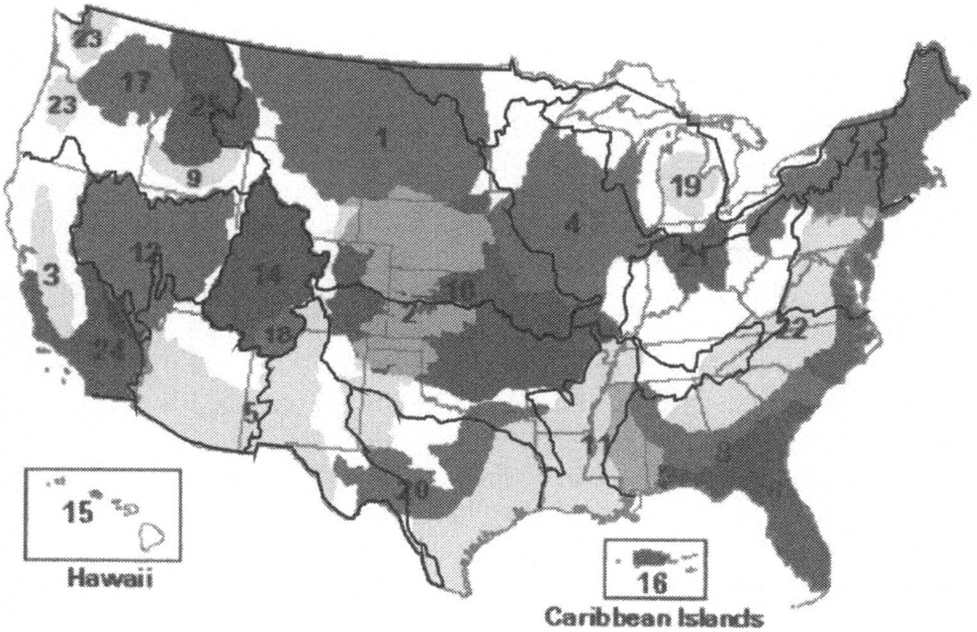

Regional Aquifers

1. Northern Great Plains
2. High Plains
3. Central Valley, California
4. Northern Midwest
5. Southwest Alluvial Basins
6. Floridian
7. Northern Atlantic Coastal Plain
8. Southeastern Coastal Plain
9. Snake River Plain
10. Central Midwest
11. Gulf Coastal Plains
12. Great Basin
13. Northeast Glacial Aquifers
14. Upper Colorado River Basin
15. Oahu, Hawaii
16. Caribbean Islands
17. Columbia Plateau
18. San Juan Basin
19. Michigan Basin
20. Edwards-Trinity
21. Midwestern Basins and Arches
22. Appalachian Valleys and Piedmont
23. Puget-Willamette Lowland
24. Southern California Alluvial Basins
25. Northern Rocky Mountain Intermontane

Major Water Resources Regions (East to West)

New England
Mid-Atlantic
Ohio
Tennessee
South Atlantic-Gulf
Great Lakes
Upper Mississippi
Lower Mississippi
Souris-Red-Rainy
Missouri
Arkansas-White-Red
Texas-Gulf
Rio Grande
Upper Colorado
Lower Colorado
Pacific Northwest
Great Basin
California

Figure 5-4. Major USGS Water Resource Regions (black outlines) Overlaid on USGS Regional Aquifers (numbered shaded areas).

The critical features used to describe hydrologic landscapes in Wolock et al. (unpublished data) were land-surface form, geologic texture, and climate. Land-surface form was used to quantify land-surface slopes and relief; geologic texture provided estimates of surficial and deep subsurface permeability, which control infiltration, the production of overland flow, and groundwater flow rates; and climate characteristics were used to approximate available water for surface and groundwater systems.

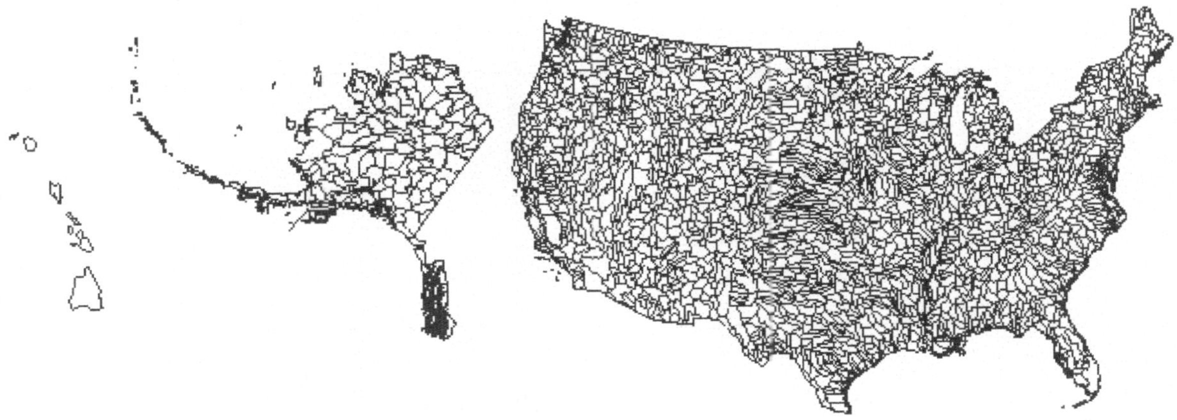

Figure 5-5. Land Surface Form, Geologic Texture, and Climate Metrics were Developed for Each of the 2244 Hydrologic Cataloging Units. (Reprinted from Seaber et al., 1987.)

The variables used to identify hydrologic settings were averaged within each of the 2244 USGS hydrologic cataloging units (Figure 5-5). This degree of spatial averaging was coarse enough to smooth the underlying data but fine enough to separate regions from each other.

5.2.2.1 Hydrologic Landscape Metrics

The metrics used to quantify land-surface form, geologic texture, and climate for the definition of hydrologic settings are described below:

- **Land-surface form** was quantified by topographic characteristics that were computed from a 1-km resolution digital elevation model (Verdin & Greenlee, 1996) analyzed over 10 km by 10 km areas and then averaged within each hydrologic cataloging unit. The land-surface form metrics included relief, and percent flatland in both upland and lowland areas.
- **Geologic texture** was described by two measures of permeability: one for surficial material and one for deep subsurface material. The surficial permeability was estimated as the percent sand based on the U.S. Department of Agriculture's STATSGO database (USDA, 1994). The deep subsurface texture was estimated by assigning permeability classes to the general lithologic groups in the USGS National Atlas of principal aquifer groups (Table 5-2). The deep subsurface permeability categories were consistent with those used to develop regions for the USGS groundwater resources program. The typical permeability is the basis for the relative rank.
- **Climate** was described by mean annual precipitation minus potential evapotranspiration (P-PET) for 1961–1990. PET was estimated from mean monthly temperature and latitude using the Hamon equation (Hamon, 1961). Negative values of P-PET were set to zero.

Table 5-2. Lithologic Groups and Permeability Classes (1 is the Least Permeable and 7 is the Most Permeable).

Lithologic group	Permeability class
Not a principal aquifer	1
Sandstone	2
Semiconsolidated sand	3
Basalt and other volcanic rocks	4
Sandstone and carbonate rocks	5
Unconsolidated sand and gravel	6
Carbonate rock	7

5.2.2.2 Identification of Hydrologic Landscape Regions Using Cluster Analysis

The land-surface form, geologic texture, and climate characteristics were combined in a cluster analysis to define hydrologic landscape regions for the 50 U.S. states. A principal components analysis first was done to reduce the number of variables (originally 13) and derive orthogonal (independent) combinations of the original variables. A clustering program then was used to assign each hydrologic cataloging unit to a cluster. The clustering program used was a hierarchical method that used a minimum variance criterion and the nearest neighbor chain algorithm (Murtagh, 1985).

The number of principal components to retain in the analysis was determined from a plot of principal component numbers versus eigenvalues. The eigenvalues were proportional to the variance in the land-surface form, geologic texture, and climate characteristics explained by the principal component. The six principal components that explained 97% of the variance in the original data were included in the cluster analysis.

The "optimal" number of clusters is the minimum number that retains a low level of within-cluster variance (within-cluster variance always increases as the number of clusters is reduced). A plot of the number of clusters versus the variance criterion indicated an increase of 31% in the within-cluster variance as the number of clusters was decreased from 17 to 16. Seventeen clusters, therefore, were used to develop the regional hydrologic landscapes to minimize the number of clusters and within-cluster variance. These clusters, which comprise the hydrologic landscapes, are shown in Figure 5-6.

5.2.2.3 Description of Hydrologic Landscape Regions

The average land-surface form, geologic texture, and climate characteristic values for the hydrologic landscapes and the geographic locations of the hydrologic landscapes were used to assign descriptive names to each landscape (Figure 5-6).

The map of hydrologic landscapes (Figure 5-6) shows areas within some of the hydrologic setting regions that are spatially contiguous. For example, Hydrologic Landscape 9 (cold arid hills) occurs only in Alaska. The areas of other hydrologic landscapes, in contrast, are located in different parts of the country. The most extreme example of this situation is Hydrologic Landscape 8 (wet mountains), which is in Hawaii, Alaska, and the conterminous United States. Hydrologic Landscape 8 consists of areas that are extremely wet and have very high topographic relief compared to other areas in the country. The large distances between the areas of Hydrologic Landscape 8 indicate that the wettest, steepest areas in the country are located far apart from each other.

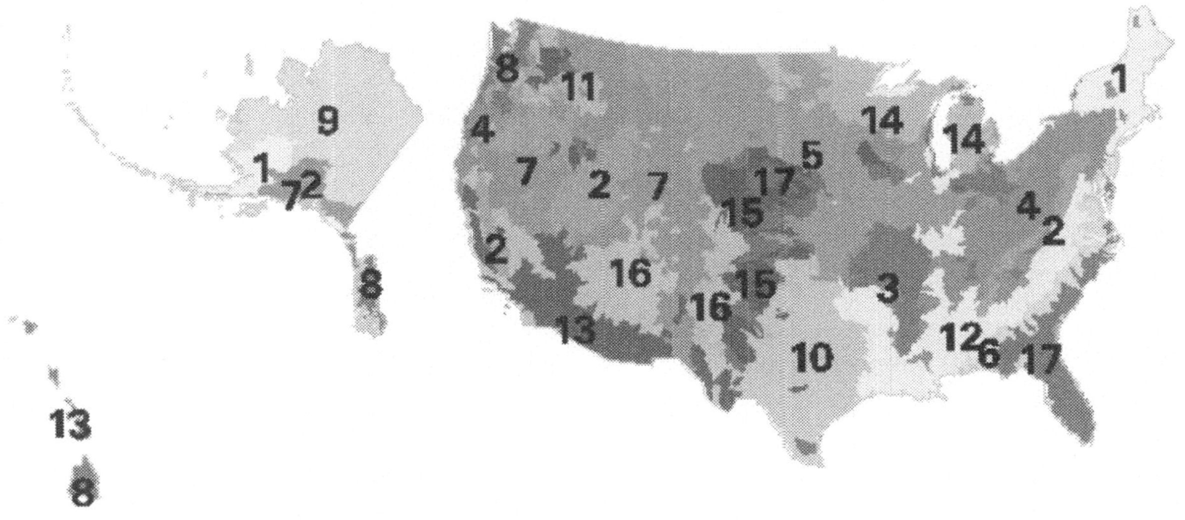

Hydrologic Setting Region Number	Name
1	Wet glaciated hills
2	Wide, steep-walled valleys
3	Permeable subsurface plains
4	Wet plateaus
5	High plains and plateaus
6	Wet inner coastal plain
7	High semiarid mountains
8	Wet mountains
9	Cold, arid hills
10	Low semiarid hills and plains
11	Plateau mountains
12	Wet hills and plains
13	Arid mountains
14	Flat, glaciated hills
15	Permeable subsurface high plains
16	Arid plateaus and hills
17	Permeable plains (lower coastal plain)

Figure 5-6. Regional Hydrologic Landscapes Derived from Cluster Analyses.

The location of areas in some hydrologic landscapes seems counterintuitive. For example, Hydrologic Landscape 2 (wide, steep-walled valleys) occurs in parts of the Appalachian Valley and Ridge, the Central Valley of California, and the Western Basin and Range. These areas are very different in terms of their average P-PET values. Other characteristics used in the cluster analysis (surficial permeability, aquifer permeability, the percent of flat land in lowlands, and the mean annual temperature), however, are similar in these areas. Therefore, the cluster analysis shows that when all the selected factors are considered, these areas are more similar to each other than they are to other areas in the country.

The particular combination of land-surface form, geologic texture, and climate characteristics for a given hydrologic landscape provides insight into the hydrologic behavior of that region. For example, Hydrologic Landscape 5 (high plains and plateaus) covers a large area of the north-central part of the nation. This area is characterized by

- moderate percentage of flatland,
- moderate to large percent of flatland in uplands,
- moderate to small percent of flatland in lowlands,
- moderate relief,
- low deep subsurface permeability,
- low percent sand at the surface, and
- moderate P-PET balance.

Surface water processes should be a prominent component of the hydrologic system in this region because of the positive balance between precipitation and PET and the low permeability (that is, low percent sand) of the surface and of the subsurface. Surface runoff probably would be slow because about two-thirds of the area is relatively flat. Poor infiltration and slow runoff could result in wetlands, mostly in the flat lowlands rather than in the flat uplands. Regional groundwater flow systems would discharge to the lowlands, resulting in a relatively steady supply of water to some areas in the lowlands. Implications for water management in this hydrologic landscape might be that agricultural and urban chemicals would be more likely to be transported into surface-water bodies rather than infiltrate to groundwater. Low-gradient streams could influence natural aquatic habitat, and flooding could be a problem because of the flat landscape and the poor infiltration.

Another example is Hydrologic Landscape 17 (permeable plains), which covers a large region of the southeastern part of the nation and part of Nebraska. This hydrologic landscape has

- the largest percentage of flatland,
- moderate to large percent of flatland in uplands,
- moderate to small percent of flatland in lowlands,
- the lowest relief nationally,
- very high subsurface permeability,
- the highest percent sand at the surface, and
- moderate to high P-PET balance.

Groundwater processes should be a prominent component of the hydrologic system in this landscape because of the high permeability of both the surface and deep subsurface soil/geology. Any generated runoff from the land surface would be slow because about two-thirds of the area is flat and sandy. This geologic framework, along with the high P-PET balance (in the southeast), should result in abundant recharge to regional groundwater flow systems in the uplands and high water tables in the lowlands. Because of this, extensive wetlands would be likely to form in the flat lowlands. Implications for water management in this hydrologic landscape might be that agricultural and urban nonpoint chemicals would be likely to infiltrate and contaminate groundwater. The close interaction of groundwater and surface water in the lowlands could result in contamination of surface water. Flooding probably would not be a problem in the uplands, but could be a serious problem in the lowlands because of the flat landscape and shallow water tables.

5.3 Watershed-based Management Units

Watersheds are a useful surface water management unit because they are easy to delineate and understand, are fairly well accepted, and have been well studied. There are a significant number of factors that have been considered in investigating hillslope or catchment hydrology: hillslope precipitation, channel precipitation, precipitation intensity and duration, evaporation, transpiration, canopy interception loss, interception and canopy storage, stemflow and drip, litter flow, litter interception loss, litter storage, depression storage, detention storage, infiltration, Hortonian overland flow, saturated overland flow, variable contributing areas, return flow, pipe flow, pipe storage, unsaturated throughflow, saturated throughflow, soil-moisture storage, seepage into bedrock, interflow in bedrock, aeration zone storage, deep seepage, baseflow, and groundwater storage (Chorley, 1978). Techniques have been developed for separating "old" water from "new" water using stable isotopes and chemical signatures (Eshlemann, 1988; Obradovic & Sklash, 1986; Rhode, 1981), and a variety of other techniques have been used to improve our understanding of hillslope and catchment hydrology (Tucker & Bras, 1998). In managing for sustainable water resources, however, this level of information is likely to be too detailed. These studies, and the information generated, however, are important in better understanding hillslope and catchment hydrology, and their relationship to the cumulative effects that occur at watershed scales.

Watershed management has been highlighted as a logical resource management strategy to integrate ecological, social, and economic factors and their interactions (U.S. EPA, 1999; NRC, 1999). Watersheds have been considered logical hydrologic management units because they nest within major drainage basins; integrate point and nonpoint source contributions; include different, and in many cases conflicting, socioeconomic sectors; contain terrestrial and aquatic ecosystems; and integrate the interactions of all of these elements. However, watersheds have shortcomings with regard to SWRM.

Watersheds are not closed systems. They are influenced by ecological, social, and economic factors outside their boundaries, and these factors and their interactions differentially influence water resources across spatial and temporal scales. Panarchy and hierarchy theory indicate there is a hierarchy of factors and interactions, with higher levels in the hierarchy constraining the response of lower levels (Gunderson & Holling, 2002; O'Neill et al., 1986). For example, larger-scale phenomena such as climate and orographic patterns constrain spatial and temporal patterns of precipitation within regions, and ultimately, watersheds. In turn, watershed recharge and runoff influence the amount of water available to local communities. Water rights doctrines are founded, in part, on these top-down constraints (see Chapter 11.0). In the precipitation-dominated eastern United States, Riparian Doctrine for water rights typically prevails. In the evaporation-dominated western United States, Prior Appropriation Doctrine for water rights generally prevails.

Another shortcoming is that watershed boundaries do not necessarily conform with groundwater aquifers, which are an integral part of the hydrologic cycle. A water budget concept can be used to determine both storage (reservoirs) of water within a watershed, and fluxes of water through a watershed (Figure 5-7). As important, the budget approach should focus on the dynamic interactions between groundwater and surface water within the watershed. Because the hydrologic cycle is not closed at the watershed scale, uncertainty and variability will be integral parts of the watershed water budget. Characterizing the amplitude, duration, and frequency of the variance is critical if water budget approaches are to be useful in assessing the sustainability of

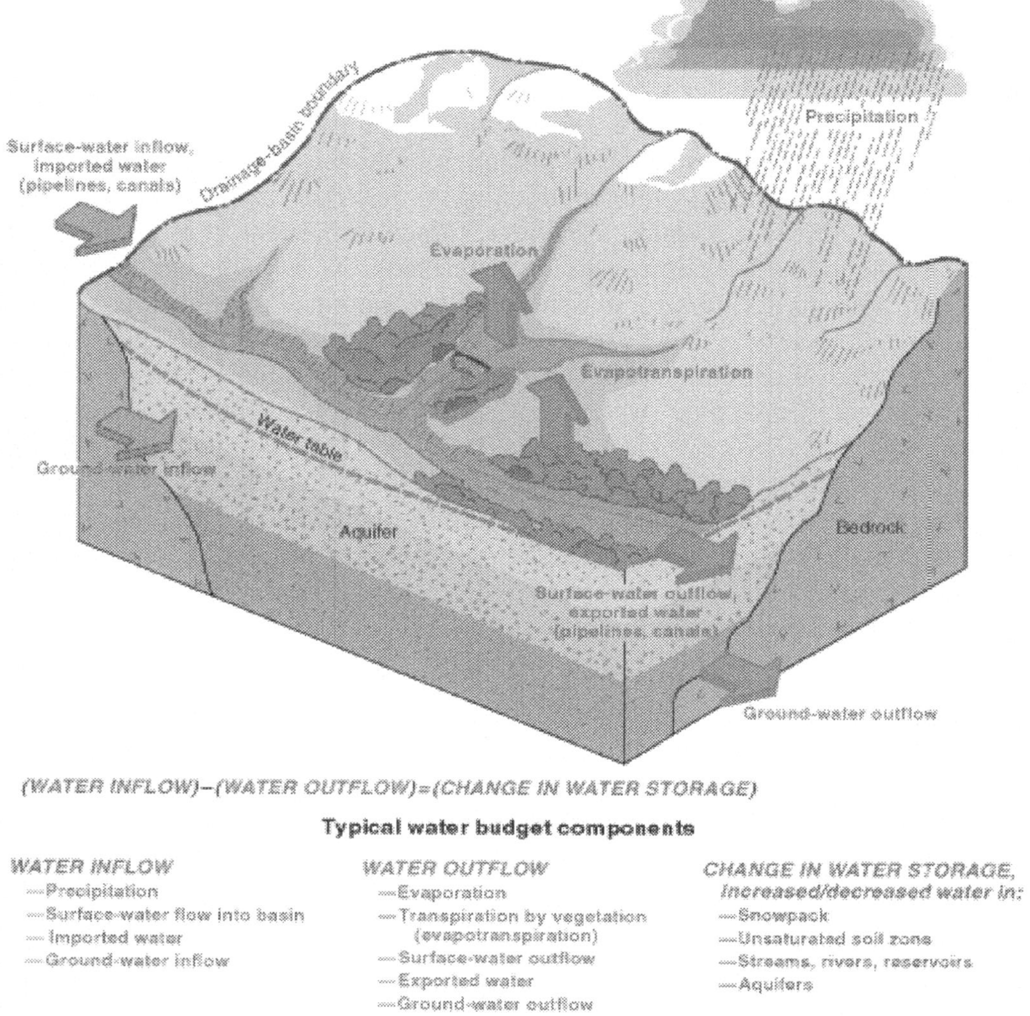

Figure 5-7. Elements of Watershed Water Budgets. (Reprinted from Barlow et al., 2002. Copyright, 2002, USGS, Reston, Va.)

water within watersheds and regions. Understanding the characteristics of these watersheds and the interaction of surface and groundwater will provide a better hydrologic perspective for developing a framework and guidelines for sustainable water resources management. For sustainable water resources, the emphasis must shift toward the hydrologic cycle *in toto*, not the independent management of its components.

CHAPTER 6.0

ELEMENTS OF SWRM: AVAILABLE WATER

Dublin Principle 1 – Freshwater is a finite and vulnerable resource
Enlibra Principle 1 – Solutions transcend political boundaries
Panarchy Principle – While there are constraints from above, it is not complete control
Panarchy Principle – Systems are dynamic, nonlinear, uncertain, and unpredictable

With an understanding of how water cycles through the environment, a logical next question might be how much of this water is available. Our discussion begins at the global scale and moves to continental and regional scales.

6.1 Available Water Quantity

There is a finite, albeit renewable, supply of freshwater in the world, but much of it is either unavailable or below the land surface (Figure 6-1). As indicated previously, about 97% of the world's water is found in the oceans with about 3% as freshwater. There is an average of about 9.2 x 10^{18} gal of freshwater on a global basis (Gleick, 2000). Of this freshwater, 69% is unavailable in glaciers, permanent snow, or ice; about 30% is present as groundwater; and less than 1% is present as surface water, soil moisture, and atmospheric vapor. An average estimate of global surface water, soil moisture, and atmospheric vapor is about 3.4 x 10^{16} gal (Gleick, 2000). These amounts, however, represent standing stocks or reservoirs, which do not consider residence time or turnover.

Understanding the replenishment of the resource and its residence or turnover time can provide a better estimate of how much water is available for use. Global and continental considerations of flow complement the estimates of standing stocks of freshwater and are needed to understand renewable quantities. Table 6-1 summarizes freshwater flows (i.e., precipitation, evaporation, runoff) at two scales—global and continental. In the table, runoff represents the portion of flows available on an annual basis. Much of this, however, is needed to maintain natural systems. Precipitation and runoff sustain both terrestrial and aquatic ecosystems and provide required freshwater input to sustain estuarine ecosystems. The annual renewable freshwater resources in the United States are only a fraction of the North American runoff reported in Table 6-1, about 30% or 1.8 x 10^6 mgd, with only a portion of this available for human use.

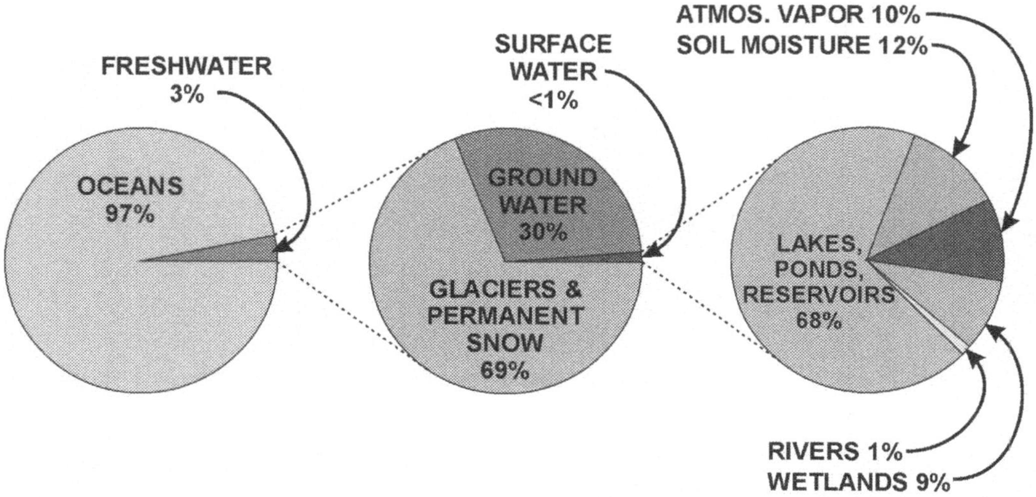

Figure 6-1. Distribution of Global Freshwater and Its Relative Availability for Use (from Gleick, 2000).

Table 6-1. Global and North American Freshwater Flows (from Gleick, 2000).

Flow (mgd)	Global	North America
Precipitation	8.6×10^7	1.8×10^7
Evaporation	5.2×10^7	0.7×10^7
Runoff*	3.4×10^7	0.6×10^7

* Includes groundwater recharge

Freshwater resources in the United States are not uniformly distributed. Figures 6-2 and 6-3 illustrate the distribution of precipitation and runoff in the United States. In general, the eastern United States is precipitation-dominated, with precipitation exceeding evaporation on an annual basis so that there is a natural water surplus (Figure 6-4). The western United States, in general, is evaporation-dominated, with evaporation exceeding precipitation so that there is a natural water deficit (Figure 6-4). Regional differences in runoff can also be seen by comparing runoff from major water resources regions (Table 6-2). The regions with the greatest mean annual runoff are the Columbia-North Pacific, the South Atlantic-Gulf, and the North Atlantic. The largest region, the Missouri River basin, has a mean annual runoff that is less than a quarter of that in the Columbia-North Pacific region. The regions with the least runoff are in the West; Great Basin, Rio Grande, and Lower Colorado.

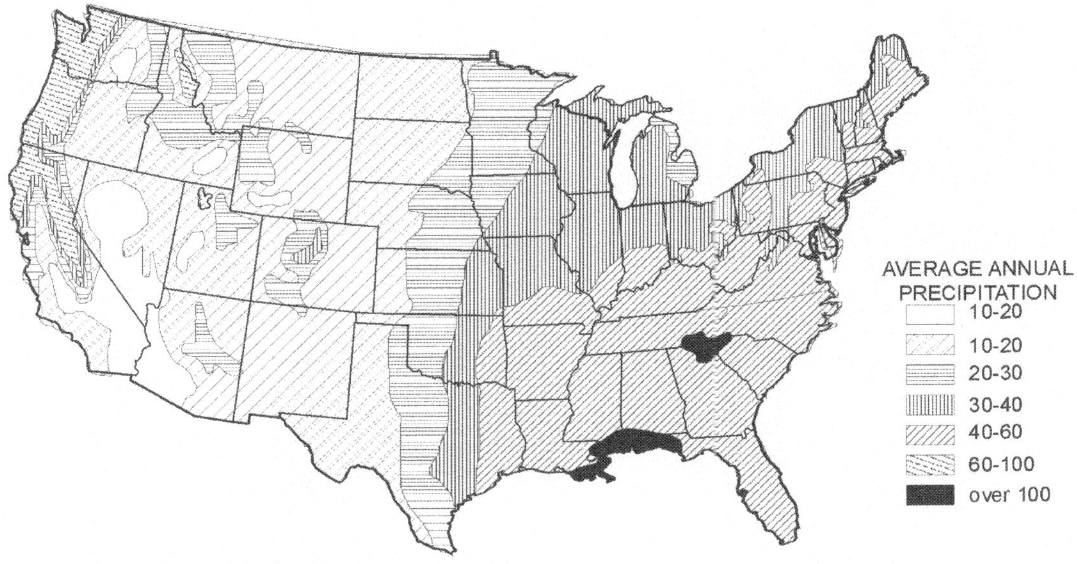

Figure 6-2. Average Annual Precipitation in the United States. (Reprinted from WRC, 1968. Copyright 1968, Government Printing Office, Washington, DC.)

Figure 6-3. Average Annual Runoff in the United States. (Reprinted from WRC, 1968. Copyright 1968, Government Printing Office, Washington, DC.)

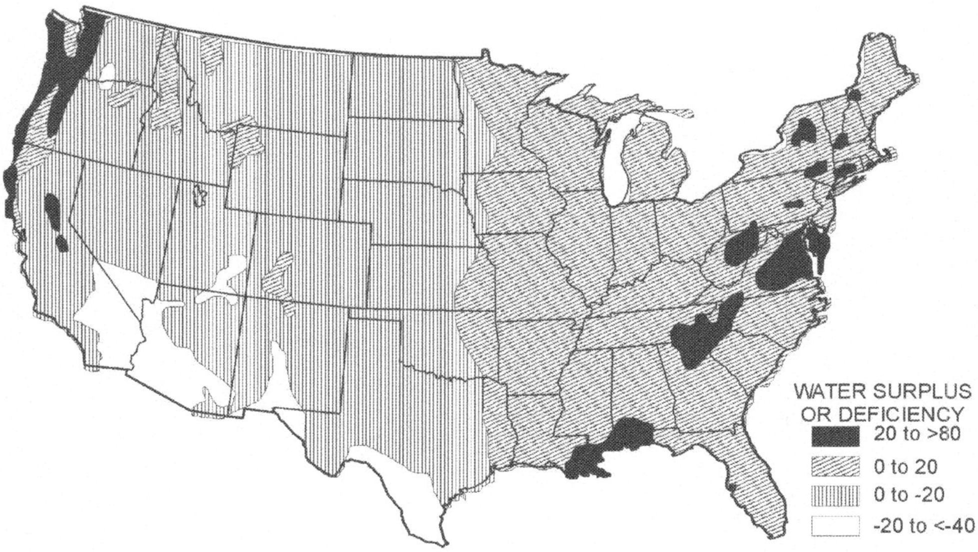

Figure 6-4. Areas of Natural Water Surplus and Natural Water Deficiency. (Reprinted from WRC, 1968. Copyright 1968, Government Printing Office, Washington, DC.)

Table 6-2 Runoff from Major Water Resources Regions. (Reprinted from WRC, 1968. Copyright 1968, Government Printing Office, Washington, D.C.)

Major Water Resources Region	Mean annual runoff (billion gallons per day)
Columbia-North Pacific	210
South Atlantic – Gulf	197
North Atlantic	163
Ohio	125
Arkansas-White-Red	95.8
California	65.1
Upper Mississippi	64.6
Great Lakes	63.2
Missouri	54.1
Lower Mississippi	48.4
Tennessee	41.5
Texas-Gulf	39.1
Upper Colorado	13.4
Souris-Red-Rainy	6.17
Great Basin	5.89
Rio Grande	4.9
Lower Colorado	3.19

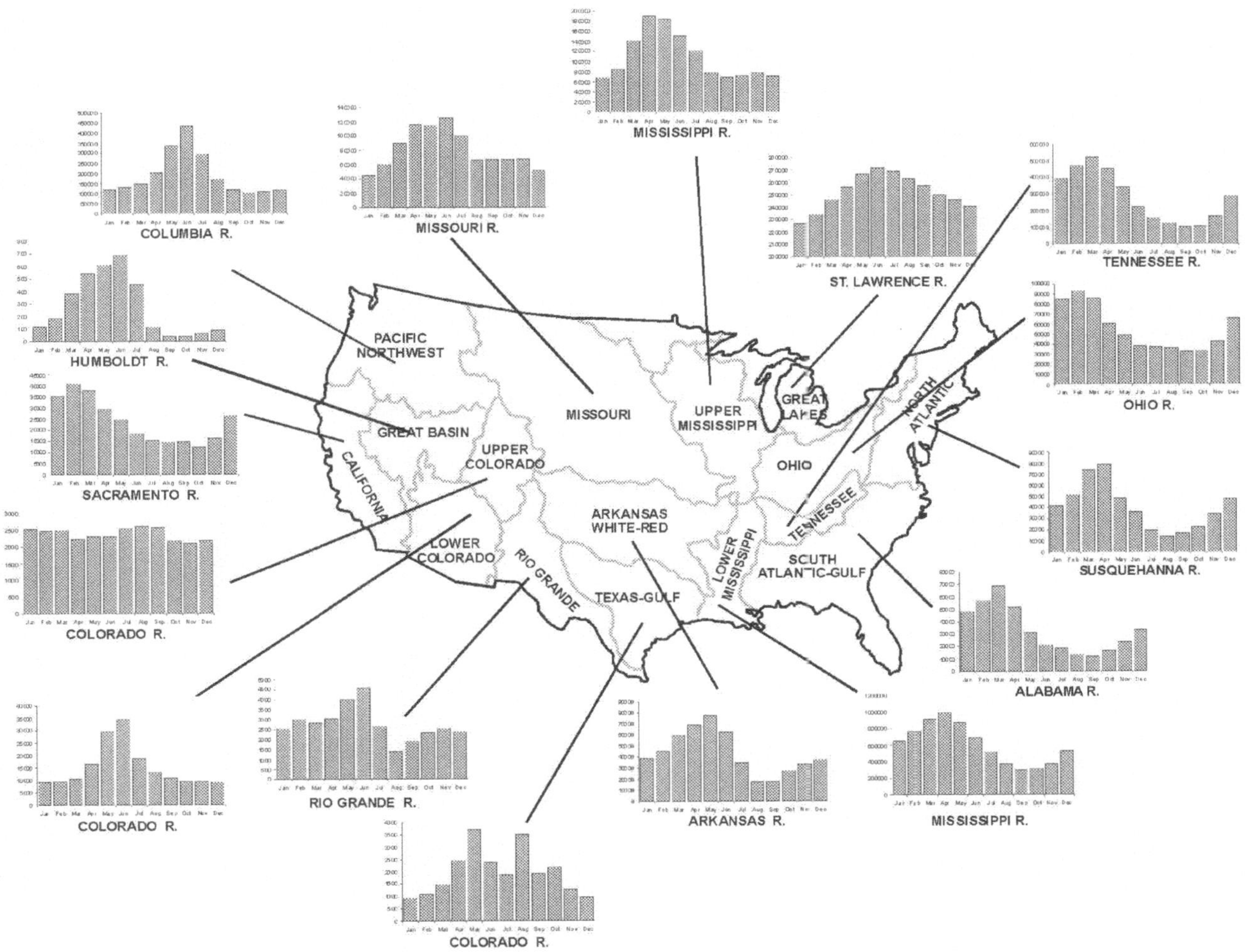

Figure 6-5. Average Monthly Flows (cfs) for Selected Rivers in Water Resources Regions of the United States (January–December).

Figure 6-5 shows average annual hydrographs for some of the rivers of the major water resources regions, which give an idea of the seasonality of water availability across the United States. These hydrographs illustrate regional variation in the peak water availability. While winter and spring are most commonly when the hydrographs peak, there are also a number of rivers where peak flows typically occur during the summer (e.g., Columbia, Lower Colorado, St. Lawrence, Missouri).

6.2 Usable Water Quality

The quality of available water determines its usability. As mentioned in Chapter 1.0, about 30,000 people a day die globally because of water scarcity (quantity) or waterborne disease (quality). In the United States, states evaluate their surface and groundwater resources every 2 years and report to the U.S. EPA, which then compiles this information into a summary of national water quality condition. Although most states have reported the results of their 2004 evaluations to U.S. EPA, the most recent national summary available on the EPA Web site, as of this writing, is for 2000. In 2000, approximately 39% of rivers and streams that had been

Table 6-3. Number of Impaired Waterbodies in United States (http://oaspub.epa.gov/waters/national_rept.control).

State Name	Effective Listing Cycle	Waters on List
Pennsylvania	2002	4298
Oregon	2002	1551
Minnesota	2002	1402
Kansas	2002	1367
Washington	1998	1327
West Virginia	2002	1152
Indiana	2002	1125
Virginia	2002	1084
New Jersey	2002	953
Tennessee	2002	863
Idaho	1998	838
Florida	2002	827
Illinois	2002	803
Massachusetts	2002	775
Mississippi	1998	734
New York	2002	718
South Carolina	2002	710
California	2002	687
North Carolina	2002	630
Wisconsin	2002	604
Montana	2002	527
Georgia	2002	445
Oklahoma	2002	436
Maryland	2002	408
Delaware	2002	381
Louisiana	1998	351
Michigan	2002	351
North Dakota	2002	257
Connecticut	2002	250
Ohio	2002	249
Texas	2000	240
Kentucky	1998	231
New Hampshire	1998	226
New Mexico	2002	206
Missouri	2002	197
Utah	2002	197
Iowa	2002	184
Alabama	2002	183
Vermont	2002	182
South Dakota	2002	176
Maine	2002	165
Rhode Island	2002	155
Nebraska	2002	128
Wyoming	2002	128
Arkansas	2002	103
Nevada	2002	85
Colorado	1998	79
Arizona	2002	52
Alaska	2002	48

assessed (approximately 19% of all streams and rivers were assessed) were evaluated to be impaired, or unsuitable for their designated uses. Approximately 45% of assessed lakes (approximately 43% of all lakes were assessed), 51% of assessed estuaries (approximately 36% of all estuaries were assessed), 14% of assessed ocean shoreline (approximately 6% of the total shoreline was assessed), and 78% of assessed Great Lakes shoreline (approximately 92% of the Great Lakes shoreline was assessed) were classified as impaired. In 2000, states reported that, for the most part, groundwater quality was adequate for supporting many uses, although there were some groundwater quality issues reported. Table 6-3 is a summary of the number of impaired waters in the United States listed by state.

CHAPTER 7.0

ELEMENTS OF SWRM: WATER USE

Dublin Principle 1 – Freshwater is a finite and valuable resource
Enlibra Principle 2 – Science for facts, process for priorities
Panarchy Principle – The complexity of systems emerges from a small number of controlling processes

The freshwater withdrawals for domestic, industrial, and agricultural uses for the United States, Canada, and Mexico are shown in Table 7-1. When expressed on a per capita basis, the United States, Canada, and Mexico use 5.4×10^4, 4.1×10^4, and 1.2×10^4 gal/yr per person, respectively, for domestic purposes. The recommended basic water requirement for domestic use is 4.8×10^3 gal/yr per person of potable water to sustain life in a moderate climate (Gleick, 2000). Although the United States, Canada, and Mexico all exceed this minimum quantity on a national basis, there are significant spatial and temporal differences in water use across these countries.

Table 7-1. Annual Freshwater Withdrawals for the United States, Canada, and Mexico, by Quantity (mgd) and Relative Use (%) (Gleick, 2000).

	Total Withdrawal (mgd)		
Use (% of Total)	**United States (3.4×10^5)**	**Canada (0.3×10^5)**	**Mexico (0.6×10^5)**
Domestic	12%	12%	6%
Industrial	46%	80%	8%
Agriculture	42%	8%	80%

Per capita use is a convenient unit for expressing water use because of the importance of population growth in increased demand on water resources. Recent global climate change research has strengthened the realization that global climate change is occurring, regardless of whether or not it is the result of anthropogenic emission influences (e.g., Vörösmarty & Sahagian, 2000). In a recent article, Vörösmarty and Sahagian (2000) projected that, in general, population growth would play a much greater role than climate change in the scarcity of global water resources over the next 25 years. However, because of the significant spatial differences in renewable and available water within countries and geographic areas, certain areas of the United States (i.e., eastern United States) are projected to have even greater water shortages because of the combined effect of population increase and climate change (Vörösmarty & Sahagian, 2000).

In addition to increasing trends in population growth, there are also increasing trends in urbanization and competition among water users, as well as growing concerns about the quality and integrity of ecological systems (Postel, 2000b). Any one of these trends can seriously challenge existing water resources management (Vörösmarty & Sahagian, 2000). In many areas of the United States, these trends are all occurring simultaneously. Projecting these trends will be important for defining sustainable approaches for water resources management, as will understanding current water uses.

Solley et al. (1998) developed a report on 1995 offstream water use in the United States for the USGS. The report included summaries of water usage by water resources region and use category, and estimates of what portion of the water use was consumptive. This information is discussed below. Hutson et al. (2004) developed a report on 2000 offstream water use in the United States for the USGS. This report summarizes water use data only by state and use category, and does not include estimates of consumptive use. Both reports also document the trends in total water use, by sector, from 1950. The USGS has announced that it will no longer support the national water use summary program, so we do not expect similar reports to be published in the future. This will impair development of water budgets necessary for management in the future.

7.1 National Water Use

Estimated offstream freshwater use for 2000 was 346,000 mgd (Hutson et al., 2004). Offstream use refers to water diverted, or withdrawn, from a surface water or groundwater source and conveyed to the place of use (Solley et al., 1998). Approximately 24% of all freshwater offstream withdrawals were from groundwater, and 76% were from surface water. Table 7-2 summarizes 2000 freshwater offstream use for each of the use categories reported by Hutson et al. (2004). Irrigation and thermoelectric power accounted for the largest percentages of freshwater use in 2000. Surface water supplied the majority of both irrigation and thermoelectric power usage. Consumptive use in 2000 was estimated only for the thermoelectric power use category. Consumptive use means the water was transpired, evaporated or incorporated into crops or products and not returned directly to a surface or groundwater source for possible further use. Approximately 12% (16,709 mgd) of water withdrawn for thermoelectric power use was estimated to be consumed in 2000 (Hutson et al., 2004).

Solley et al. (1998) provided a source-use-disposition network that summarizes freshwater offstream use within the United States during 1995 (Figure 7-1). This network analysis indicated that about 78% of all the freshwater offstream withdrawals in the United States were from surface water sources, with 22% of the withdrawals from groundwater sources. Almost half of all the surface water withdrawals were for thermoelectric cooling use, with almost a third of the surface water and two-thirds of the groundwater withdrawals being used for irrigation. In addition to irrigation being one of the primary users of freshwater, about 60% of the irrigation withdrawals were consumed in the process with only 40% return flow to receiving systems. In contrast, almost 98% of the thermoelectric withdrawals were returned to receiving systems.

About 10% of the surface water and 20% of the groundwater withdrawals were for public water supply. Almost 90% of the public water supply was used for domestic or commercial purposes. Over 80% of the domestic and commercial withdrawals were returned to receiving systems. Industrial-mining withdrawals represented the smallest user of freshwater with about 85% of these withdrawals returned to receiving systems. About 30% of all freshwater offstream withdrawals in 1995 went to consumptive uses.

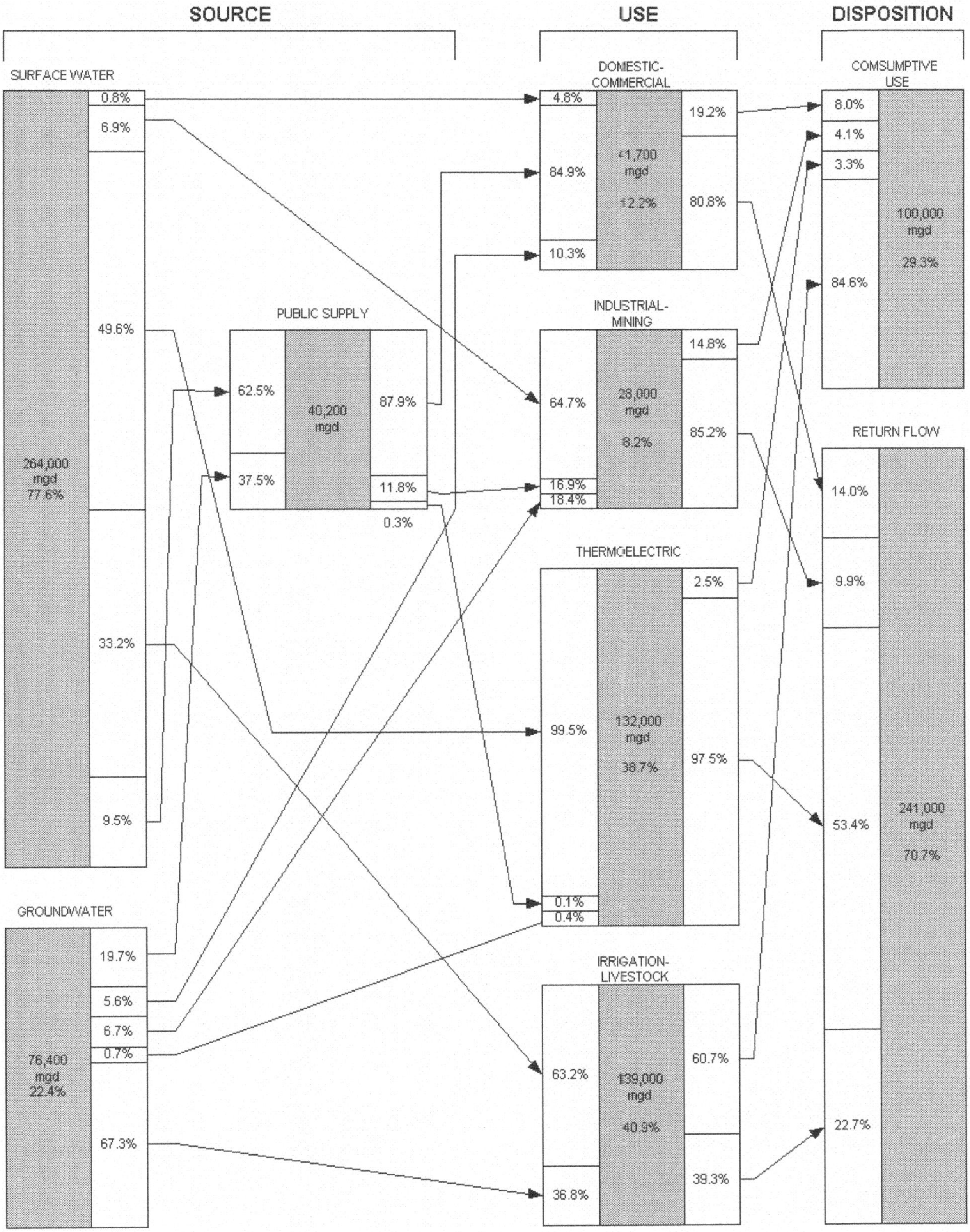

Figure 7-1. Source, Use, and Disposition of Offstream Freshwater in the United States in 1995 (from Solley et al., 1998). While detailed, this figure provides information on the quantity of surface water and groundwater withdrawn, and its fate.

Table 7-2. 2000 Freshwater Offstream Uses for the United States by Category (from Hutson et al., 2004).

Water Use Category	Freshwater Usage (mgd)			Percent of Total		
	Ground-water	Surface Water	Total	Ground-water	Surface Water	Total
Public Supply	16,000	27,300	43,300	19	10	13
Domestic (self-supplied)	3,660	58.9	3,720	4	0	1
Irrigation	56,900	80,000	137,000	68	31	40
Livestock	1,010	747	1,760	1	0	1
Aquaculture*	1,060	2,640	3,700	1	1	1
Industrial (self-supplied)	3,570	14,900	18,500	4	6	5
Mining	767	1,240	2,010	1	0	1
Thermoelectric Power	409	135,000	136,000	0	52	39

*19 states only, these states accounted for 94% of aquaculture withdrawals in 1995.

7.2 Regional Offstream Water Use

Although the national perspective is interesting and useful, it is also instructive to evaluate freshwater offstream withdrawals by major water resource region (Figure 7-2). Withdrawals for 2000 are not summarized by water resources region, so the discussion below is for 1995 freshwater offstream withdrawals (Solley et al., 1998). Some regional patterns emerge from these analyses:

- The regions with the greatest offstream freshwater withdrawals were the California, South Atlantic–Gulf, and Mid-Atlantic regions.
- In the eastern regions, thermoelectric use accounted for the majority of the freshwater withdrawals while in the western regions, irrigation use accounted for the majority of the freshwater withdrawals. Only in the Texas–Gulf region did irrigation use represent less than 68% of the freshwater withdrawal for western regions (31%). In the Upper Colorado region, almost 95% of all freshwater offstream withdrawals were for irrigation. The largest irrigation withdrawals occurred in the California, Pacific Northwest, and Missouri regions (Solley et al., 1998). These regional differences were also reflected in the water consumption estimates.
- In the eastern regions, freshwater consumptive use was about 12% of the offstream freshwater withdrawals (Solley et al., 1998). In the western regions, freshwater consumptive use was about 47% of offstream freshwater withdrawals. This higher consumptive use in the western regions was because of the large irrigation withdrawals (Solley et al., 1998).

7.2.1 Public Supply

Total public supply surface and groundwater withdrawals, including the public uses and losses for 1995, are shown in Figure 7-1. Public uses and losses represent the difference between the quantity of water withdrawn by public suppliers and the quantity of water delivered to all users (Solley et al., 1998). Losses are not estimated for 2000 (Hutson et al., 2004). The 1995 public uses and losses include losses in the distribution systems, filter back-flushing, and public use (water for firefighting, swimming pools, street washing, municipal buildings, and parks).

Some regions export water whereas others import it. For example, the Great Lakes region exports about 37% of its water to other regions, such as the Upper Mississippi region, which imports about 32% of its water (Solley et al., 1998). In general, the public uses and losses are relatively consistent among most regions, averaging about 10–15% of public supply.

Public water supply provides water to both domestic and commercial sectors (Figure 7-1). Supply delivery was not estimated for 2000 (Hutson et al., 2004). The average per capita domestic water supply delivery in 1995 was 101 gal/day (Solley et al., 1998). This was very similar to the per capita delivery rate in 1990 and 1985. Withdrawals for the population served by their own water systems averaged about 80 gal/day per person. Gleick (1996) provided estimates of 1984 per capita domestic water use for Sweden and the Netherlands of 56 and 27 gal/day, respectively. Based on an estimate of 13 gal/day per person as the total basic recommended water requirement per person (Gleick, 1996), United States domestic consumption was about eight times the basic minimum requirement.

The South Atlantic-Gulf and Mid-Atlantic regions had the largest self-supplied withdrawals for domestic use in 1995, representing 23% and 15%, respectively, of the public supply deliveries (Figure 7-2). The South Atlantic-Gulf and Mid-Atlantic regions, along with the California region, had the largest total domestic withdrawals and deliveries (Solley et al., 1998). This was expected because these three regions also had the largest populations.

7.2.2 Irrigation

Irrigation water use includes all water artificially applied to farm and horticultural crops as well as self-supplied water used to irrigate public and private golf courses (Solley et al., 1998). Total acres irrigated are reported for three types of irrigation—sprinkler (e.g., center pivot and traveling gun); micro (trickle and drip); and surface flooding, furrowing, or ditching (Solley et al., 1998; Hutson et al., 2004). Conveyance losses of irrigation water for 1995 are estimated at 19–20% with consumptive use estimated at 61% and return flow estimated at 20% for the United States (Solley et al., 1998). Almost 60% of all 1995 irrigation withdrawals occur in the Missouri, Pacific Northwest, and California regions (Figure 7-2). The 1995 conveyance losses for these three regions are estimated at 32%, 33%, and 6%, respectively (Figure 7-3). Conveyance losses in the California region are significantly lower than other western regions (Solley et al. 1998). However, 1995 consumptive use in the California region represents 80% of the irrigation withdrawals.

In general, conveyance losses, and consumptive uses are difficult to obtain or to estimate. Irrigation withdrawal estimates also vary greatly, from theoretical estimates of plant requirements for water to accurate records of application rates (Solley et al., 1998). Therefore, although there were 2% less withdrawals in 1995 compared to 1990 and 1% more irrigated acres, these differences are probably not statistically significant.

In the Lower Mississippi, Arkansas-White-Red, and Texas-Gulf regions, groundwater withdrawals for irrigation were nearly equal to surface water withdrawals for other uses (Figure 7-4). Irrigation in the Gulf coastal states was a major user of water in 1995. In the Missouri region, groundwater withdrawals were about 50% of the surface water withdrawals for irrigation during 1995. The largest users of groundwater during 1995 were located in the California region (Figure 7-4). Seventy-five percent of the groundwater withdrawals in the California region during 1995 were used for irrigation (Solley et al., 1998).

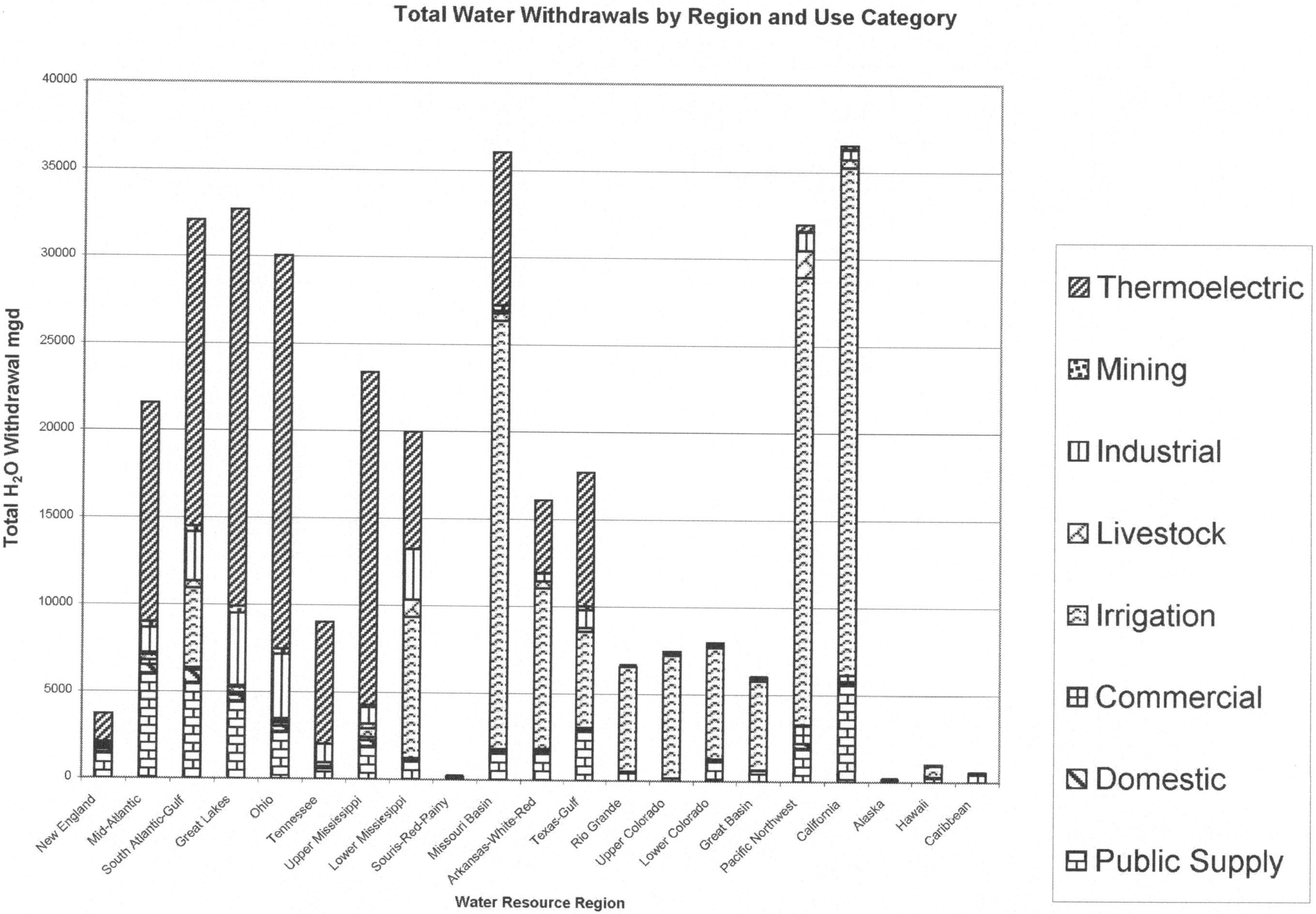

Figure 7-2. Total Offstream Water Withdrawals by Major Water Resources Regions and Use Category During 1995 (after Solley et al., 1998).

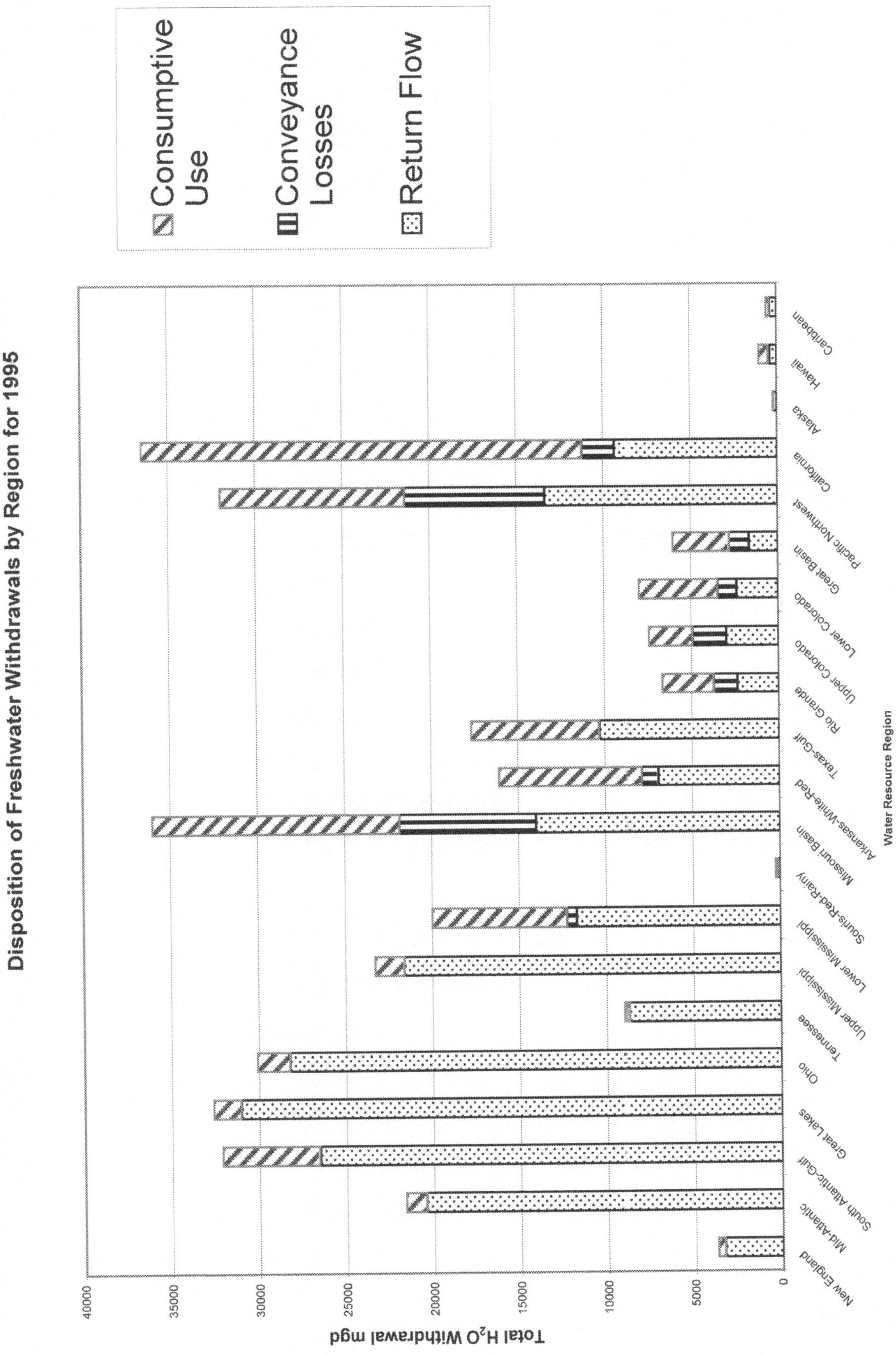

Figure 7-3. Disposition of Offstream Freshwater Withdrawals by Major Water Resources Regions for 1995 (after Solley et al., 1998).

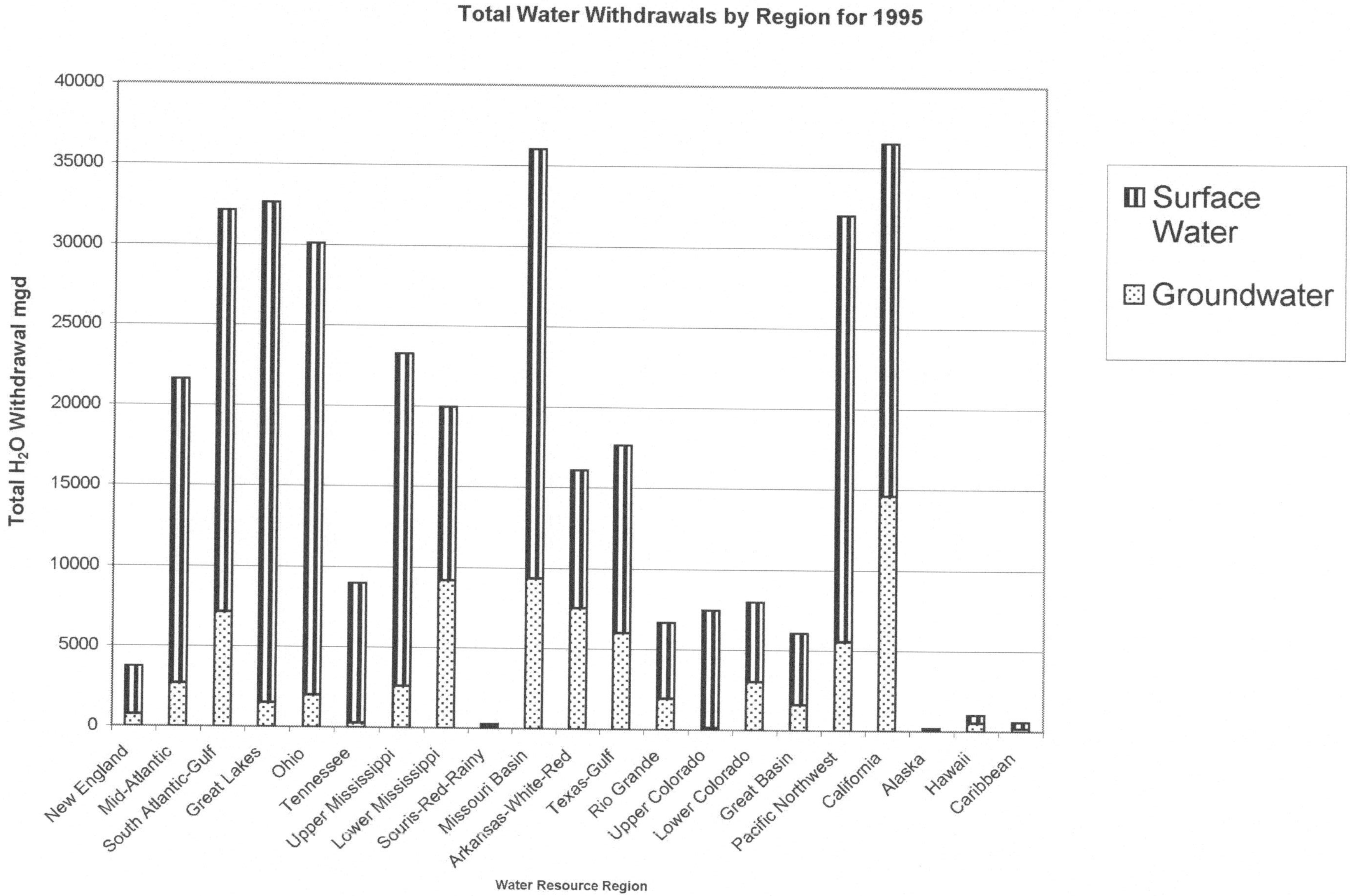

Figure 7-4. Total Offstream Water Withdrawals by Major Water Resources Regions for Surface Water and Groundwater for 1995 (after Solley et al., 1998).

7.2.3 Industrial

Industrial withdrawals for 1995 were primarily from self-supplied sources (82%) compared with public supplies (20%) (Solley et al., 1998). Public supply to industrial water users was not estimated for 2000 (Hutson et al., 2004). Surface water comprised 80% of the 1995 self-supplied industrial freshwater withdrawals, compared to 20% from groundwater sources. Industrial uses consumed about 13% of the freshwater that was withdrawn in 1995. The two largest regions for industrial withdrawals in 1995 were the Ohio and the Great Lakes regions, which together accounted for 36% of all industrial uses (Figure 7-2).

7.2.4 Thermoelectric Power

Withdrawals for thermoelectric power generation, primarily cooling water, accounted for 38% of all the 1995 freshwater withdrawals in the United States. The 1995 once-through cooling water or cooling pond use made up the 98% of the withdrawals accounted in return flows. The 2% consumptive use probably was more reflective of loss in the system than actual consumptive use. In addition, 99% of the 1995 freshwater withdrawals for thermoelectric power were from surface water sources (Solley et al., 1998).

The greatest use of water for thermoelectric power generation in 1995 was in the eastern regions. Eight eastern regions accounted for 75% of the 1995 withdrawals for thermoelectric power use. The Great Lakes, Ohio, and Upper Mississippi regions together accounted for almost 50% of all the 1995 thermoelectric power withdrawals in the United States (Figure 7-2) (Solley et al., 1998).

The majority of offstream freshwater withdrawals for thermoelectric power during 1995 (almost 100%) were made from surface water sources (Figure 7-1). However, groundwater withdrawals in some regions were almost equivalent to the surface water withdrawals during 1995 (Figure 7-4).

7.3 Regional Instream Water Use

Hutson et al. (2004) did not estimate instream water use for the United States. Hydroelectric power generation was the only instream use tracked by the USGS (Solley et al., 1998). The total volume of water used to generate hydroelectric power in 1995 was 2.6 times the average annual runoff in the conterminous United States (Graczyk et al., 1986). Hydropower generation can exceed the average annual runoff because the same volume of water can be used multiple times as it passes through hydroelectric dams on rivers. This estimate was confounded by pump-storage facilities that generate during peak demand periods when electricity prices are higher, and pump water from a lower storage facility back into the upper reservoir during off-load periods when electricity is less expensive (Solley et al., 1998).

Fresh surface water provides essentially all the water for hydroelectric power generation reported for 1995. Almost 40% of all the hydroelectric power generated in the United States was generated in the Pacific Northwest region, primarily in the Columbia River system. The use of water for hydroelectric power generation in the Pacific Northwest region during 1995 was more than triple that use in the Great Lakes region, the region with the next largest water use for hydroelectric power (Solley et al., 1998). Interestingly, although the Pacific Northwest generated the largest quantity of hydroelectric power of any region in the United States during 1995 (140,000 million kWh), the next largest generator of hydroelectric power was the California region (47,000 million kWh) even though the California region was eighth in water use for hydroelectric generation. Multiple generating facilities are located along California region rivers, taking advantage of the large head generated as streams travel from the mountains to the Pacific Ocean.

7.4 Use versus Supply

Groundwater depletion represents the prime example of use exceeding supply, where groundwater is being depleted because the withdrawal exceeds recharge. Unfortunately, regional groundwater depletion and recharge data are not readily available. With the exception of a few regional aquifer studies, most of the groundwater information is collected at the state level (Barlow et al., 2002). Areas of significant groundwater depletion in the United States include the High Plains aquifer from Nebraska to Texas, the Sparta aquifer in the southeast, and the Chicago-Milwaukee area (USGS, 1998). Galloway et al. (1999) described land subsidence studies conducted in areas where groundwater depletion has occurred, but these studies represent site-specific information and did not describe the regional extent of groundwater depletion.

In general, coordinated groundwater monitoring has not been done for aquifers that cross state boundaries (USGS, 1998). The most recent report we were able to find on regional and national groundwater depletion was prepared by the U.S. Council on Environmental Quality (1981).

7.5 Historical Trends

Hutson et al. (2004) summarized trends in total water use from 1950 through 2000 (Figure 7-4). Some of the water use categories for recent data were changed to match those used for older data in the trend analysis. For example, self-supplied domestic withdrawals were combined with livestock and aquaculture withdrawals into a rural-use category (Hutson et al., 2004). Commercial water use was not estimated for 2000, so it was not possible to estimate a comparable value for "Other industrial use" (Hutson et al., 2000).

The public supply and rural categories are the only two that consistently show a increase over the entire period, corresponding with the increase in population (Figure 7-5). Thermoelectric power generation has withdrawn more water than any other category since 1965. Irrigation withdrawals are the second largest use category.

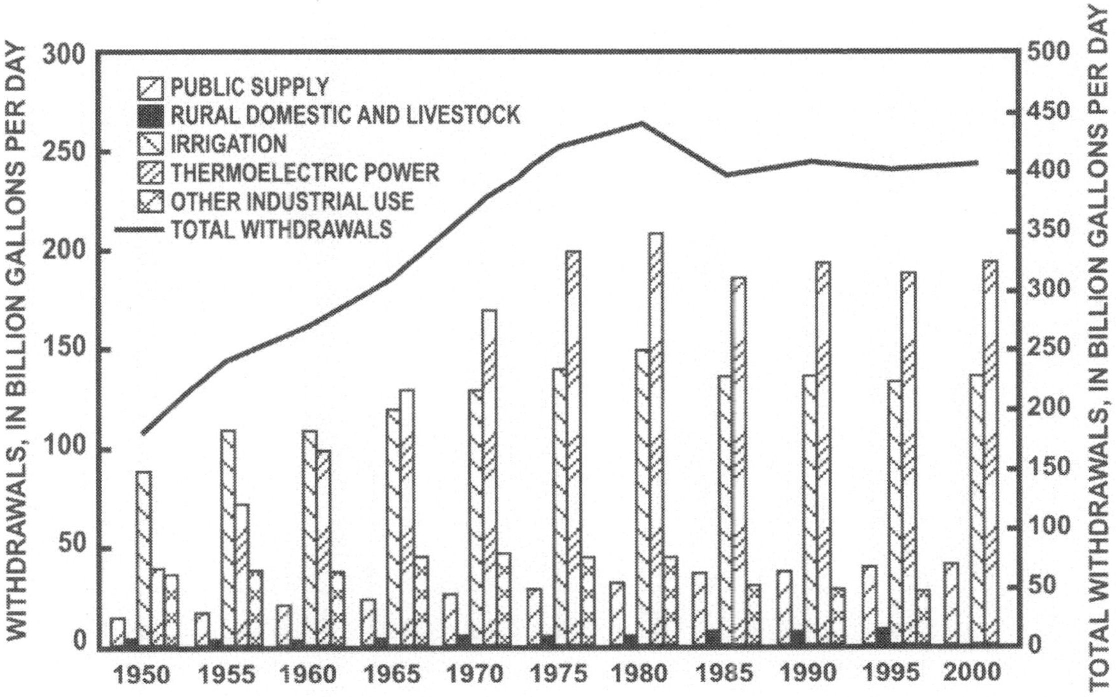

Figure 7-5. Estimates of Water Withdrawals by Use Category Over Time. (Reprinted from Hutson et al., 2004.)

Total , surface, and groundwater withdrawals (Figure 7-6) and irrigation, thermoelectric, and industrial use categories (Figure 7-5) all increased until 1980, then decreased and have remained relatively constant from 1985 to 2000. This decreased water withdrawal and use has resulted from improved technologies that are more water efficient, new industries that use less water, recycling, and conservation programs (Solley et al., 1998; Hutson et al., 2004). Irrigation withdrawals have decreased because of improved irrigation systems and technologies, conservation acreage programs, increased dry land farming in the West, and the sale of water rights to municipalities (Solley et al., 1998).

Although the trends in total water withdrawals for the United States are interesting, information on regional trends in water use would be more useful for local water managers.

7.6 Future Trends

Brown (2000) has projected U.S. freshwater withdrawals for both the United States and for six aggregated water resources regions from 2000 to 2040. These six aggregated water resources regions are

1. Northeast (New England, Mid-Atlantic),
2. North Central (Ohio, Great Lakes, Upper Mississippi, Souris-Red-Rainy),
3. Southeast (Tennessee, South Atlantic-Gulf, Lower Mississippi),
4. Great Plains & Texas-Gulf (Missouri, Arkansas-White-Red, Texas-Gulf),
5. Southwest (Rio Grande, Upper and Lower Colorado, Great Basin), and
6. Pacific Coast (Pacific Northwest, California).

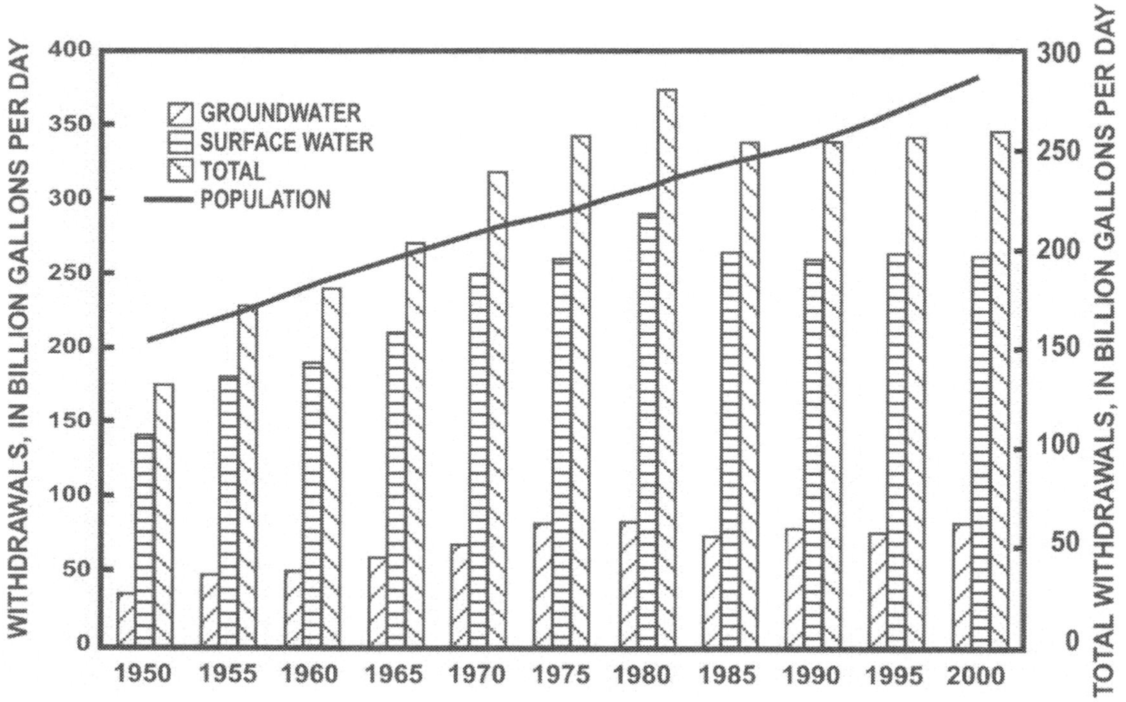

Figure 7-6. Estimates of Water Withdrawals by Source Over Time. (Reprinted from Hutson et al., 2004.)

The projected withdrawals were based on historical trends calculated for these basins using the data provided in Solley et al. (1998). Assumptions were made concerning the factors influencing five water use categories—livestock, domestic and public supply, industrial and commercial, thermoelectric, and irrigation. Use of historical trend data and an understanding of the determinants of water use was in deference to Shabman's admonition of the dangers of forecasting future water use without this information (Shabman, 1990). Population was a factor affecting all categories but irrigation. The 1996 Census Bureau's U.S. projections were used to estimate changes in population, assuming the middle series projections (Census Bureau provides low, middle, and high population projection estimates) for population. Other factors that affected water withdrawals for various categories included per capita withdrawal, income per person, total kilowatt hours per person, irrigated acres, and withdrawal per acre.

All six regions showed overall increases in population (Table 7-3). With the increased population, there were also projected increases in livestock, domestic, and public supply uses. Regions with the greatest population increase were projected to experience the greatest increase in industrial and commercial withdrawals (Table 7-3). The Northeast and North Central regions, with smaller projected population increases, were projected to have decreased industrial and commercial withdrawals.

Thermoelectric withdrawals also were projected to increase in all regions. The Southwest and Pacific Coast regions were projected to have the largest percentage population increase, but only moderate net increases in withdrawals. This is because one of the assumptions was no new hydroelectric projects (i.e., no new dams) would be built. The western regions rely on hydroelectric power generation so increased power demand would have to be met by increasing efficiency, adding turbines at existing facilities and/or building thermoelectric generation

Table 7-3. Change in Population, Irrigated Acreage, and Withdrawal from 1995 to 2040 for Aggregated Regions, Assuming the Census Bureau's Middle Population Projection Series. (Reprinted from Brown, 2000. Copyright 2000, American Geophysical Union, Washington, D.C).

	Northeast	Southeast	North Central	Great Plains and Texas Gulf	Southwest	Pacific Coast
Population, millions	17 (31)*	25 (51)	20 (30)	14 (41)	7 (64)	22 (52)
Irrigated acres, millions	0.1 (12)	3.7 (40)	0.8 (40)	-1.4 (-6)	0.2 (4)	-1.7 (-11)
Withdrawal, billion gallons per day						
Livestock	0.0 (31)	0.8 (51)	0.1 (30)	0.4 (41)	0.1 (64)	1.0 (52)
Domestic and Public	1.9 (31)	3.0 (51)	1.9 (30)	2.1 (41)	1.2 (64)	3.1 (52)
Industrial and Commercial	-0.2 (-5)	1.3 (14)	-0.5 (-4)	0.3 (8)	0.3 (26)	0.6 (12)
Thermoelectric	0.5 (3)	6.2 (20)	1.2 (2)	2.4 (11)	0.1 (33)	0.7 (120)
Irrigation	-0.0 (-5)	5.4 (41)	0.4 (35)	-3.8 (-10)	-0.3 (-1)	-5.7 (-10)
Total	2.2 (9)	16.6 (27)	3.2 (4)	1.3 (2)	1.5 (5)	-0.2 (0)

*Percent change from 1995 to 2040 is indicated in parenthesis.

projects. The need for thermoelectric generation in these regions resulted in the moderate increase in withdrawals

Finally, irrigation was projected to increase only in the Southeast and North Central regions, where irrigation is used to supplement precipitation (Brown, 2000). The number of irrigated acres was projected to decrease in the other regions, resulting in a decrease in irrigation withdrawals compared to 1995 (Brown, 2000). Overall, total water withdrawals were projected to increase in all regions except the Pacific Coast, where withdrawals will remain relatively constant. The greatest increase was projected for the Southeast, following the greatest projected net change in population.

Riebsame et al. (1997) and Case and Alward (1997) discussed changes in land use and demographics, respectively, in the West and its implications for water. Case and Alward (1997) noted that the West has been experiencing dramatic demographic changes over the past 15 years and the population is projected to increase by more than 30% by 2020, if current population trends persist. This growth has occurred primarily in the "urban archipelagos" of Denver, Salt Lake City, Boise, Missoula, Portland, Phoenix, Albuquerque, Dallas, Houston, and Seattle (Case & Alward, 1997). Exurban commercial and residential land development occurred in a ring out from the urban centers. Riebsame et al. (1997) noted this development pattern puts pressure on agricultural land, and the associated water rights, for conversion to urban/residential uses. Some of the projections of changes in irrigation withdrawals might be overestimated as agricultural areas are taken out of production. However, withdrawal projections for livestock and domestic and public water might be underestimated. It is uncertain if decreased irrigation withdrawals can compensate for increased domestic and public supply withdrawals.

CHAPTER 8.0

ELEMENTS OF SWRM: ECOLOGY

Dublin Principle 1 – Water is a finite and vulnerable resource
Dublin Principle 4 – Water has an economic value and should be recognized as an economic good
Enlibra Principle 2 – Science for facts, process for priorities
Enlibra Principle 4 – Recognition of benefits and costs
Enlibra Principle 6 – National standards, neighborhood solutions

Ecology has been defined as the study of the relationships of organisms or groups of organisms to their environment, the science of the interrelationships between living organisms and their environment, or the study of the structure and function of nature (Odum, 1971). As such, ecological studies would include hydrologic components.

8.1 Hubs and Corridors

The hubs and corridors planning framework, often used in transportation, is also being used in ecological conservation. The framework consists of a network of large protected areas (hubs) connected by narrow linear protected areas (corridors), much like cities are connected by roads. This type of network of protected lands can be considered "green infrastructure" (www.greeninfrastructure.net). Well-planned and maintained systems of natural resources are as important to society as well-planned and maintained transportation or water supply distribution systems.

Hubs in the green infrastructure protect those species that require large areas of habitat, and ecological processes that require large spatial units. For example, large areas of contiguous forest can protect and/or maintain bird species that live in interior forest habitats, and large protected groundwater recharge areas can protect groundwater quality.

Corridors facilitate movement of resources between hubs, and further protect the hubs by protecting resources that contribute to the hubs. Corridors provide potential migration routes between hubs for both animal and plant species. This helps to prevent populations from becoming genetically isolated, and to maintain the genetic robustness of species. In the case of a stream flowing between hubs, protecting a corridor along the stream helps protect the water quality of the stream, which contributes to the quality of the habitat in the hub.

The Greenprint Program in Maryland provides an example of how a hubs and corridors conservation network can be designed and implemented. Detailed descriptions of methodology are available at the Maryland Department of Natural Resources Greenprint Program Web site (www.dnr.state.md.us/greenway/greenprint). Maryland used a three-step process in the Greenprint Program: identify, link, and save.

The first step in the Maryland process was to identify hub areas (Figure 8-1). These areas included existing protected areas, such as state parks, federal lands, national parks, etc.; lands that would be beneficial to have set aside as hubs; and lands that would be beneficial to add to existing hubs. Although there are large "hubs" of forest in the Mid-Atlantic, it is also readily apparent from Figure 8-1 that these forests are being fragmented (white pixels). As part of the identification activity, land parcels were assigned a priority based on their importance in meeting the goals of the program.

The second step of the Maryland process was to link the hubs with corridors. Streams and ridge tops were commonly identified for corridors in Maryland's Greenprint Program (Figure 8-1). In a similar program in Pennsylvania, abandoned railroad beds and existing hiking trails were also considered as corridors. A 10-mile-wide forest buffer along streams (dark gray in Figure 8-1) would be an option for identifying corridor areas to protect.

The last step in the Maryland process is to "save" the lands identified for inclusion in the network as hubs or corridors. This might include acquiring land, enrolling land in conservation or set-aside programs, purchasing easements, encouraging land owners to manage lands so they contribute to the goals of the program, or restoring lands to a state that would enable them to achieve the goals of the program.

A hubs and corridors network can be used to preserve landscape processes and habitat functions needed for sustainable water resources, while accommodating other land uses (e.g., farming, timber, towns). A large-scale perspective is needed to assess the functions and processes at work (e.g., groundwater infiltration, recreation, migration routes, farming) and which areas in the region are critical for the continuation of those functions (top graphic in Figure 8-1). With a regional network developed, it would be possible to know the best course of action to pursue with land resources planning to move toward sustainable water resources. Having an integrated plan for an entire region would allow coordination of the activities of all the agencies, organizations, and stakeholders working in the region, regardless of their scale of interest (e.g., continental, regional, state, watershed, county, city, or parcel). The bottom graphic in Figure 8-1 illustrates with greater resolution where a county could purchase land or easements to reconnect the isolated forest patch near the center of the panel with the larger forest hub to the right.

A hubs and corridors network can be developed to meet a range of goals at multiple scales. In Maryland, the hubs and corridors network is developed by the Department of Natural Resources, at a state scale. In Pennsylvania, however, the network is being developed at the county scale. Each county is responsible for developing their own network of hubs and corridors to meet the goals of the state program, as well as goals at the local scale. Both the Maryland and Pennsylvania hubs and corridors networks are being used to protect natural resources (including both plant and animal species), protect and improve air and water quality, offer increased recreational opportunities, protect cultural resources, improve quality of life, and improve economic climate. A discussion of expected benefits from a "green infrastructure" is available at www.greeninfrastucture.net.

Although the above example of hubs and corridors is for the Mid-Atlantic, the concept and approach is also applicable and appropriate for the western United States. Protecting riparian and recharge areas in arid regions is critical for sustainable water resources.

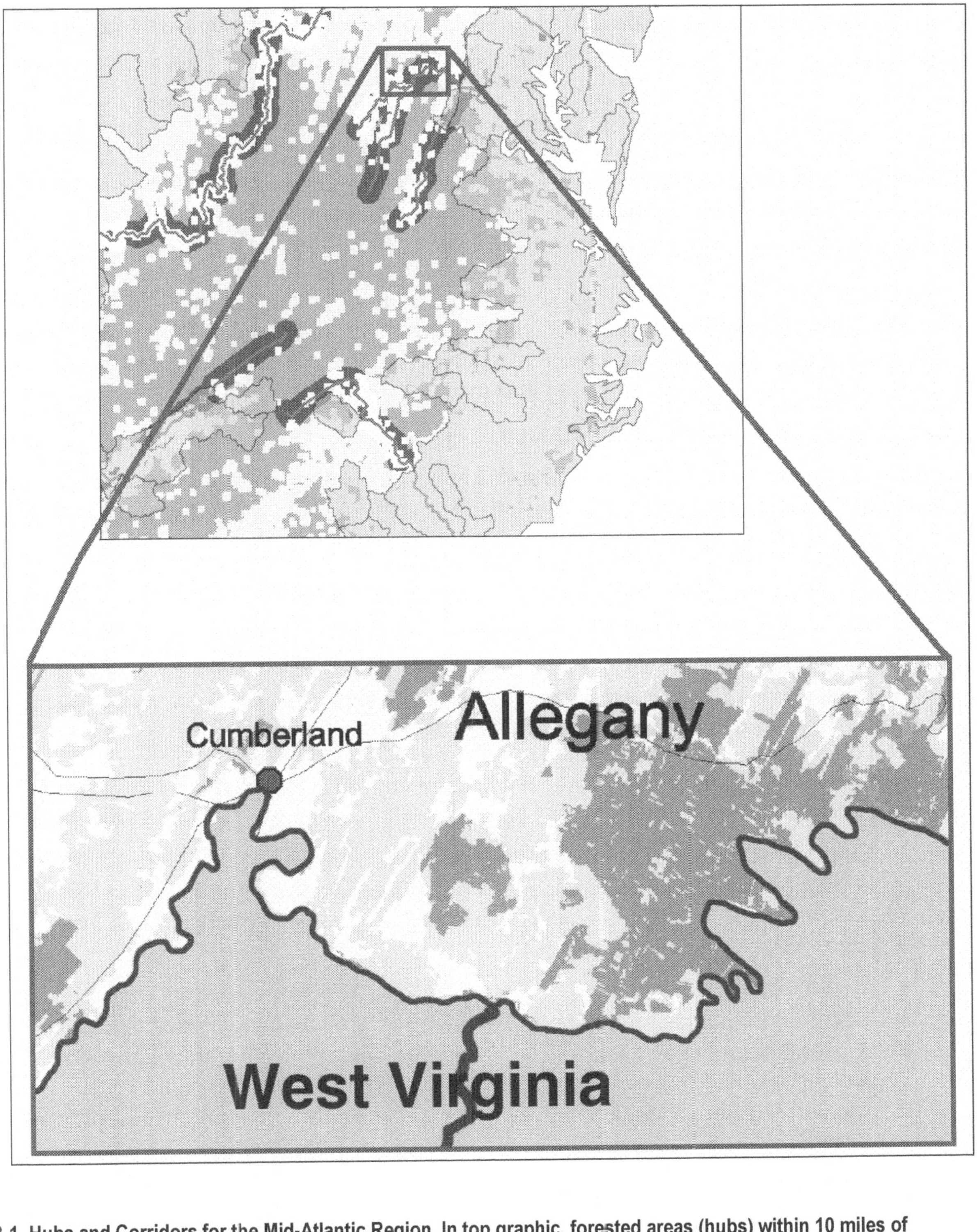

Figure 8-1. Hubs and Corridors for the Mid-Atlantic Region. In top graphic, forested areas (hubs) within 10 miles of major rivers (corridors) are shown in dark gray, and forest areas outside the 10-mile buffer are shown in medium gray (other land covers are shown in lighter colors). The bottom graphic shows Greenprint Program areas (dark gray) within one of the 10-mile buffers in Allegany County, Maryland.

The San Pedro River Basin in Arizona and Mexico is using a hubs and corridors approach to protect groundwater recharge areas and sustain summer low flow in the San Pedro River. The case studies presented in Chapter 20.0 discuss this in greater detail.

Southern California is using Natural Community Conservation Planning to identify hubs and corridors to protect biodiversity (California Biodiversity News, 1997). The Natural Community Conservation Planning is a joint effort among local, state, and federal agencies working with landowners and developers to protect and maintain habitat corridors between habitat hubs so these hubs do not become isolated islands. Some of the hubs and corridors in southern California are shown in Figure 8-2. Although the emphasis of Natural Community Conservation Planning is protecting biodiversity, retaining and maintaining these hubs and corridors also contributes to water resources management. These hubs and corridors protect riparian habitat, reduce erosion and sedimentation, provide infiltration and ponding for stormwater, reduce flash flooding, and improve overall stormwater quality.

8.2 Ecological Goods and Services

Ecological services are the functions and processes through which natural ecosystems, and their associated species, sustain and fulfill human life (Daily, 1997). Ecological services maintain biodiversity and the production of ecological goods, such as seafood, forage and grains, timber, biomass fuels, natural fiber, and many pharmaceuticals, industrial products, and their precursors (Daily, 1997). Goods typically have a market value. Ecological services also have value, but not as estimated through traditional market approaches.

8.2.1 Ecological Goods

Ecological goods from freshwater systems that have market value include water for multiple uses (e.g., domestic and commercial water supply, irrigation, thermoelectric cooling water), commercial shell and fin fisheries, recreational fisheries, tourism, recreation (e.g., boating, swimming), and similar marketable commodities. The market value of these goods can be estimated directly.

8.2.2 Ecological Services

Ecological services, in general, do not have direct market value. However, the services provided are essential for sustaining aquatic and human ecosystems and are of value. Examples of ecosystem services include

- Air and water purification;
- Floods and droughts mitigation;
- Wastes detoxification and decomposition;
- Soil/soil fertility generation and renewal;
- Crops and natural vegetation pollination;
- Agricultural pests control;
- Seeds dispersal;
- Nutrients translocation;
- Biodiversity maintenance;
- Pharmaceutical, industrial, and commercial raw materials;
- Protection from solar ultraviolet radiation;
- Temperature, wind, and wave moderation;
- Diverse human cultural support; and
- Aesthetic beauty and spiritual renewal (after Daily, 1997).

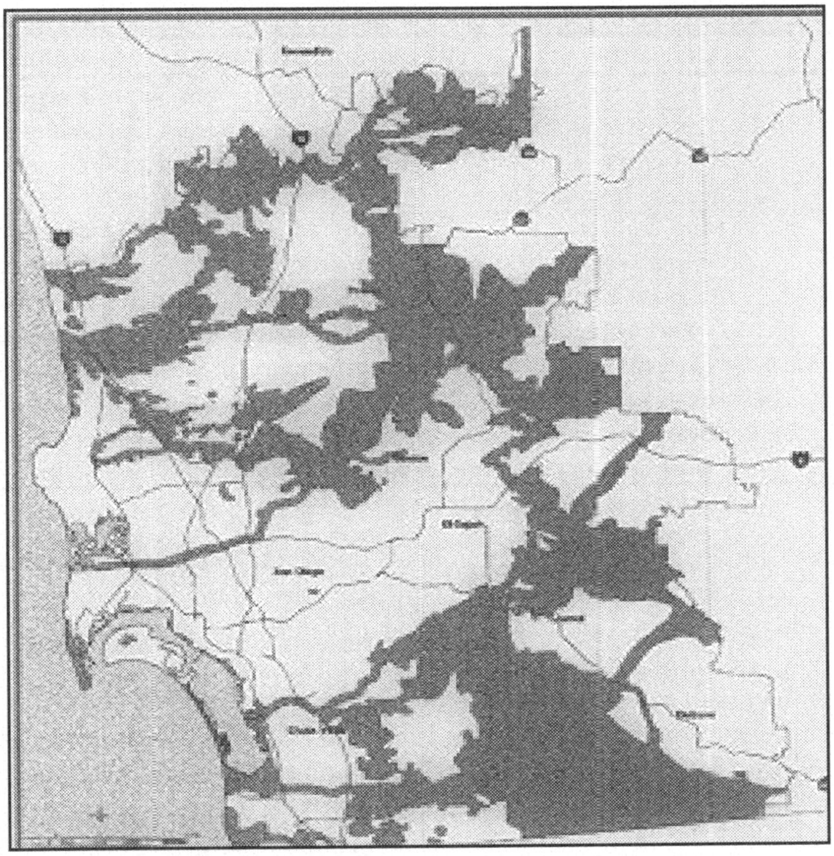

Figure 8-2. Hubs and Corridors throughout San Diego County, California, Resulting from Natural Community Conservation Planning. (Reprinted from California Biodiversity News, 1997. Copyright 1997, CERES, Sacramento, Calif.)

Sustainable water resources require that all these services be sustained. What happens on the land affects the water—surface water and groundwater, freshwater and saltwater. Despite their importance to human society, ecological services are difficult to consider in evaluating management alternatives because they are difficult to describe in terms of monetary value. As a result, they are difficult to include in common evaluation methods, such as benefit–cost analyses. This issue, and possible methods for including ecological services in management analysis, are discussed in detail in Chapter 10.0.

8.2.3 Shifts in Ecologic State

There is increasing evidence that human induced loss of resilience can result in sudden changes in ecosystem states (e.g., crossing a condition threshold) when triggered by events like storms or fires (Table 8-1). These changes in ecological state can result in the loss of ecological goods and services provided by these ecosystems. For example, water clarity in lakes might change little with additional nutrient loading until a critical threshold is passed, at which point lake water abruptly turns from clear to turbid (Carpenter et al., 1999; Scheffer et al., 2001). With increased turbidity, submerged macrophytes disappear, followed by changes in the fish and macroinvertebrate diversity (Carpenter et al., 1999; Scheffer et al., 2001). Lake restoration is possible, in some instances, through reduced external and internal nutrient loading, but at considerable expense.

Table 8-1. Examples of Documented Shifts in States in Different Kinds of Ecosystems.

Ecosystem Type	Alternative State 1	Alternative State 2	References
Freshwater Systems	Clear water	Turbid water	Carpenter, 2001
	Benthic vegetation	Blue-green algae	Scheffer et al., 2001
	Oligotrophic macrophytes and algae	Cattails and blue green algae	Gunderson, 2001
Marine Systems	Hard coral	Fleshy Algae	Nystrom et al., 2000
	Kelp forests	Urchin dominance	Estes & Duggins, 1995
	Seagrass beds	Algae and muddy water	Gunderson, 2001
Rangelands	Grass structure	Shrub structure	Walker, 1993
Forests	No pest	Pest outbreak	Holling, 1986
	Hardwood plants dominate	Pine trees dominate	Peterson, 2002
Arctic Systems	Grass dominated	Moss dominated	Zimov et al., 1995

As another example, resilience of rangelands depends on maintaining water infiltration, water storage capacity, nutrient cycles, and vegetation cycles. Rangeland shifts between grass dominance and woody shrub/tree dominance can be driven by fire and grazing pressure under variable rainfall (Walker, 1993). Persistent high grazing pressure precludes fire from mobilizing soil nutrients, with the result that woody shrub/tree vegetation becomes dominant, and the rangeland becomes less productive for grazing. Even if the animals are removed, there may be insufficient grass fuel to allow fires to occur. It can take decades for the woody plant community to become susceptible to fire pressure (Carpenter et al., 2001). Other examples are listed in Table 8-1, with references. Sudden shifts in ecosystem state have effects beyond ecological consequences, including socioeconomic and institutional consequences.

8.3 Green Water and Blue Water

The terms "green water" and "blue water" are used in discussions of the relationships between land use and hydrology (Falkenmark, 2003). Green water represents evapotranspiration and infiltration in the terrestrial ecosystem whereas blue water represents withdrawals and return flows to the aquatic ecosystems.

Land use is one of the primary forcing functions on water, both directly and indirectly (Figure 8-3). Land use and land cover management affect both water quantity and quality. The effects of agriculture, urban areas, mining, forests, and other land uses on water quality are well documented. Less attention has been given to the effects of land use and land cover on water quantity, particularly green water.

Terrestrial ecosystems play a fundamental role in runoff generation, infiltration, and groundwater recharge. Terrestrial ecosystems "consume" about two-thirds of continental precipitation as green water (Falkenmark, 2003). Forests consume the greatest proportion of this green water, returning over 50% of the continental precipitation to the atmosphere through evotranspiration (Table 8-2). See Chapter 5.0 for additional discussion of land use effects on water quantity.

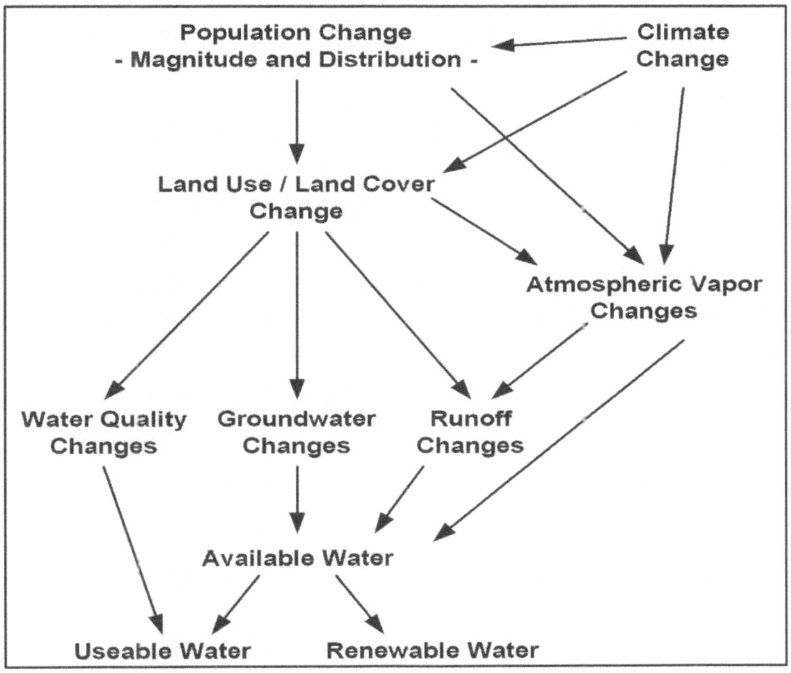

Figure 8-3. Forcing Functions on Water.

Table 8-2. Terrestrial Ecosystem Consumption of Green Water (after Falkenmark, 2003).

Ecosystem	Green Water ET (ac-ft/yr)
Temperate and tropical forest, woodlands	3.2×10^{10}
Temperate and tropical grasslands	1.2×10^{10}
Croplands (including weeds, periphery)	5.5×10^{9}
Bogs, fens, marshes, swamps	1.1×10^{9}
Tundra and desert	4.6×10^{9}
Other systems	1.6×10^{9}
Total	5.7×10^{10}

The cumulative effects of land use management and change, especially at the basin and regional scale, need to be better documented and considered in water quantity management. These cumulative effects include changes in total available water quantity, shifts in frequency and timing of hydrologic events, and shifts in the spatial distribution of available water.

8.4 Minimum Needs

As human demand for water increases, it becomes more important to consider the water required to maintain ecosystems on which human society depends for essential goods and services (Section 8.2). Assessing ecological requirements for water could begin with a consideration of how much water is available to satisfy human uses and ecological requirements.

Recent estimates indicate that the global long-term average annual runoff is about 3.2×10^{10} ac-ft/yr, with estimated global withdrawals of 3.2×10^{9} to 4.0×10^{9} ac-ft/yr (Fekete et al., 1999; Shiklomanov, 2000). Therefore, 10–15% of the average annual runoff is withdrawn for human use, which might indicate there is sufficient available water to satisfy ecological needs. However, not all of this runoff is accessible for human use. About 31% of global runoff is accessible to society; 35% of this 31% is withdrawn for human uses, and 19% is maintained for instream uses, leaving 45% to run into the sea unused by humans (Postel et al., 1996). Over half of the accessible water, then, has already been committed. Spatial and temporal differences in the demand for and availability of water are not considered as part of these global budgets. In addition, accessibility and availability do not ensure that water has satisfactory quality to support human uses or aquatic life. Both quantity and quality determine if water is usable. Water quantity and quality are both important for sustainable water resources.

Two questions that arise are

1. What is needed to sustain aquatic ecosystems?
2. How do aquatic ecosystems interact as part of the local, regional, and global hydrologic cycles?

These two questions are addressed in the following subsections. Because ecosystems are conceptual, without clear space and time boundaries, it is useful to discuss global, regional, and local ecosystems in a context similar to that used to discuss the hydrologic cycle. Aquatic systems, by definition, depend on hydrologic processes, but they also influence hydrologic processes and, therefore, need to be considered part of the global hydrologic cycle.

8.4.1 Global Ecosystems

The Pilot Analysis of Global Ecosystems (PAGE), a precursor to the ongoing Millennium Ecosystem Assessment, evaluated the condition of five global ecosystem categories: agroecosystems, forest ecosystems, grassland ecosystems, freshwater ecosystems, and coastal ecosystems (Revenga et al., 2000). The approach was to assess each ecosystem with respect to the goods and services it provides in sustaining humans and other living organisms. This integrated ecosystem approach attempted to consider the entire range of goods and services provided by the ecosystem. As such, it faced the same methodological challenges that were faced in the development of guidelines for sustainable water resources. Unlike a living organism that can be determined to be healthy or unhealthy, but not both simultaneously, ecosystems can be in good condition to provide some goods and services, but in poor condition for providing other goods and services. The PAGE conclusions and trends are summarized in Table 8-3. Additional information can be obtained for any of the individual ecosystem reports at www.wri.org/wr2000.

Global trends similar to those identified in PAGE (Table 8-3) have been noted by others in projected water scarcity (Gleick, 1998a,b; Postel, 2000a,b; Postel et al., 1996; Vörösmarty & Sahagian, 2000), land use transformation (Houghton, 1994; Naiman & Turner, 2000; Vitousek et al., 1997, Vörösmarty & Sahagian, 2000), forest fragmentation (Laurance et al., 1998; Riitters et al., 2000), agricultural intensification (Matson et al., 1997), irrigation (Postel, 1999), and loss of biodiversity (Loreau et al., 2001; Myers et al., 2000; Pimm & Raven, 2000; Revenga et al., 2000).

Table 8-3. Pilot Assessment of Global Ecosystems Conclusions and Trends (after Revenga et al., 2000).

1. Overall, there are global signs that the ecosystem capacity to continue to produce many of the goods and services on which societies depend is declining.
2. Human activities have significantly disturbed the global water, carbon, and nitrogen cycles.
3. Agriculture, industry, and human settlements have permanently converted extensive natural areas and contributed to fragmentation, pollution, and increased incidence of pest attacks, fires, and nonnative invasive species.
4. Irrigated agriculture has contributed to water logging, and salinization, depletion and chemical contamination of surface and groundwater supplies.
5. Dams, diversions, and other engineering works have altered the quantity and location of freshwater available for human use and for sustaining aquatic ecosystems. Water engineering, however, has also significantly improved living standards by providing flood control, fresh and safe drinking water, and water for irrigation, energy production, and transportation.
6. Globally, from 1900 to 1995, water withdrawals increased six-fold, more than double the rate of population increase. Surface and groundwater resources are being depleted in many parts of Asia, North Africa, and North America.
7. River modification has affected the flow of rivers on every continent. The average residence time of river water in regulated basins has tripled to over one month worldwide. Reservoirs trap about 30% of global suspended sediments.
8. Freshwater wetlands have been reduced by as much as 50% worldwide.
9. Almost 40% of the world's population experiences serious water shortages, and water scarcity is expected to increase dramatically over the next several decades.
10. Surface water quality has improved for many pollutants in developed countries, including the United States, but nutrient runoff from agriculture is a serious global problem. Phosphorus concentrations in U.S. waterways have improved, while nitrate concentrations have remained relatively stable from 1980 to 1990.
11. In North America, Europe, and Russia, inland fisheries landings have declined. The greatest global threat to sustainable long-term fisheries is habitat loss and degradation.
12. Forests currently store about 40% of all the carbon held in terrestrial ecosystems. Forests in the northern hemisphere are slowly increasing their storage capacity after historic clearing, but this gain is more than offset by deforestation in the tropics.
13. Global biodiversity has significantly declined over the past century, primarily as a result of habitat destruction. Freshwater ecosystems overall appear to be the most severely degraded, with an estimated 20% of freshwater fish species becoming extinct, threatened or endangered in recent decades. Physical alteration, habitat loss and degradation, water withdrawal, pollution, overexploitation, and nonnative species introductions are all contributing to these declines.
14. Although wetland birds are at high risk of imperilment in many regions of the world, there has been a 66% increase in wetland birds in the United States and parts of Canada.

Although the condition of freshwater ecosystems in the United States, in general, is better than in many other areas of the world, activities that affect global ecosystems can also affect U.S. ecosystems through teleconnections. As indicated in Chapter 5.0, deforestation and land use change alter interactions among soil moisture, evaporation, and precipitation, which can affect sea surface temperatures, and subsequently climate and hydrologic patterns around the globe, including the United States. Altering the residence time of water on land (e.g., reservoir storage) also affects evaporation and precipitation, as well as the timing, magnitude, and duration of freshwater discharge to downstream freshwater and estuarine ecosystems. Similar changes in timing, magnitude, and duration of storm flow occur with the loss of wetlands. Wetland loss also results in altered water quality and loss of biodiversity. Monitoring global trends must be an integral part of evaluating sustainable water resources in the United States. Knowing these trends can help us anticipate potential changes and begin managing for them. In addition, this knowledge provides insight into what changes may be irreversible and require adaptation as the management strategy.

8.4.2 Regional Ecosystems

The concept of regional ecosystems or ecoregions has been developed by Omernik (1987, 1995) and Bailey (1976, 1995). Ecoregions are geographical areas within which the biotic and abiotic components of terrestrial and aquatic ecosystems exhibit relatively homogeneous patterns compared to the patterns between ecoregions. Ecoregions can provide a framework for managing ecosystems because ecosystems within the same ecoregion are expected to respond similarly to land use changes or other anthropogenic stressors, thereby permitting similar management practices to be implemented to control impacts. The attributes considered important for management of a specific resource will determine the specific ecoregional delineations. Comparison of Bailey and Omernik ecoregions reveals differences, as well as similarities between the two approaches.

Omernik ecoregions are based on delineating regions of the United States that have greater among than within variance in geology, physiography, vegetation, climate, and soils than contiguous regions (Omernik, 1987, 1995). Omernik has developed three classification schemes—Level II, III, and IV—based on increasing level of detail (Level II is the coarse scale, Level IV is fine scale). U.S. EPA has aggregated some of the Level III ecoregions to delineate areas in which nutrient criteria for lakes, streams, and estuaries might be expected to be similar (Figure 8-4). The U.S. EPA Office of Water uses Omernik's ecoregions in their regional and national evaluations and assessments of surface water condition in the United States.

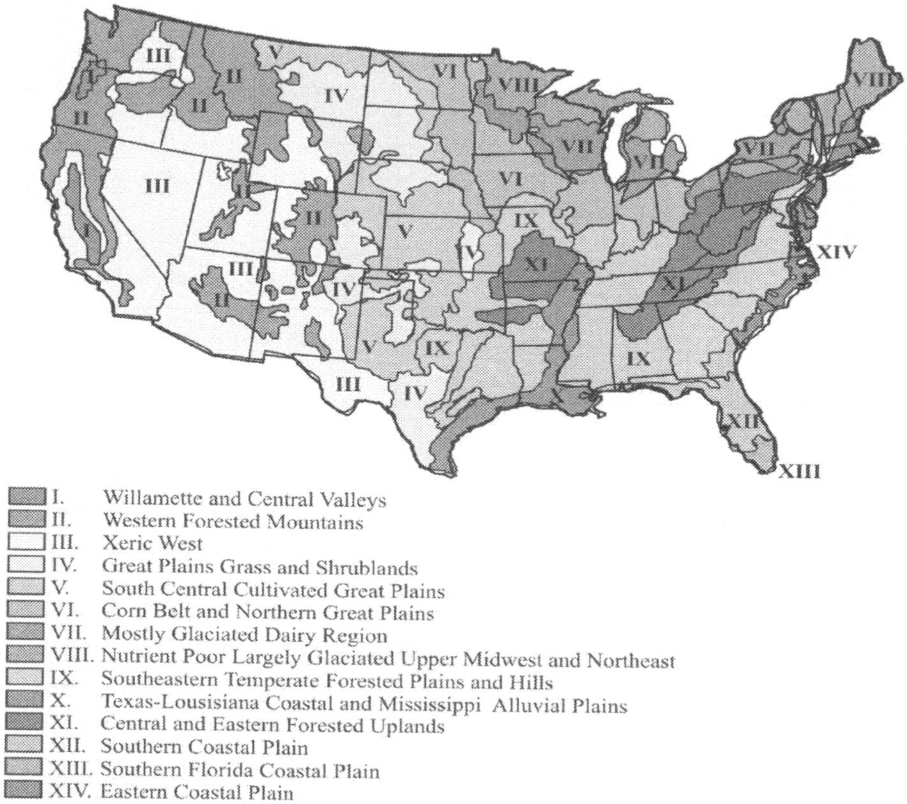

Figure 8-4. U.S. EPA Aggregations of Level III Ecoregions Expected to be Similar with Regard to Nutrient Criteria.

Bailey ecoregions are based principally on differences in macroclimate and prevailing plant formations among regions (Bailey, 2002). Bailey ecoregions are used by the USDA Forest Service to evaluate and assess the effects of stressors and management practices within regions where the forests are expected to respond similarly to natural climatic factors and disturbance.

Although ecoregions delineate areas with similar ecologic attributes, these boundaries typically do not correspond with watershed boundaries. Watersheds and their associated aquatic ecosystems represent the primary management unit for many local, state, and federal agencies.

8.4.3 Watersheds and Aquatic Ecosystems

Just as surface water and groundwater cannot be considered independent, neither can aquatic ecosystems be divorced from their watersheds (Hynes, 1975). Management practices that do not consider the watershed in conjunction with the aquatic ecosystem provide ephemeral improvements, at best, or at worst are both scientifically and economically ineffective. In discussing the restoration of aquatic ecosystems, the NRC (1992) states that stream and river restoration should begin with improved land management practices that permit natural restoration of the stream or river to occur. The NRC recommends that watershed-scale restoration projects that integrate lakes, streams, and wetlands are needed (NRC, 1992). For sustainable water resources, the lake, stream, and/or wetland must be considered in context with the watershed and groundwater system.

8.4.3.1 Aquatic Ecosystem Needs and Requirements

Much of what is known about aquatic ecosystems and the role of water in supporting these ecosystems is fragmented and piecemeal (NRC, 2000b). Even for individual species, water requirements are not well known (NRC, 1992). Instream flow techniques have been developed to provide information on minimum flow requirements, but these requirements are typically for individual sport or game fish species (Bovee et al., 1998). These techniques have been criticized as being overly simplistic in their treatment of complex physical-biotic interactions (Gore & Nestler, 1988; Stanford, 1994; Williams, 1996).

Minimum downstream flow requirements for aquatic life use have been negotiated for almost every major reservoir in the United States, and minimum instream flows have been established for many streams. Most interstate streams also have negotiated minimum flow requirements. Although there are still streams and some major rivers (e.g., Colorado River) that go dry during low flow periods, minimum low flow requirements for aquatic life use are recognized even in states with Prior Appropriation water rights. Better procedures and techniques need to be developed to establish sustainable flow volumes for aquatic ecosystems, rather than individual species.

Although minimum flow or quantity requirements are clearly important to aquatic ecosystems, recent studies have shown that flow dynamics are as important, if not more so (Dynesius & Nilsson, 1994; Richter et al., 1997; Rosenberg et al., 2000). Changes in the hydrologic regime affect water quality and ecosystem functioning. For example, hydrologic alteration affects aquatic processes such as nutrient cycling and primary productivity (Pringle, 1997; Rosenberg et al., 1997), biodiversity (Master et al., 1998; Richter et al., 1998; Rosenberg et al., 1997, Wilcove et al., 1998), assimilative capacity for point and nonpoint source pollutants (Gillian & Brown, 1997; NRC, 1992), and sediment transport, which affects downstream deltas and wetlands (NRC, 1990; Rosenberg et al., 1997). Hydrologic variability is key to sustaining riverine ecosystems because dynamic fluvial processes maintain the mosaic of habitat structures need to sustain biotic diversity (Gosselink et al., 1990; NRC, 1992; Sparks. 1995; Standford et al., 1996). Preserving or restoring river ecosystems requires this natural flow variability. The frequency, duration, and timing of flow is as important as the total volume of water (Kingsolving & Bain, 1993; Richter et al., 1996; Walker et al., 1995).

The extent of flow fragmentation and flow regulation is documented by Dynesius and Nilsson (1994) and monitored in PAGE (Revenga et al., 2000). Sixty percent of the 227 largest rivers in the world are strongly or moderately affected by either river channel fragmentation from dams, or flow regulation resulting from reservoir operation, interbasin diversion, or irrigation. Fragmented rivers account for 90% of the total volume flowing through all 227 rivers. Every large river in the United States has been fragmented (Graf, 1992, 1999). There are around 75,000 dams in the United States that have a height of at least 6 feet and storage of at least 50 ac-ft (Graf, 1999; U.S. Army Corps of Engineers, 1996). Every major river, most smaller rivers, and many streams have been altered hydrologically, either through damming, or through channel modifications such as dredging, snagging, channelization, clearing, or similar geomorphological changes. Sustaining aquatic ecosystems requires more than ensuring minimum flows are maintained. In many cases, it also requires habitat restoration, restoration of a more natural flow regime, and geomorphic improvements (NRC, 1992).

Restoring aquatic ecosystems as part of water resources management has received considerable attention over the past decade (e.g., NRC, 1992; Richter et al., 1997; Rosgen, 1996; Trush et al., 2000). Trush et al. (2000) provides a list of 10 attributes of alluvial rivers to help managers identify desired processes and prescribe changes necessary to move closer to alluvial river integrity (Table 8-4). Derived from field observations and experimentation, these attributes provide a minimum checklist of critical geomorphic and ecological processes for restoring and preserving alluvial river ecosystems. Natural stream channel design incorporates many of these attributes in restoring streams to a stable channel configuration (Federal Interagency Stream Restoration Working Group, 1998; Rosgen, 1996). Trush et al. (2000) state that these attributes can help guide

- restoration of alluvial processes below an existing dam without resorting to extreme measures such as decommissioning the dam; and
- preservation of alluvial river integrity below proposed dams.

Richter et al. (1997) have developed an approach for assessing long-term streamflow records to evaluate hydrologic variability. Their goal is to use long-term streamflow data and statistical analyses of natural flow variability in setting ecosystem based management targets for river management decisions (NRC, 1992). In applying this approach, either measured or synthetic daily stream flows from a period with negligible anthropogenic perturbation are selected for analysis (Richter et al., 1996). The streamflow record is characterized by 32 different hydrologic parameters (e.g., annual minima 1-day means, annual maxima 1-day means, annual minima 3-day means, annual maxima 3-day means, number of high pulses per year) (Richter et al., 1996). These hydrologic parameters are then used to establish guidelines or bounds for desired hydrologic variability in the regulated river flow. Targeting can be used to identify candidates for restoration or improvement. Some systems can be improved, but not restored. By monitoring and modifying the guidelines or bounds, this approach can be integrated into an adaptive management framework (Walters, 1990).

The influence of flow dynamics and geomorphology extend beyond the stream ecosystem to riparian vegetation and wetland ecosystems. Naiman and Decamps (1997) indicate that the ecological diversity of riparian zones is directly related to variable flood regimes,

Table 8-4. Attributes of Alluvial River Ecosystems. [After Trush et al. (2000). Copyright 2000, National Academy of Sciences, Washington, D.C.]

	Attribute	Importance
1.	The primary geomorphic and ecological unit of an alluvial river is the alternate bar sequence.	Dynamic alternating bar sequences are the basic structural underpinnings for aquatic and riparian communities in healthy alluvial river ecosystems.
2.	Each annual hydrograph component accomplishes specific geomorphic and ecological functions.	Annual hydrograph components (including winter storm events, baseflows, snowmelt peaks, and snowmelt recession limbs) collectively provide the impetus for processes that shape and sustain alluvial river ecosystems. These components are uniquely characterized by year-to-year variation in flow magnitude, duration, frequency, and timing.
3.	The channel bed surface is frequently mobilized.	Coarse alluvial channel bed surfaces are significantly mobilized by bankfull or greater floods that generally occur every 1 to 2 years.
4.	Alternate bars must be periodically scoured deeper than their coarse surface layers.	Floods that exceed the threshold for scouring bed material are needed to mobilize and rejuvenate alternate bars. Alternate bars are periodically scoured deeper than their coarse surface layer, typically by floods exceeding 5- to 10-year annual maximum flood recurrences. Scour is generally followed by redeposition, often with minimal net change in the alternating bar topography.
5.	Fine and coarse sediment budgets are balanced.	River reaches export fine and coarse sediment at rates approximately equal to sediment input rates.
6.	Alluvial channels are free to migrate.	During lateral migration, the channel erodes older flood plain and terrace deposits on the outside bend whereas it deposits sediment on the bar and flood plain of the inside bend. Although outer and inner bend processes may be caused by different hydrograph components, the long-term result is maintenance of channel width.
7.	Flood plains are frequently inundated.	Flood plain inundation typically occurs every 1–2 years. Flood plain inundation attenuates flood peaks, moderates alternate bar scour, and promotes nutrient cycling.
8.	Large floods create and sustain a complex mainstem and flood plain morphology.	Large floods – those exceeding 10- to 20-year recurrence events – reshape and/or redirect entire meander sequences, avulse mainstem channels, rejuvenate mature riparian stands to early successional stages, form and maintain side channels, scour flood plains, and perpetuate off-channel wetlands, including oxbows.
9.	Diverse riparian plant communities are sustained by the natural occurrence of annual hydrograph components.	Natural interannual variability of hydrograph components is necessary for woody riparian plant life history strategies to perpetuate early and late successional stand structures.
10.	Groundwater in the valley bottomlands is hydraulically connected to the mainstem channel.	When flood plains are inundated, a portion of surface runoff from the watershed is retained as groundwater recharge in the valley bottomlands.

geomorphically unique channel processes, and upland influences on the fluvial corridor. Riparian vegetation zones and wetland ecosystems serve as the interface between aquatic and terrestrial ecosystems (Karr & Schlosser, 1978).

Water quality, in addition to water quantity, is important in protecting, restoring, and sustaining aquatic ecosystems. Over the past three decades, there has been significant improvement in the quality of point source discharges into the nation's waters (NRC, 1992, 2001). Water quality criteria related to physical, conventional chemical and biological, toxic, and radiological pollutants have been established for all waters of the nation. Sediment, nutrients, combined sewer overflow, toxic chemicals, endocrine disrupters, pathogens, and other pollutants are also contributed by nonpoint sources. Currently, the greatest water quality challenge is addressing nonpoint pollutant sources, and habitat loss and degradation (NRC, 1992, 2000b, 2001). Future improvements in technology for point source discharges are unlikely to result in significant improvements in water quality (NRC, 2001). For sustainable water resources, watershed management practices must be implemented to control nonpoint source pollution.

Aquatic ecosystems have requirements for water quantity and quality just as do human communities. Aquatic ecosystems, however, also provide some of basic goods and services that are used to sustain human communities. Satisfying the ecological requirements for water quantity and quality, therefore, contributes to satisfying human needs and desires.

8.4.3.2 Watershed Influences

Watersheds, by definition, play an important role in the hydrologic cycle. Global climate change research has highlighted the importance of changes in watershed land use/land cover in affecting not only global climate, but also the global water cycle (Houghton et al., 1995; Subcommittee on Global Change Research, 1999; See Section 8.3). Chase et al. (2000) indicate that land use change will be a critical factor in availability of freshwater in the future.

The general effects of land use change and practices on freshwater ecosystems, particularly streams, are illustrated in Figure 8-5 (Naiman & Turner, 2000). These changes occur in both ecosystem structure and function, and range from individual species effects to alterations of energy and water balances. These changes are complex, and have cumulative effects on downstream communities and habitats. Some of the specific effects of deforestation, agricultural intensification, and wetland loss are listed in Appendix C (Calder, 1992). Other land use changes of concern include desertification and soil erosion in agricultural areas. Desertification can occur in semi-arid regions when overgrazing or similar intense land use denudes and dries the soil, which results in a net loss of soil water storage, reduced evaporation, and increased storm runoff (Vörösmarty & Sahagian, 2000). Soil erosion in agricultural areas alters the relation between surface topography and the groundwater table (Vörösmarty & Sahagian, 2000). Erosion in uplands and deposition in valleys and lowlands can change hydraulic gradients and alter baseflow and net water storage. As with desertification, this can result in net loss of soil, loss of water storage, lower evaporation, and increased storm runoff. The cumulative effects of land use changes such as deforestation of tropical rainforests, and fragmentation of other large, intact forests, can significantly alter precipitation recycling, the amount of water vapor that is released to the atmosphere, and the timing, magnitude, and frequency of storm events and runoff (Eagleson, 1986).

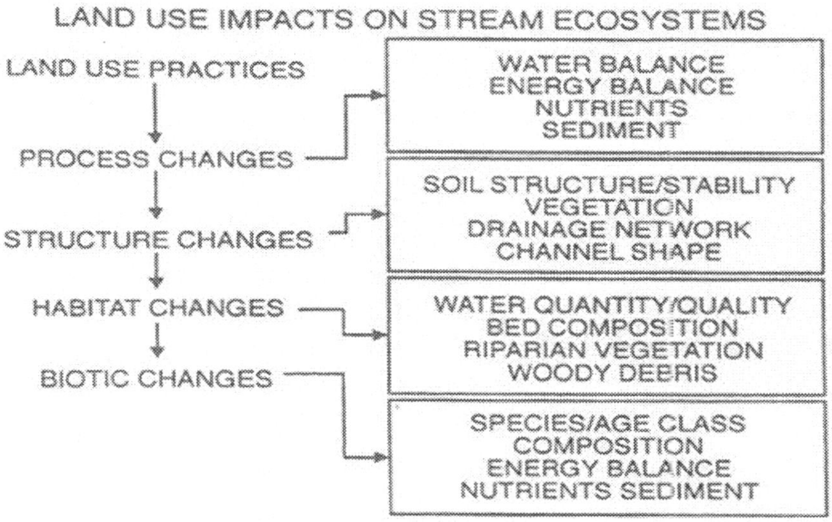

Figure 8-5. Cascading Effects of Land Use Practices and Land Use Change on Stream Ecosystem Structure and Function. [Reprinted with permission from Naiman & Turner (2000). Copyright 2000, Ecological Society of America, Washington, DC.]

Watershed land use change includes loss of wetlands and riparian habitat, which are both important components of sustainable water resources. Effects of wetland loss are further confounded through wetland mitigation because, in many instances, the created wetland does not have the same integrity as the wetland being lost. No net loss of wetlands means more than just maintaining extent of wetlands; it includes creating wetlands with similar functions and conditions (Hurby, 1999).

8.5 Working River Concept

This concept is the basis for a water management approach being used in Australia. Under this concept a healthy "working river" is a river that is being managed to maintain community determined levels of societal and environmental uses (Whittington, 2002). Under this approach it is accepted that the river is not in a natural or pristine state, does not function like a pristine river, and is not likely to ever achieve that state or function again. The community decides on the desired balance of naturalness and societal work, and the river is managed to sustain this agreed upon balance.

In the United States, decisions have already been made with regard to desired levels of water resources condition and use through the assignment of designated uses to all water bodies. Designated uses can be used as a basis for managing to maintain ecosystem services, while accepting that it is unlikely that waterbodies can be managed at, or restored to, a pristine state. Use attainability analyses can be conducted to change waterbody uses when the current designated use(s) cannot be attained (U.S. EPA, 2004b). Sustainable water resources management must consider not only ecological factors, but also societal desires for water resources.

8.6 Working Lands Concept

We think the Malpai Borderlands Group has developed a "working lands" concept similar to the Working River concept of Australia. The Malpai Borderlands Group is a nonprofit organization consisting of ranchers in the great Chihuahuan and Sonoran deserts of Arizona and New Mexico, just north of the U.S.–Mexican border. Ranchers in this area were concerned about the demise of cattle ranching as a way of life and the increase in development. Some of the families have been ranching in the Malpai region for six generations. In 1994, the Malpai Borderland Group, a nonprofit organization, was formed to protect not only the ranching life style, but also the landscape. This group of private ranchers protects 800,000 acres while also sustaining working cattle ranches.

The Malpai Borderland Group has protected about 42,000 acres through conservation easements and invented a new form of conservation easement called grass banking. During serious droughts, cattle ranchers can transfer some of their herd to a ranch owned by a nonprofit organization that has returned about 137,000 acres to extraordinary grassland health. This is the grass bank. Cattle ranchers that use the grass bank agree to convey land use easements that prohibit development on their property to the nonprofit organization.

The Malpai Borderland Group has both an active science and an active habitat restoration programs to protect endangered species. Cooperators in this effort include the USDA Forest Service, Natural Resource Conservation Service, U.S. Fish and Wildlife Service, U.S. Bureau of Land Management, various state Soil and Water Conservation Districts, state Game and Fish Departments, Arizona State University, New Mexico State University, University of Arizona's Desert Laboratory, University of New Mexico, and The Nature Conservancy (www.malapaiborderlandsgroup.org). The Malpai Borderlands Group represents a model for implementing "working lands" concepts in water resources management.

CHAPTER 9.0

ELEMENTS OF SWRM: SOCIETY

Dublin Principle 2 – Water development and management must be participatory
Enlibra Principle 3 – Collaboration, not polarization
Enlibra Principle 8 – Change a heart, change a nation
Panarchy Principle – Ecological, economic, social, and institutional systems are interlinked

Water is not only a basic element for human survival, it is also an integral part of the social and cultural fabric of society. Because water is essential for survival of all living organisms, both natural and social systems have evolved approaches for protecting and conserving water so it is available to satisfy basic needs. These approaches range from life cycle adaptations to storage mechanisms in nonhuman organisms and natural systems, to legislation, water policies, and economic markets for human societies. Water also contributes to our spiritual, emotional, mental, and physical well-being, as individual humans, communities, and global societies.

9.1 Rights, Needs, and Desires

9.1.1 Rights

An ongoing debate in international circles is whether water is a basic human right (Gleick, 2000). Clearly, water is a basic element required for human survival, and as discussed above, an integral part of every society. A number of international declarations, covenants, and agreements on human rights have been issued, beginning with the 1948 Universal Declaration of Human Rights up to the 1989 Convention on the Rights of the Child (Gleick, 2000), but none of these made explicit statements about water as a right. According to Gleick (2000) explicitly acknowledging water as a fundamental human right is important for the following five reasons:

1. Encourages governments to renew their efforts to meet the basic water needs of their populations,
2. Encourages governments to legislate or promulgate specific laws and regulations to protect this right,
3. Highlights the deplorable state of water management around the world,
4. Helps resolve conflicts over the use of shared water by identifying minimum water requirements and allocations among all basin parties, and
5. Ensures human rights to water take precedence over other water allocations or investment decisions.

In 2002, the United Nations Economic and Social Council passed General Comment No. 15, that defines the, "human right to water as indispensable for leading a life of human dignity" and a "prerequisite for the realization of other human rights" such as the right to food. It concludes, "States parties have to adopt effective measures to realize without discrimination, the right to water, as set out in this General Comment" (United Nations, 2002). Acknowledging that access to, and use of, water is a basic human right leads to the question of how much water is needed to "lead a life of human dignity."

9.1.2 Needs and Desires

The per capita water use in the United States is about 5.4 x 10^4 gal/yr per person (see Chapter 7.0), or more than an order of magnitude greater than the minimum domestic use of 4.8 x 10^3 gal/yr per person to sustain life in a moderate climate suggested by Gleick (1996, 2000). The need for water, however, includes more than just sustaining basic domestic water requirements.

As the 1974 Technical Committee of Water Resource Centers of Western States considered water issues, they asked a basic, but broad question, "How does water contribute to society?" The Committee decided they first needed to identify society's primary social needs and then determine how water contributed to satisfying these needs (Holdren et al., 2001). Building on this initial list of social needs, Klessig and Hagengruber (1999) developed a list of conditions necessary for social sustainability (shown in Table 9-1). Water contributes either directly or indirectly to nearly all of the 11 conditions needed for social sustainability. Jonch-Clausen (2004) developed a table (Table 9-2) illustrating specifically how water resources management affects the social issues that are the focus of the Millennium Development Goals (United Nations, 2001). This table shows how sustainable water resources can contribute to social sustainability.

Every individual, community, and society has needs and desires that they are motivated, or strive, to satisfy. There have been numerous theories on what factors influence or motivate individuals to satisfy these needs and desires, whether these are hierarchical (Maslow, 1943, 1954) or concurrent (Alderfer, 1969, 1972), whether it is satisfaction or dissatisfaction (Herzberg, 1968), needs versus desires (McClelland, 1987), or goals that are the drivers. As listed in Table 9-1, societal needs and desires range across the sphere of human perspective from security to economics to cultural to spiritual.

Because water is essential for life, sustainable water resources contribute to satisfying basic human/societal physical and physiological needs. However, sustainable water can also contribute, either directly or indirectly, to simultaneously satisfying multiple factors related to social sustainability (Table 9-2). For example, stream restoration can improve both aquatic habitat and water quality. Stream restoration can use green construction technologies, which can contribute to the development of a restoration industry. The use of economic incentives and incubator programs can help promote inner city restoration-related businesses. Urban stream restoration can contribute not only to improved riparian habitat and water quality, it can create jobs during stream restoration (natural stream design and construction, bank stabilization, revegetation) and afterward through recreational activities (fishing, rafting, etc.). In addition, stream restoration can contribute to improved urban and exurban community aesthetics and community sense of well-being. This can result in inner city economic development that provides jobs, employment, community programs, increased community pride, and appreciation for environmental amenities such as parks, water features, and green infrastructure. Thus, the process of stream restoration can contribute directly to satisfying the aesthetic, collective, economic, emotional, and environmental security, individual safety, recreational, and spiritual aspects of social sustainability (Table 9-1).

Table 9-1. Conditions for Social Sustainability (from Klessing & Hargengruber, 1999).

Condition	Examples
Aesthetic Opportunity	Natural beauty/landscapes Attractive buildings/cities Soothing interior spaces
Collective Security	International peace National defense Protection from internal disorder
Cultural Opportunity	Music Art Heritage
Economic Opportunity	Good wages/job security Return on investment Efficient production
Emotional Security	Family bonds Friendships Sense of community
Environmental Security	Clean water Clean air Biodiversity
Individual Freedom and Variety	Private property rights Consumer choices Unrestricted speech/media
Individual Safety	Health care Police protection/prisons Fire protection
Recreational Opportunity	Outdoor sports Indoor leisure activities Relaxation time
Spiritual Dimension	Humility before Supreme Being Awe of nature Set of moral principles

If we assume that individuals, communities, societies, and cultures are striving to satisfy their needs and desires, then it is critical that we know and understand what these specific needs and desires are and how the individual and social sectors perceive reality with respect to the attainment of these needs and desires.

9.2 Social Mindscapes and Mental Models

Mental models are psychological representations of reality that individuals form in their minds. The idea of the mind constructing small-scale models dates back to 1943 (Craik, 1943). The concept and properties of mental models have developed extensively over the past 60 years (Johnson-Laird, 1983; Garnham & Oakhill, 1996; Leighton & Sternberg, 2003; Rogers et al., 1992).

Table 9-2. Water and the Millennium Development Goals. [Reprinted with permission from Jonch-Clausen (2004). Copyright 2004, Global Water Partnership, Stockholm, Sweden.]

Issues Addressed by Millennium Development Goals	Contribution of Improved Water Resources Management and Access to Water Supply and Sanitation
Poverty	• Water is a factor of production in agriculture, industry, and other economic activities • Investments in water infrastructure/services as a catalyst for local/regional development • Reduced vulnerability to water-related hazards reduces risks in investments and production • Reduced ecosystems degradation makes livelihood systems of the poor more secure • Improved health increases productive capacities, reduces burden on those who care for the sick
Hunger	• Water is a direct input to irrigation for expanded grain production • Reliable water for subsistence agriculture, home gardens, livestock, tree crops • Sustainable production of fish, tree crops, and other foods gathered in common property resources (also affects poverty when such goods are sold for income) • Reduced urban hunger because of cheaper food prices • Healthy people are better able to absorb the nutrients in food than those suffering from water-related diseases, particularly worms
Primary Education	• Improved school attendance from improved health reduced water-carrying burdens, especially for girls • Having separate sanitation facilities for girls and boys in schools increases girls' school attendance
Gender Equality	• Community-based organizations for water management improve social capital of women • Reduced time, health, and care-giving burdens from improved water services give women more time for productive endeavors, adult education, empowerment activities, leisure • Water sources and sanitation facilities closer to home put women and girls at less risk for sexual harassment and assault while gathering water and searching for privacy • Higher rates of child survival are a precursor to the demographic transition toward lower fertility rates; having fewer children reduces women's reproductive responsibilities
Child Mortality	• Improved quantities and quality of domestic water and sanitation reduce main morbidity and mortality factor for young children • Improved nutrition and food security reduces susceptibility to diseases
Maternal Mortality	• Improved health and reduced labor burdens from water portage reduce mortality risks • Improved health and nutrition reduce susceptibility to anemia and other conditions that affect maternal mortality • Sufficient quantities of clean water for washing pre- and postbirth cut down of life-threatening infections • Higher rates of child survival are a precursor to the demographic transition toward lower fertility rates, and fewer pregnancies per woman reduce maternal mortality
Major Disease	• Better water management reduces mosquito habitats • Better water management reduces incidence of a range of other waterborne diseases • Improved health and nutrition reduce susceptibility to/severity of HIV/AIDS and other major diseases
Environmental Sustainability	• Improved water management, including pollution control and water conservation, is a key factor in maintaining ecosystems integrity • Development of integrated management within river basins creates situation where sustainable ecosystems management is possible and upstream-downstream effects are mitigated • Biodiversity conservation, combating desertification furthered by sound water management

Individuals' mental models are formed through experience, education, and interaction with others and the environment, and include the prejudices and unwritten assumptions that define how they believe the world functions. As with any model, mental models are simplified representations, or processes, used for dealing with and solving problems. Mental models are used by individuals to understand, explain, and predict how things happen in their lives and the world. They also act as filters through which the world is seen. There is evidence most mental models only represent what is true, not what is false (Goldvarg & Johnson-Laird, 2000). Differences in mental models explain, in part, why different individuals describe the same event differently. Part of the reason the descriptions differ is that the individuals' mental models, which define their perspective of reality, differ.

Mental models have been studied primarily with respect to individual perception and behavior, but mental models are applicable not just for individuals. Recently, studies have been initiated to address social mindscapes, which are mental models associated with communities or social groups, and how these social mindscapes affect the perception, attention, and significance of events, activities, and actions within the social group (Zerubavel, 1998). A great deal of what individuals think and how they respond to actions depends on the society or culture in which they reside or of which they are part.

So what does this have to do with sustainable water resources? Mental models and social mindscapes are exceptionally powerful in influencing or affecting individual decisions. If information is presented that does not conform with an individual's mental model or the social mindscape, in most cases it is ignored, refuted, or rejected, regardless of its validity. In fact, mental models can, in some instances, serve as a barrier to "wiser" environmental agreements (Bazerman et al., 2000). Win–win situations created in negotiating environmental agreements can be ignored or dismissed because of the mental models of the participants.

In large, we have not engaged sociologists, behaviorists, or psychologists in water resources management to help us determine what mental models exist in the community, watershed, and region. Without this understanding, it will be difficult to determine what information needs to be provided, and in what format, to affect a change in these communities. Identifying the mental models that exist is the first step, and one of the most difficult steps (Easum, 1998).

Ross et al. (1998) described an approach used to document the mental models of farmers, scientists, and extension agents used in managing rangeland. The research was conducted to bridge rural communication gaps among farmers, extension agents, and agricultural research scientists. The approach included site visits to two different properties, in-depth interviews in the field and following the site visits, and analysis using two software packages, *Influence* and *Vensim*, to identify the mental models of these three groups (Ross et al., 1998). The individual mental models identified were more complex than originally hypothesized, and did not necessarily conform with other individual models for the same group (e.g., farmers versus researchers), although there were similarities within groups. In general, farmers relied on learning from observation, whereas researchers used theory to integrate their observations (Ross et al., 1998). A similar approach could be used to identify and formulate the mental models associated with water resources management. In fact, Thornton and Laurin (2005) proposed this premise and approach to move toward more effective lake management in the United States.

We need to begin identifying and specifically codifying the mental models that exist within local communities, watershed associations, and similar groups if we are to move toward effective, SWRM. Documented property owner, community, or watershed association mental models about water resources management simply do not exist. We need to begin formulating these models.

But, once we identify these models, how could we possibly change them?

9.3 Changing Mental Models and Social Marketing

Lasting change in attitude or belief will not occur without a change in the underlying mental models. There are a number of social mindscapes that exist in communities that will have to be changed if water resources are to become sustainable, such as:

- Water is an infinite resource.
- New water sources will become available.
- Technology will save us.
- It isn't safe to drink recycled wastewater.
- We can forecast what will happen to water in the future.

How do you go about changing mental models? To change a mental model or social mindscape, a new model must be presented, thoroughly explained (multiple times), and proven to be true. This is a difficult process and one that takes time, unless there is a crisis situation that demands a different mental model for survival. Unfortunately, water shortages or water contamination events are crises only during the period when they are occurring, and take on diminished importance as time passes. Costa and Kallick (1995) state that growth and change are typically found in crises, not status quo. Order is built out of chaos, learning takes place, understandings are forged, and gradually, the organization begins to function more efficiently. This is the Ω through α process described in Panarchy Theory (Chapter 3.0).

Changing mental models is not a linear process, but there are generally several steps in the process that must occur (Easum, 1998). The first step, which is also the hardest, is to recognize that we have mental models and document that these models are part of the reason for the conflict or lack of effectiveness in addressing issues. The next step is to decide if changing the mental model is really something that we want to do, even though we recognize that it is contributing to the problem or issue. Third, if the decision is made to change our mental model, then we must decide if the pain of change is greater than the existing pain (i.e., cost is less than the potential benefits of change). Unless the pain (economic, social, personal, institutional, etc.) associated with the issue is greater than the pain associated with changing the mental model, it typically doesn't happen (Easum, 1998). Fourth, the new mental model must be formed in our minds and become reality. The new model must be the image or picture we see when we make decisions about that issue. Finally, the new mental model must change our behavior. If our behavior doesn't change then we have not fully accepted this new mental model, regardless of what we say. How many of us know we are not eating or exercising properly, accept we need to change, go on a diet and exercise, lose weight and improve our physical condition, and then, one year later, find we are in the same shape we were before we started the diet. We did not sustain our behavioral change, implying we did not fully change our mental model.

One of the approaches used in the public health fields to change individual and community mental models is social marketing. Social marketing is the planning and implementation of programs designed to bring about social change using concepts from commercial marketing (Kotler et al., 2002). The concept is that if commercial marketing can encourage people to buy certain products, similar approaches might be used to encourage people to adopt behaviors that will improve not only their own lives, but also the lives of their fellow citizens (communities, societies) (Andreasen, 1995). The social marketing campaigns against smoking and for designated drivers both have had national implementation and national success. Similar successes have been attained for other issues at the state and local levels, as well (Weinreich, 1999).

Social marketing, in contrast to commercial marketing, involves changing intractable behaviors in complex socioeconomic and political climates, which is much more difficult to achieve and document than product purchases (MacFadyen et al., 1999). Both commercial and social marketing focus on consumer orientation, but the perspective is different. In social marketing, the question is often asked, "What is it we don't understand about our target audience that is preventing us from getting the message through?" (Kotler et al., 1996). Too often, as scientists and engineers, our question is, "What is wrong with these people, why don't they understand?" The mental model of an engineer describing a water resources management issue is likely not at all similar to that of a property owner or businessperson. One of the first steps in social marketing, then, is to determine the "consumers" mental model and how it is different from our own. This involves identifying the community mental models, as discussed in the previous section, and how they have become defined as they are.

With an understanding of these models, exchanges can then begin occurring among water resources managers, scientists, engineers, public officials, businesses, and the public. The exchange is an exchange of resources or values among these sectors with an expectation of some benefits. These benefits might be an exchange of goods for money, improved education for a fee, a vote in return for lower taxes, or improved community pride because of better environmental stewardship. The emphasis in social marketing is on voluntary behavior. Therefore, water resources managers must ensure that people are getting something they really want (and that they believe they really want it) because the benefits must outweigh the costs of making a sustained change in their behavior (i.e., mental model). Although changing our mental models is not an easy task, it is also not an intractable task. We can implement and maintain water resources management practices – if the right disciplines are included in the management team as players.

One of the keys to marketing is that it is a continuing process. So is water resources management. We must plan for the long term. Water resources management must be an adaptive management process with an iterative cycle of monitoring → assessment → adaptation → monitoring, and so on. This includes continuous interaction among the management team and the stakeholders. Without long-term and continuous reinforcement of the water resources management messages and practices, there will be no long-term improvement in sustainable water resources. Adaptive management is discussed in Chapter 15.0.

Changing the mental models for water resources management is a critical component of sustainability. It requires a major educational effort. Because the mind thinks in pictures (Aristotle), it is useful to have figures, schematics, and illustrations that support the alternative mental model. These alternative mental models should emerge from, or help form the basis for, alternative futures (see Chapter 13.0).

CHAPTER 10.0

ELEMENTS OF SWRM: ECONOMY

Dublin Principle 4 – Water has an economic value
Enlibra Principle 4 – Recognition of benefits and costs
Enlibra Principle 5 – Markets before mandate
Panarchy Principle – Ecological, economic, social, and institutional systems are interlinked

Historically, water resources have been managed for uses such as public drinking water supplies, power generation, irrigation, transportation, recreation, fisheries, flood control, and waste disposal/assimilation. Increasingly, however, additional services provided by water resources, such as habitat and aesthetics, are being recognized, leading to new conflicts in water management. Tradeoffs become necessary as the uses for which water resources are managed increase and water supplies become stretched.

Estimates of market and nonmarket values associated with changing the management of water resources can help decision-makers (and the public) understand the nature of these tradeoffs and allocate water resources across competing societal demands. For example, a water authority may want to know how a specified water price increase will affect the amount of water used by agriculture or other sectors, or the broad range of both costs and benefits from changes in the prices it charges for water or in the rules or processes it uses for allocating water among competing users. A land use planner might want to know about the potential benefits and costs of stream corridor restoration, wetland conservation, and best management practices to compare alternative strategies for protecting aquatic ecosystem services. A regulatory agency might want to know about the costs and benefits of alternative policies for reducing discharges of pollutants into streams. These examples demonstrate the importance of economic factors in water resources management decisions.

Economics focuses on the allocation of resources among alternative, competing ends, or the appropriate use of available means to achieve desired ends (Daly & Farley, 2004). Economics is generally discussed in terms of macro- or microeconomics. Macroeconomics focuses on growth as maximizing throughput or the flow of natural resources from an infinite source (i.e., the environment) through the economy and back to an infinite sink (i.e., the environment). Clearly, the environment is neither an infinite source of raw material nor an infinite sink for waste material. Therefore, sustainable growth might be a misnomer or oxymoron. Although sustainable growth might not be possible, sustainable development is theoretically possible (Holling, 2004). Daly and Farley (2004) define sustainable development as an increase in the quality of goods and services, as defined by their ability to increase human well-being, provided a given throughput. The analogy provided is that as adults we have reached our limits of growth, but not our limits for continued development of intellect, interests, skills, spirituality, or similar traits. Macroeconomics addresses the economy as a whole, at the national and international scale. Macroeconomics typically considers policy interventions such as fiscal policy (government spending and taxation) and monetary policy (money supply and interest rates) to promote sustainable development (Daly & Farley, 2004).

In a global society, both macroeconomics and microeconomics are important for sustainable water resources. Because of the emphasis in this report on communities, watersheds, and regions, microeconomic concepts and principles are the focus of this chapter. Macroeconomic concepts and principles are discussed in subsequent chapters on policies and adaptive management tools.

10.1 Microeconomics

Microeconomics specifically focuses on the efficient allocation of scarce resources among alternatives to achieve desired goals. Water clearly represents a resource that must be efficiently allocated among alternative, competing ends to achieve desired goals. Unfortunately, the full cost and value of water and the effective use of economic principles and approaches for allocation have rarely been determined. Economists and ecologists both are aware of the problems associated with subsidized prices for water, neglect of external costs and benefits, and the political unwillingness to respect the basic notions of scarcity and opportunity costs associated with water (Daly & Farley, 2004). Yet, there are microeconomic approaches for allocating water among competing uses.

Water policy changes usually entail both benefits and costs. Benefits refer to changes that positively affect individuals, whereas costs refer to changes that negatively affect individuals. Measuring the benefits and costs relies on some way of describing preferences among combinations of goods and services (e.g., by estimating the amount of other goods and services a person would need in compensation for, or would be willing to pay for, a given change). Dollars are convenient value measures for decision-makers to use.

The dollar value of a change that improves an individual's economic welfare (benefit) is the maximum amount of income that could be taken away from the individual after the change such that the individual would be no better off than before. This is referred to as the person's willingness to pay (WTP) for the new higher level of well-being. The dollar value of a decline in welfare (cost) is the amount that must be given to an individual after a change so that the individual's original level of well-being is restored. This is referred to as the person's willingness to accept (WTA) compensation. This approach assumes that individual judgments are the appropriate basis for determining the payment or compensation needed.

Both WTP and WTA measures can be applied to the same change in water resources, however, WTP generally is smaller than WTA. Estimates from WTP methods are accepted more widely, partly because they provide a lower bound on value. However, choosing the correct measure depends on which baseline is relevant: do the affected individuals start with property rights to the lower level of well-being (so that WTP is appropriate), or with property rights to the higher level of well-being (so that WTA is appropriate)? For clarity, our discussion is in terms of WTP, but the WTA perspective also can be used.

There are several steps in analyzing the economic benefits and costs of a water management decision (Wilson & Carpenter, 1999). The first step is to identify the physical and economic impacts of the management decision. The next steps are to monetize and then aggregate impacts over individuals and time, for the specific community, watershed, or region.

10.1.1 Identify Impacts

Water resource impacts are generated through a chain of actions and reactions. These include changes in prices, wages, profits, the use of land and other resources, commodity production, environmental quality levels, and direct and indirect human uses of the environment. Impacts can be quite diverse, affecting different people and places over varying time periods. Some impacts involve interactions within the economy, within the environment, and between the environment and the economy. The interactions we explore below are outlined in Figure 10-1, using the example of pricing agricultural irrigation water. This example is relevant because, in the United States and globally, agriculture is the largest user of water withdrawals and a leading cause of water quality problems.

The impacts of an increase in water price depend on how farmers respond. Possible responses for individual farmers include going out of business, relocating the farm business, changing the mix of outputs produced, and changing production and irrigation practices. Given adjustment costs (including learning-by-doing about new practices), farmers' short-run decisions generally will differ from their long-run decisions. In addition, the higher water price could induce technical changes that bring about long-term changes in the structure and location of agricultural production, further affecting the demand for irrigation water.

Economic linkages to input and output markets can generate additional economic welfare consequences. In the irrigation pricing example, if the number of acres (or producers) directly affected is sufficiently large, then the prices of inputs (e.g., local farm labor or farmland) or farm products could be affected because farmers change their demand for inputs. For instance, farmers might use more labor (or some other input) to compensate for using less water. This would mean benefits to input suppliers, and would modify the costs in the target farm population if prices for those inputs rise. If those farmers facing higher costs can increase their output prices, this would yield higher costs for consumers. A similar chain of events can be described if the increase in water prices discourages some farmers so that they use less labor or other inputs, putting downward pressure on those input prices. Changes in input or output prices also would lead to benefits or costs to farmers who are not directly affected by the local increase in water price.

There could also be environmental impacts from the price increase. Reduced irrigation use can yield environmental benefits. Smaller irrigation return flows (that carry fertilizers, salts, pesticides, and sediments) generally will improve water quality in the receiving aquatic habitat. Improved habitat can increase biodiversity and the filtering and flood control capacity of the aquatic ecosystem.

Changes in water quality will lead to changes in the use of water resources, and to other economic responses that influence the benefits of water quality improvements. For example, improved water quality in a lake that improves fishing conditions might enhance the experience of current anglers and lead to an increase in their use of the resource. It may also lead to use of the resource by individuals who had been selecting other sites. It is important to account for these behavioral responses when estimating benefits of the water quality improvement. Because there can be many individuals making decisions and many individuals affected, it is important to look at the statistical (or probabilistic) outcomes, rather than impacts on particular individuals or results from the decisions of particular individuals (Smith, 1997).

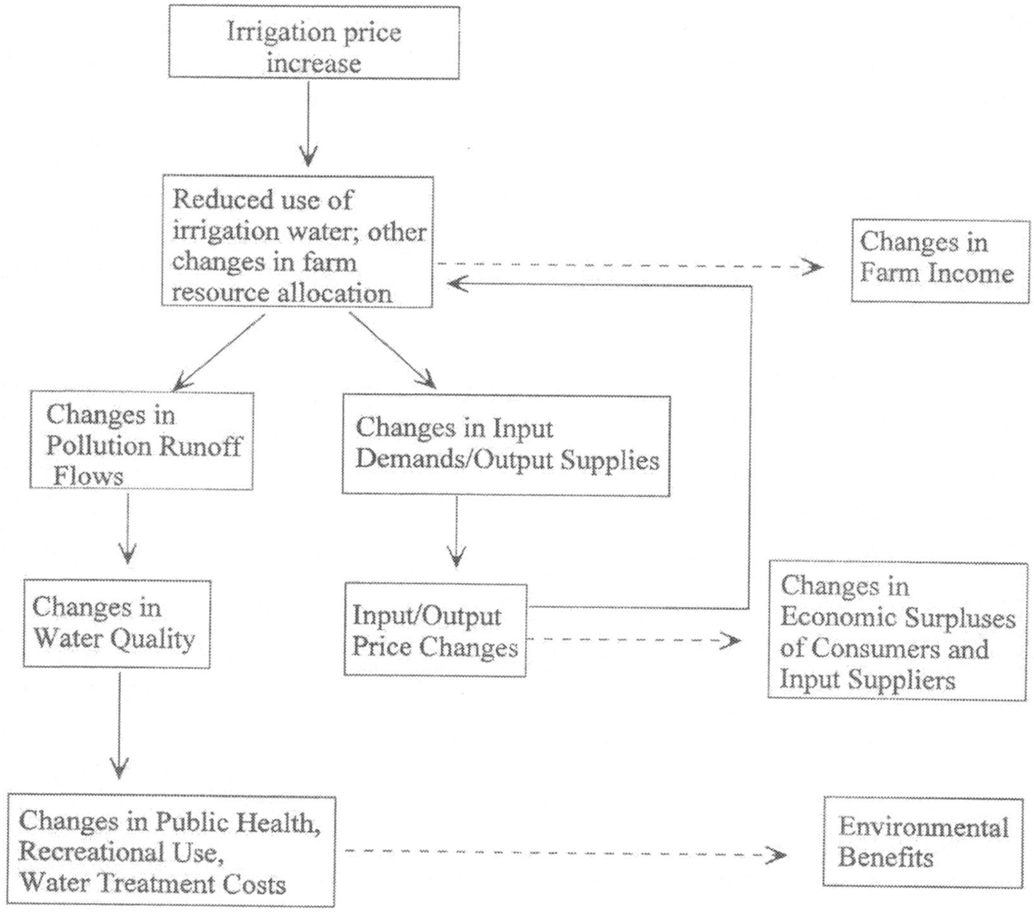

Figure 10-1. Benefit–Cost Linkages Associated with the Price of Irrigation Water.

A fundamental contribution of economics to policy analysis is describing and forecasting such behavioral responses to policy initiatives. Several techniques are available for estimating the demand for water resources in agriculture and other sectors, and nonmarket uses of water resources, such as recreational fishing and boating. These techniques are discussed in Section 10.1.3.

10.1.2 Water Resources Uses Values

Ecological entities are valued for a wide variety of reasons and from a number of perspectives, including habitat value, scarcity value, option value, existence value, cultural value, historical value, biodiversity, intrinsic value, and bequest value. Table 10-1 shows that water resources can be viewed as providing use values and nonuse values. Values for water resources services can be related to the categories shown in Table 10-1. *Use values* are values of goods and services that people use or consume. Many, but not all, of these goods are bought and sold in markets, where the prices and quantities sold reflect the demand for, and supply of, these goods. Examples of freshwater use values that can be estimated in markets include food, building materials (e.g., gravel), drinking water supplies, electric power generation, transportation of coal, and tourism. Watershed use values that often are not traded in markets include recreational fishing, swimming, boating, hunting, bird-watching, sight-seeing, flood control, mitigation of drought, stormwater treatment and/or retention, water purification, and cycling of nutrients and minerals. *Nonuse values* (also called existence values or passive-use values) are values that people hold for a resource without using or visiting that resource. For example, many Americans value the existence of the Everglades whether or not they ever plan to visit this ecosystem. Thus, damage to the Everglades would reduce their well-being.

10.1.3 Valuation Methods

Methods for obtaining value estimates are based on the principle that commodities have value because of their attributes. For example, a car has value because of specific characteristics such as size, color, comfort, body style, handling, gas mileage, and price. People generally have preferences among these attributes and are willing to make tradeoffs among them. Economic methods are designed to measure the underlying preference relations in terms of people's willingness to make such tradeoffs.

Market data can be used to estimate values of commodities that are bought and sold in ordinary markets. Alternative nonmarket valuation methods must be used to estimate use values for goods and services not sold in markets as well as for goods and services that have nonuse values—as is the case for many ecological services. Note that the distinction among categories is not always perfect. For instance, recreation often is in the nonmarket category, even though some recreation is purchased in markets. Nonuse values comprise a sizable portion of total economic value associated with aquatic ecosystems (Wilson & Carpenter, 1999).

Several methods are available to estimate the value of changes in watersheds, including changes in ecosystem services. Each method has a different set of assumptions, data requirements, and features, and thus each has a different applicability. The amounts of technical expertise and data collection effort required also vary among methods. The choice of methods depends on the importance of the decision for which the ecological assessment is to be used; more effort makes sense when the wrong decision would have substantial adverse consequences. The subsections below summarize techniques for valuing both market and nonmarket goods.

10.1.3.1 Market Goods Valuation Methods

Market price methods estimate the economic value of products or services that are bought and sold in commercial markets, or that are sold by government agencies (e.g., water sold through irrigation districts or municipal water authorities, hydroelectric power sold by private utilities or government agencies, and fish or shellfish sold in markets). These methods can be used to value changes in either the quantity or quality of a good or service.

The standard market price method is the estimation of changes in consumers' surplus and producers' surplus using market price and quantity data. The consumers' surplus for a good that is sold at a price is defined as the total amount of money that demanders of the good are willing to pay for the quantity of the good that they consume at the going price, less the amount actually paid. Producers' surplus is the amount producers receive from selling a quantity of a good over and above the variable costs of production. To illustrate the use of this market price method, suppose that the price of water charged by a municipal water supplier is increased. The increase will lead to a reduction in water consumption by households and businesses. Estimates of the water demand functions can be used to estimate the reductions in use. The economic cost to these water consumers is the reduction in their consumers' surplus.

Table 10-1. A Typology of Services from Freshwater Resources.

Value Class	Benefit Category	Benefit Subcategory (example)
Use	Instream	Recreational (water skiing, fishing, swimming, boating, waterfowl hunting) Commercial (fishing, transportation)
	Withdrawal	Public drinking water supplies Power generation Irrigation for agriculture, golf courses Snow making Building materials (sand, gravel)
	Aesthetic	Recreation support (duck hunting) Near-water recreation (hiking, picnicking, photography) Routine viewing (commuting, office/home views)
	Ecosystem	General ecosystem support (biodiversity, food chain) Landscape diversity Water quality protection Regulation of water flows Maintenance of gene pool Weather moderation Pest control Erosion control Storm protection, flood control, regulation of hydrologic cycles Nutrient cycling Sequestration and transportation of materials
Nonuse	Vicarious Consumption	Significant others (relative, close friends) Diffuse others (general public)
	Stewardship	Inherent (preserving remote wetlands) Bequest (family, future generations)

Productivity methods are used to estimate the economic value of ecosystem products or services that contribute to the production of commercially marketed goods. This approach is applied in cases where the products or services of an ecosystem are used, along with other inputs, to produce a marketed good. For example, water quality affects the productivity of irrigated agricultural crops, the productivity of fisheries, and the costs of purifying municipal drinking water. Thus, the economic benefits of improved water quality can be measured by the increased returns from greater agricultural or fisheries productivity, or the decreased costs of providing clean drinking water. Statistical models and economic-engineering models are commonly used for estimation of these benefits (Adams & Crocker, 1991).

10.1.3.2 Nonmarket Techniques

Market price and productivity methods are limited to cases in which prices are available, either from actual markets, simulated markets, or administrative pricing. However, nonmarket valuation techniques are required for the many cases in which prices are not available for services provided by water resources and aquatic ecosystems. One set of value estimation methods is based on how people behave when faced with real-world tradeoffs, for instance between ecological services and other goods. These techniques are called *revealed-preference* approaches. Revealed-preference approaches include the Hedonic property value method, which determines values for ecological attributes based on nearby property values; and the travel-cost method, which determines values for resources based on the travel costs users incur to visit them.

However, revealed-preference approaches cannot yield estimates when there simply is no observed behavior related to a particular ecological service, or when the observed behavior does not permit separating the value of ecological services from other values. Survey techniques, called *stated-preference* approaches, have been developed to obtain value estimates in such circumstances.

Stated-preference approaches use surveys to determine preferences. The two most common stated-preference approaches are contingent-valuation and stated-choice. In the contingent-valuation method, respondents are asked to place a value on different levels of services (e.g., potential changes in fish habitat and population) (Heberling, 2000). In the stated-choice method respondents perform a series of judgment tasks to express their preferences among attributes that include specific services (e.g., fish catch, protecting an endangered species) (Heberling, 2000). Stated-choice methods are currently preferred because of their ability to develop values for individual attributes (e.g., water quantity, water quality, wetland condition) as well as for commodities as a whole (e.g., the watershed), and because they encourage respondents to think about their preferences in terms of trade-offs. All stated-preference approaches risk introducing potential hypothetical bias because we observe what respondents say they would do, rather than what they actually do (Schkade & Payne, 1994).

10.1.4 Challenges in Translating Ecological Value to Economic Value

Conditions are ideal for economic analysis of ecological change when it is possible to describe or predict the ecological change accurately, the nature of the ecological good or service that is lost/gained is understood, and the importance of the change can be quantified (e.g., monetized). However, these ideal conditions seldom are met. Three major challenges for translating ecological value to economic value are discussed here:

1. uncertainty,
2. irreversibility and cumulative incremental effects, and
3. issues of fairness, such as intergenerational equity, discounting, and environmental justice.

10.1.4.1 Uncertainty

Uncertainty in translating ecologic value to economic value arises from ecological sources of uncertainty, economic-ecological sources of uncertainty, and economic sources of uncertainty. One ecological source of uncertainty is in the cause of impact. Another is lack of understanding of how ecosystems respond to stressors (Frost et al., 1999). Spatial and temporal scale affect uncertainty in measuring ecological change. It is generally easier to predict or measure change at smaller scales (Meyer & Wallace, 2001) and in shorter time frames (Firth & Fisher, 1992).

Economic-ecological sources of uncertainty result from the fact that our understanding of how people depend on ecosystems for life-support and fulfillment is incomplete. There are large gaps in the nonmarket benefits category because of limited understanding about how ecological processes help with flood control, stormwater treatment, climate control, pollution mitigation, cycling of nutrients and minerals, flow of energy, and some life support functions (such as provision of oxygen). Similarly, problems arise in describing nonuse values. It often is not clear just how a particular ecosystem change affects religious, ethical, spiritual, or cultural concerns, or how to measure and describe the resulting impacts on these values.

Economic sources of uncertainty result from the fact that monetary methods of quantification sometimes are inappropriate or impossible. The typical economic approach is to add up the changes in ecosystem values over all individuals, over all time periods, and over all affected ecosystem services. However, the relevant population over which to aggregate is often unknown because individuals far from an ecosystem can hold values for the ecosystem, especially if it is unique. In addition, the time period over which to aggregate often stretches far into the future, but the preferences of future generations are unknown and thus assumptions must be made to incorporate values from future generations.

10.1.4.2 Irreversibility and Cumulative Incremental Effects

The value of changes in an ecosystem is not simply the value of changes in a few plant and animal populations or communities, although the valuation process may determine the marginal value for each of these as part of estimating a value of all changes in the system. Furthermore, the key aspect of change in an ecosystem, as opposed to the aggregated value of its parts, could be the value of changes in its embodied interrelationships—especially if crossing a threshold leads to very large changes.

Some ecosystem impacts from individual small actions aggregate to large cumulative impacts. Sometimes effects are reversible, so that restoration is a potential option. In other instances, or beyond some point, even "nickel-and-dime" decisions can chip away at ecosystems and accumulate so that effects become irreversible. The challenge is to identify such instances in time for conscious decisions about whether such degradation is in society's long-term best interest.

Furthermore, what appear to be "small" changes in one ecosystem in essence are magnified if similar changes are happening in many places at the same time. For example, the draining of one particular wetland to create a shopping mall may have modest local impacts because of similar nearby wetlands. But if several wetlands in the large river basin are drained at about the same time, this could affect the viability of that type of wetland as well as the ecosystem services they provide (such as flood control, water purification, and recreational fishing).

10.1.4.3 Fairness: Intergenerational Equity and Discounting, Intragenerational Equity, and Environmental Justice

Future generations are likely to experience significant effects from watershed management policies made today. Intergenerational concerns include several issues:

- Future generations are not "given a vote" in current policy-making.
- Future preferences are dynamic and unknown.
- The common practice of discounting eliminates the preferences of future generations.

Although economics can analyze efficiency, equity issues fall to policy makers and social processes for decision-making. Note however, that economic measures can indicate the efficiency costs of alternative actions based on equity considerations. Individuals and policy makers either implicitly or explicitly discount future benefits and costs when making decisions. The most common example is the need for a bank to pay interest so that savers will give up the use of their money for some time period. With an interest rate of 5%, the promise of $1.05 a year from now is discounted to a present value of $1.00.

The literature on discounting includes ethical and empirical considerations of whether or how to discount future values (Norgaard & Howarth, 1991; Harvey, 1994). Variations in valuation estimates based on choices made with respect to the discount rate, the functional form for discounting, and the possibility of different discount rates for different types of values all can outweigh variations based on choices of value estimation methods or choices of which ecosystem service flows to analyze. Any discount rate over a few percentage points used in conventional discounting functions essentially eliminates from consideration any events more than a few decades into the future. This can be a concern when the ecosystem impacts will not become evident for a long time, or when they will last a long time. Newell and Pizer (2002) argue that uncertainty about future discount rates leads to a higher valuation of future benefits (compared with ignoring this uncertainty).

The structure of preferences is difficult to ascertain for current populations, much less for future generations. Making policy decisions in the present inherently affects future generations' well-being, but also can affect the structure of their preferences. Any attempt to model dynamic preference structures requires assumptions about human behavior that may have little theoretical or empirical foundation (Bowles, 1998). The dilemma is that the alternative approach of assuming stable preferences also has little theoretical or empirical basis.

Intragenerational issues (including environmental justice) also are likely to arise in choices over water resources, especially when different groups have significantly different preferences for ecosystem services. Because individuals with low income have a smaller economic "vote," impacts on subsistence users tend to have little weight in traditional monetization and benefit–cost analysis. Likewise, spatial distributional impacts can be relevant, especially when ecosystem functions that fall within specific spatial boundaries generate nonuse values for distant populations, whereas they may generate primarily use values for the local population.

10.2 Economic Criteria for Water Resources Management

Mechanisms for managing water resources can be judged by economic and social criteria. Social criteria include fairness, opportunities for public participation, the degree of local control, and orderly conflict resolution (Young & Havemann, 1985). Because there is less controversy about economic criteria, we focus on them, in particular

- economic efficiency,
- transaction costs associated with water resources management, and
- the incentives created for technology changes.

Having good information on the economic aspects of water resources management can show the efficiency changes associated with meeting social criteria.

10.2.1 Economic Efficiency

The first criterion for water resources management mechanisms is economic efficiency, exemplified by the Equimarginal Principle. According to this principle, the most efficient allocation of resources occurs when the gain in benefit that would be achieved from allocating resources among different uses is the same for all uses. Economic efficiency has been highlighted as one of the cornerstones of sustainable development (environmental integrity and equity being the other two cornerstones) (Simonovic, 1996). Economic efficiency, however, is difficult to evaluate when the true costs and benefits of water are unknown and often does not adequately address social equity. The Equimarginal Principle is manifested differently in water quantity and water quality management.

10.2.1.1 Water Quantity

In water quantity management, an allocation of water among uses is economically efficient if it maximizes the economic net benefits to society from the water use. The Equimarginal Principle can be manifested as either static or dynamic economic efficiency. Static efficiency deals with allocating water use efficiently across a set of competing uses during a specific period of time (e.g., a month or year). In this case, the Equimarginal Principle is met when the marginal benefit of water in any use, net of conveyance and other supply costs, is equal to its opportunity cost. The marginal benefit of water in a particular use is the additional benefit that would accrue from a small increase in the amount of water used (e.g., an increase in expected profits from increased crop yields resulting from an additional cubic foot of irrigation water). Opportunity cost is the value of a resource in its best alternative use.

To illustrate this concept, suppose that water in a particular watershed is fully appropriated so that additional consumption in one use, say industry, requires diminished use in another, say agriculture. If there are only these two uses, then the opportunity cost of industrial water use is its value in the forgone agricultural use. Suppose the benefit from another cubic foot of industrial water use is $10, while the (opportunity) cost of taking the water from agriculture is only $6. In this case there would be a net benefit of $4 from the transfer. As long as there are differences in the marginal benefits and opportunity costs of water use in different sectors, there are potential net benefits from water transfers. No further efficiency gains are possible when marginal benefits and opportunity costs are equal for all uses.

Dynamic efficiency deals with allocating water use over time. This concept is important in the management of groundwater supplies. When groundwater withdrawals exceed recharge rates, supplies will diminish over time, increasing pumping costs as the water table falls, until the resource is exhausted. Depletion produces benefits for current users but reduces opportunities for future users. Conversely, saving the resource provides benefits for future users but requires sacrifices in the present. Dynamically efficient allocation balances the benefits from current use against the forgone benefits to future users. The latter is referred to as the *user cost* of groundwater. Dynamic efficiency is characterized by equality of the marginal benefits of current use with the *marginal user cost*.

10.2.1.2 Water Quality

In water quality management, we are concerned with the economic efficiency of pollution control. The economic efficiency of pollution control can be judged based on cost-effectiveness or based on a benefit–cost analysis. Cost-effectiveness is related to allocation of pollution control costs among pollution sources. Benefit–cost analysis looks at the cost of an improvement in water quality related to the value of the improvement.

Optimization theory leads to a convenient rule for a cost-effective allocation: an allocation of load reductions is cost-effective for a given water quality goal when the marginal costs of water quality improvements are equalized across sources. With stochastic environmental outcomes, useful cost-effectiveness analysis must consider variations in the ambient impacts of different sources and the natural variability of nonpoint loadings (Milon, 1987; Shortle, 1987; Segerson, 1988; McSweeny & Shortle, 1990; Byström et al., 2000; Shortle, 1990; Beavis & Walker, 1983; Lichtenberg & Zilberman, 1988; Lichtenberg et al., 1989; Shortle et al., 1999).

Most studies of cost-effectiveness in water pollution control sidestep formal consideration of stochastic elements. Effectiveness of pollution control is evaluated on the basis of estimated changes in long-term average or expected effluent flows, and control costs are defined over these flows. Several studies demonstrate that planning based on this deterministic approach could lead to poor environmental and/or economic outcomes (e.g., McSweeny & Shortle, 1990; Shortle, 1990; Shortle et al., 1999; Horan et al., 2002a,b). These studies also present numerical optimization techniques that can be used to compute cost-effective allocations for alternative types of probabilistic water quality constraints.

The concept of cost effectiveness is independent of the water quality goal. Water quality goals can be subjected to benefit–cost tests to determine whether the economic benefits of achieving a particular goal warrant the economic costs. Going a step further, an economically optimal water quality goal would be selected to maximize the (positive) difference between benefits and costs. This leads to a second equimarginal principle: a water quality program is economically efficient when the *marginal benefit* of water quality improvement is equal to the marginal cost of the improvement.

As with the cost-effectiveness concept, when there is uncertainty about costs or benefits because of stochastic elements that affect ambient conditions, or uncertainty about underlying economic values, the rule must be modified. Horan et al. (2002a) illustrates these principles by analyzing the economically optimal allocation of nutrient pollution loads between point and nonpoint sources, and the economically optimal total nutrient load reduction in the Susquehanna River basin. The study also illustrates numerical methods that can be used to compute the economic optimum.

The practical application of economic optimization to the choice of water quality goals is limited by the high cost of measuring the economic benefits of environmental improvements, and the limited political acceptance of benefit–cost analysis for determining environmental goals (Hanley et al., 1997; Hahn, 2000). However, although society may choose not to use benefit–cost analysis for selecting water quality goals, benefit–cost analysis still has an essential role because it forces consideration of the economic rationality of water quality decisions. Although society may want more from its water quality programs than economic rationality, programs in which the costs are grossly out of line with the benefits either because the means of achieving the water quality goals are not cost effective, or because the goals are out of line with what citizens are willing to pay, are unlikely to be sustainable. Benefit–cost analysis can help to avoid choosing excessively costly means for achieving water quality goals, or choosing goals that are either too lax or too strict, judged by the difference between what citizens are willing to pay and the actual costs. Executive Orders 12291 (Reagan, 1981) and 12866 (Clinton, 1993) both emphasize the importance of having the benefits from government regulations justify the costs.

10.2.2 Transaction Costs

Another economic criterion for evaluating water resources management mechanisms is transaction costs. These include the cost of gathering information and determining strategy, and the costs of the bargaining, negotiating, arbitration, judicial, or other processes by which decisions are made and enforced (Randall, 1975). Other things being equal, lower transaction costs are preferable to higher costs. However, because good decisions require good information and analysis, lower cost processes are not necessarily better processes (Colby, 1995).

10.2.3 Long-Run Incentive Effects

The final economic criterion discussed here is whether the program provides incentives that will lower costs associated with water quantity or water quality management. Lower costs could result from changes in technology that would conserve water, reduce the costs of obtaining water, or reduce the costs of water quality protection. Lower costs could also result from shifting activities to less costly locations (e.g., moving water intensive activities to regions with relatively abundant water supplies, or locating water pollution intensive activities in regions where the environmental costs are less).

10.3 Full Cost, Full Value

The Global Water Partnership Technical Advisory Committee Report No. 2 discusses water as both a social and economic good (Rogers et al., 1998). Two general principles are identified for assessing the economic value of water and costs associated with its provision (Rogers et al., 1998). First, both direct and indirect costs of providing water need to be determined. Second, a value for water needs to be assigned based on the desired use of the water. This value will be influenced both by the reliability of the supply and the water quality. For sustainable water, the value and costs of water should balance (Rogers et al., 1998).

The general elements of the full cost of water are briefly described below:

- Full Supply Cost – costs associated with supplying water to users. This cost is the sum of the Operations and Management Costs and Capital Charges.
    - Operations and Management Costs – costs associated with running and maintaining the water supply system (infrastructure). Typically these costs are easy to identify and quantify.
    - Capital Charges – capital consumption (depreciation charges) and interest costs associated with water supply infrastructure, including reservoirs and treatment plants. There is some disagreement as to whether these charges should be calculated based on repaying investments or providing for replacement/upgrade.
- Full Economic Cost – the sum of the Full Supply Cost, the Opportunity Cost of the water, and Economic Externalities associated with the water use.
    - Opportunity Cost – cost associated with the allocation of water, based on the differences in value placed on water by different users (i.e., the difference between the value of the water to the current user and the value of the water to alternative users, including the environment). Zero only when all uses of water are fully supplied (i.e., no shortage).
    - Economic Externalities – costs or economic benefits to users that result from others' use of water (e.g., cost of treating surface water polluted by upstream users, investment savings that result from raising the water table by allowing irrigation water to recharge an aquifer).
- Full Cost – the full cost of water consumption is the Full Economic Cost plus Environmental Externalities.
    - Environmental Externalities – costs associated with public health or environmental impacts to users that result from others' use of water. It is difficult to assign economic values to these externalities, but possible.
- The general elements of the full value of water are briefly described below:
- Economic Value – the sum of the Value of Water to Users, Net Benefits from water use, and Adjustments for Societal Objectives.

- Value of Water to Users – based on the marginal value of product for industrial and agricultural users, based on willingness to pay for domestic users.
- Net Benefits from Return Flows – improvement in groundwater level or surface water flows resulting from return flows from uses (e.g., irrigation, public water supply).
- Net Benefits from Indirect Uses – benefits associated with concurrent water uses that also have value (e.g., in India, irrigation canals provide groundwater recharge, hydropower, and habitat for wildlife).
- Adjustment for Societal Objectives – value added to water, over and above the value to the user, that reflects the value of the water use in meeting societal objectives (e.g., poverty alleviation, employment, and food security). It is determined based on the best available methods to give real gains to society from price differentials among sectors.

♦ Full Value – the full value of water is the Economic Value plus the Intrinsic Value.

- Intrinsic Value – the value people place on the presence of good quality water, irrespective of any real use that can be gotten from it (e.g., a water view, the Everglades). This value is difficult to define and estimate.

Microeconomic practices and procedures are suited for contributing to the efficient allocation of water among desired uses, when the full value and cost of water are known. It is important that all the cost and value elements be quantified.

10.4 Summary

Part of sustainability is being able to pass on "stocks" so future generations have options. Diversity as a stock, therefore, is a critical aspect of sustainability. However, it is important to understand that diversity is *not* achieved by sustaining specific economic sectors or specific ecosystems in their current state. The approach must be to sustain economic and ecological viability, rather than specific sectors or ecosystems. For example, Pennsylvania's Allegheny County (Pittsburgh) is experiencing economic decline, but neighboring Butler County is experiencing economic growth, so the economic viability of the region is sustained. Change will occur and specific sectors or systems might not be sustainable over time, even with enormous inputs of capital and legislation. The status quo does not necessarily define sustainable systems. Sustainability is dynamic, including both positive and negative changes. The Panarchy cycle from α to Ω, with r and K phases, is equally applicable for economic systems as for ecological and social systems (see Chapter 3.0).

CHAPTER 11.0

ELEMENTS OF SWRM: LAWS, REGULATIONS, AND POLICIES

Dublin Principle 2 – Water development and management must be participatory
Enlibra Principle 6 – National standards, neighborhood solutions
Enlibra Principle 7 – Reward results, not programs
Enlibra Principle 8 – Change a heart, change a nation
Panarchy Principle – Ecological, economic, social, and institutional systems are interlinked

11.1 Water Law

Statutes, regulations, and regulatory agencies are integral to water resources management. In general, there are three purposes of water law:

1. Promote desirable behavior and discourage undesirable behavior;
2. Resolve conflicts in an orderly fashion; and
3. Define and protect water rights with adequate security to encourage investment and development (Thompson, 1999).

Water law is distinct from water policy. The objective of water policy is to achieve the maximization of benefits derived from available water resources and their rational management (Wouters, 2001). The primary purpose of water law is to promote and facilitate attaining policy objectives through a system of regulatory and institutional measures and mechanisms (Wouters, 2001). Water law is not limited to water legislation, although legislation usually provides the foundation for water law. Water law can be found in a variety of sources, including environmental law, constitutional law, land law, mining law, administrative, civil, and criminal legislation, and, in some instances, customary law (Wouters, 2001). Water law, doctrine, policy, regulations, and agencies vary considerably from state to state and region to region. There are growing concerns that current water law is incompatible with management approaches required to move toward sustainable water resources.

11.1.1 Federal Laws

At the national level, water is generally considered the property of the state held in trust for its citizens, with overall responsibility for resource-based activities vested in the state. However, there are more than 200 sets of rules, regulations, and laws that provide a legal framework for water resources. Some of the federal regulations and laws relating to ecosystems, and ecosystem processes, related to water are shown in Appendix D.

Conflicts among states on water rights or allocation are adjudicated by the U.S. Supreme Court. The Supreme Court has both original and exclusive jurisdiction over conflicts between states (Original Jurisdiction, *U.S. Code,* 2005). In a review of Supreme Court cases on water law,

Sherk (2000) concludes that litigation is the least favorable mechanism for protecting individuals' interests. Individuals' interests are better served through negotiated settlements.

There are a number of international treaties that have been negotiated between the United States and Canada and the United States and Mexico that deal with international water resources. The cornerstone of international water law is "equitable and reasonable utilization." The rules of international water law apply to sovereign nations, but it is primarily the nations themselves that ensure compliance with international commitments. There is no supra authority to enforce international rules, except for specific circumstances such as a threat to international peace or security (Wouters, 2000). In general, the procedures and rules established through treaties work toward prior notification or exchange of information and consultation concerning any planned measures that might adversely affect other watercourse nations. This has been recently demonstrated through the bilateral U.S.–Canadian International Joint Commission report recommending against the bulk transfer of water from the Great Lakes, which also sets forth criteria for the removal of water (Dendauw, 2000). With increased need for water, similar situations are likely to arise in the future.

There is concern that many of the laws and institutions that are relied on for water resources management were developed in the 19th century and are no longer applicable, particularly in the West (Gillian & Brown, 1997; NRC, 2000b).

11.1.2 Regulatory Agencies

Federal, state, and local agencies and organizations all have specific laws, regulations, and policies that relate to water use and allocation. At the federal level, 22 federal executive agencies in 10 cabinet departments, including the Executive Office of the President, are involved with policies and programs to protect and manage water quantity and quality (NRC, 2000b). Each agency approaches water from its legislatively mandated perspective. In Congress, there are 12 standing committees in the House and Senate that have direct jurisdiction over various components of federal water resource policy, and these are in addition to the Appropriations Committees, Budget Committees, the Ways and Means and the Finance Committees, the House Government Reform and Oversight committees, and the Senate Government Affairs Committees, which all have ongoing interests in water. The federal water management responsibilities also include the competing interests for water among their constituency.

Major federal agencies involved with water quantity and quality issues include:

- U.S. Army Corps of Engineers – responsible for public water resources development and management as well as regulatory functions for waterways and wetlands.
- U.S. Bureau of Reclamation – responsible for providing water in the 17 western states.
- U.S. Environmental Protection Agency – responsible for protecting the quality of the nation's water and wetlands.
- Federal Emergency Management Agency – responsible for responding to natural disasters such as floods and hurricanes, and coordinating activities to reduce future damage.
- U.S. Fish and Wildlife Service – responsible for protecting, conserving, and enhancing fish and wildlife and their habitats.
- U.S. Geological Survey – responsible for providing hydrologic information on surface and groundwater. The USGS provides information on both water quantity and quality.

- National Oceanic and Atmospheric Administration – responsible for describing and predicting changes in the Earth's environment, and managing the nation's coastal and marine resources to ensure sustainable economic opportunities.
- Tennessee Valley Authority – responsible for managing the Tennessee River basin and its water resources including flood control, power generation, navigation, water supply, and recreation.

Many of the federal water management responsibilities are mirrored at the state level, with multiple agencies and organizations involved in water quantity and quality management. State agencies that complement the federal agencies listed above include Departments of Conservation or Natural Resources, Environmental Quality or Environmental Protection, Health, and Emergency Preparedness. Most decisions related to water are made at the state level according to a variety of common law and state statutory approaches that vary from state to state. One of the purposes of resource laws in a civil society is to help ensure that the common good is achieved while protecting individual rights. Local agencies and organizations also have laws and regulations related to the management of water resources, from the safety of drinking water in public water supplies, to water pricing, and restrictions on water use. Allocating water among competing uses is one of the purposes of laws and regulations, and is an area where economics and law interface. Water pricing and demand management, are primarily concerned with the efficiency of water use, which is part of the broader topic of water allocation within and among water systems.

Many of the agencies currently involved in water resources management were instituted in the 19th century to respond to 19th century issues (Gillian & Brown, 1997). In addition, there is currently little coordination among federal and local state agencies, in managing water resources (NRC, 2000b).

11.1.3 Regional Laws and Regulations

Water law and water rights doctrines are established on a state-by-state basis, but regional compacts can, and have been, developed among states (e.g., Chesapeake Bay Compact). As stated earlier, all interstate water conflicts that go to adjudication are referred directly to the Supreme Court, so the legal precedents for addressing interstate conflicts are established (Sherk, 2000).

11.2 Water Rights

Water rights are usufructuary, which means the individual has the right to *use* the water, in contrast to property rights, where the individual *owns* the property (i.e., water).

11.2.1 Open Access Allocation

In some regions of the United States there are no formal rights to water, particularly for groundwater resources. If there are many actual or potential users, and no security of rights, individuals will have little incentive to conserve because the water they save for future use may well be used by someone else. In this situation, users tend to ignore marginal user costs, resulting in an excessive rate of depletion. Note though, that open access is not a sufficient condition for excessive depletion rates. For example, the lateral movement of water in some aquifers is so small that individual users, for all practical purposes, have unique supplies (Beattie, 1981; Nieswiadomy, 1985).

11.2.2 The Riparian Doctrine

Brought as English Common law to the humid eastern United States, the Riparian Doctrine holds that water rights are tied to the ownership of land bordering a natural watercourse. "Reasonable" use is allowed and transfers are prohibited. This doctrine works reasonably well when water is plentiful and population densities are small enough so access to water can be gained easily. However, nontransferability becomes limiting as the amount of land adjacent to water becomes scarce and transfer to nonriparian lands becomes desirable.

11.2.3 Prior Appropriation Doctrine

The Prior Appropriation Doctrine evolved in the western United States in response to the limitations of the Riparian Doctrine. The Prior Appropriation Doctrine assigns water rights according to the first date of beneficial use. It is essentially a queuing system under which those with earlier (senior) rights have prior claim to those with later (junior) rights. Initially, rights could be traded without governmental restriction. However, as the frontier developed, negative externalities (e.g., disturbance of instream uses because of changes in flow regimes) caused by changes in the location of withdrawals associated with water transfers created increasing conflicts between alternative users. Review processes to limit damage to third parties and restrictions on the types of uses to which water rights could be transferred were imposed by the federal government and state governments beginning in the 1860s (Miller, 1995). The merits of these restrictions are the subject of much debate. However, it is apparent they were excessive in some cases, with significant harm to the efficiency of water allocation (Anderson, 1982; Frederick, 1985). The high costs of these restrictions have led several western states, most notably California, to undertake reforms to allow for freer market allocation (Rosegrant & Gazumuri, 1994). At the same time, water laws in the eastern and southern United States are evolving in the direction of the Prior Appropriation Doctrine (Howe et al., 1986).

11.2.4 Comparison of Riparian and Prior Appropriation Doctrines

Thompson (1999) summarized five differences between Riparian and Prior Appropriation Doctrines:

1. In Riparian Doctrine, the right is derived from and tied to ownership of the riparian land; in Prior Appropriation, the right is given to someone who diverts the water and applies it to a beneficial use.
2. Water rights are never lost under Riparian Doctrine as long as the land is owned, although nonuse or abandonment can eliminate water rights under Prior Appropriation Doctrine.
3. Water can be used only on adjacent land within the watershed under Riparian Doctrine, while under Prior Appropriation water can be used anywhere, although the right is appurtenant to a specific land parcel.
4. There are no fixed water quantities assured riparian owners under Riparian Doctrine, while specific flow or quantities are based on seniority date under Prior Appropriation.
5. All water users suffer equally in times of water shortage under Riparian Doctrine, while a seniority system determines allocation during times of shortage under Prior Appropriation.

11.3 Water Policy Decisions

The objective of water policy is to achieve the maximum benefits that can be derived from available water resources and their rational management (Wouters, 2001). Lindblom (1959,

1992) contrasts a "rational" approach for policy decisions with a "muddling through" approach. Lindblom (1959) initially identified "muddling through" as the process by which effective decision-makers determined which policies to implement.

The rational, or scientific, approach to establishing and evaluating policy would be to identify the desired goals; establish objective, quantitative criteria for evaluating the policy options; conduct the analyses; and select the policy with the highest cumulative quantitative criteria score. Although this approach can be rationally described, in practice, it only works for simple, and usually trivial, policy decisions (Lindblom, 1992). Over the last decade, ecological systems have been described and depicted using chaos and complexity theory, mathematically modeled using nonlinear equations, and analyzed using nonlinear systems analysis, with the understanding that simple, linear relationships rarely exist in ecosystems (Berkes & Folke, 1998; Holling, 2001; Levin, 1999).

Adding humans to these equations increases the complexity, which is reflected in the evaluation of policy alternatives. The attempt to approach water policy decisions from the "rational" approach has contributed to the distrust between scientists and policy analysts because the scientists desire to pursue an approach using quantitative, objective criteria and the policy analysts recognize the futility of this objective approach in terms of complexity, information, time, and funding requirements.

"Muddling through," or successive limited comparisons, is an alternative approach for reaching policy decisions (Lindblom, 1992). Some of the characteristics of this approach are listed below:

- Selection of value goals and empirical analyses of needed action are closely intertwined and not considered independent.
- A succession of comparisons reduces or eliminates theoretical constructs or assumptions.
- The emphasis is on marginal or incremental values.

Rather than consider the entire universe of policy alternatives, a few alternatives are selected for comparison. Because each of the policy alternatives being considered would make an incremental advance in achieving the desired end, the distinctions among alternatives are the marginal, or incremental, benefits that accrue. The decision on which policy to implement, therefore, is based on which policy has the greatest marginal benefits (Lindblom, 1959).

The advantages of this successive limited comparison are three-fold. First, the number of policy alternatives for which information and analyses are needed is small and finite. This saves both time and money in evaluation. This approach is fully consistent with alternative futures approaches (see Chapter 13.0). Second, in many cases, it is possible to get agreement on the policy even when different groups can agree on neither why the policy change is needed or the ultimate endpoint. For example, in the recent debate over health care reform, many groups, from American Association of Retired Persons to pharmaceutical companies agreed on the need for revised healthcare policies for prescription drugs even though they differed dramatically in their rationales and desired endpoint for healthcare reform. Third, policy decisions are considered to be interim, rather than final, steps toward a desired end product, and are therefore, fully consistent with adaptive management.

Given the issues associated with SWRM, "muddling through" is a logical and consistent approach for making water policy decisions to further sustainability of water resources.

11.4 Policy Coordination

Water demands, supplies, and quality typically are influenced by public policies that pursue other objectives. Major examples are irrigation water subsidies, agricultural price and income supports, and land use policies. Many of these policies incorporate macroeconomic principles to promote sustainable growth rather than sustainable development (see Chapter 10.0).

11.4.1 Irrigated Agriculture and Water Subsidies

Irrigated agriculture is the leading societal use of water. Globally, about 70% of the water drawn from rivers, lakes, and aquifers is used in agricultural production, and irrigation's comparatively low return flow ratio accounts for about 70% of total consumptive use. The amount of agricultural land under irrigation has increased tremendously during the past 100 years. The United States has the third largest land area under irrigation, 52.8 million acres in 1997, representing 12% of the nation's cropland. The irrigated land is found mostly in the arid West (Gleick, 2000).

In addition to being a big water user, irrigated agriculture is also a major cause of water-related environmental problems. Reservoirs constructed to supply water to agriculture and other sectors have destroyed significant natural assets by inundation. Dams and diversions present physical barriers to fish migration, alter stream flow regimes and water temperatures, and trap sediments. Consequences of these effects include severe degradation of aquatic and riparian habitats, with significant threats to aquatic species. Surface return flows and drainage from irrigated agricultural lands carry salts, fertilizers, pesticides, and other pollutants into surface waters, causing harm to fish and wildlife, and impairing water for human uses. Presently, scores of fish species are threatened or endangered in western rivers where habitat has been degraded by large-scale irrigation projects (Moore et al., 1996). U.S. EPA's 1990 water quality inventory reports that irrigated cropland accounted for 89% of the water quality-impaired river mileage in 11 states in the West (U.S. EPA, 1992). Irrigated agriculture was listed as the source of impairment for 40% of the impaired lakes in the United States in 1990, and was the most common cause of pollution in wildlife refuges in 1990. Aquifers used for drinking water are also contaminated by agricultural pesticides and nitrogen when irrigation water percolates through the ground into the saturated zone.

Subsidies for irrigation water increase the volume of water used in agricultural production. The U.S. Bureau of Reclamation charges farmers less than the full cost of water, encouraging excessive use and resulting environmental problems (U.S. EPA, 2001). Weinberg and Kling (1996) show that full cost pricing of irrigation water in California would improve the efficiency of water use and reduce the environmental harm associated with irrigation return flows. They find that coordinated policies for addressing both salinity and water overuse problems offer the greatest rewards.

Two economic tools have emerged to ameliorate tendencies to overinvest in dams and overuse water in irrigation, benefit–cost analysis, and water marketing. Benefit–cost analysis has illuminated the inefficiency of many water projects before construction, thereby slowing the further appropriation of water from natural systems. In recent decades, water marketing has provided a vehicle for reallocating water from irrigation, where it has a relatively low value, to growing urban and industrial demand, where water has a relatively high value (e.g., water transfer agreement between San Diego County Water Authority and Imperial Irrigation District, www.sdcwa.org/manage/awmp.phtml). Water marketing has even fostered transfers of water rights to instream uses (e.g., Oregon's Instream Leasing Program, Oregon Administrative Rules

Ch 690, Div. 77; and Arizona Instream Water Use Permit, www.water.az.gov/watermanagement/content/waterrights/default.htm).

11.4.2 Agricultural Price and Income Supports

Agricultural price and income policies also affect water quality and the performance of water policies. Since the 1930s, the federal government has intervened in agricultural markets to support farm income. There is mounting evidence that agricultural price supports, input subsidies, and other agricultural policies influence the nature, size, and spatial distribution of agriculture's impacts on water quality through their impacts on crop selection, the scale and location of production, and farming practices and structure (e.g., Abler & Shortle, 1992; Just & Antle, 1991; Lichtenberg & Zilberman, 1986; Platinga, 1996; Swinton & Clark, 1994; Liapis, 1994; Weinberg et al., 1993a,b). For example, policies that increase producer prices without restricting output (e.g., price floors, output subsidies, import restrictions) encourage farmers to increase production. Adverse environmental impacts occur insofar as these policies induce farmers to produce on environmentally sensitive lands or to make more intensive use of environmentally harmful inputs (e.g., pesticides, fertilizer, irrigation water, fossil fuels). Agricultural policies that increase livestock production also increase the volume of livestock wastes. Similarly, input subsidies can have adverse impacts when they encourage the use of potentially harmful agricultural inputs.

Adverse environmental impacts of price and income policies are clearly at odds with environmental goals. They are also at odds with programs intended to improve the environmental performance of agriculture. This situation has stimulated considerable interest in coordinating environmental policies with other policies, and in policy innovations that eliminate conflicts [Organization for Economic Cooperation and Development (OECD), 1989, 1993, 1998].

In particular, there is growing interest in replacing traditional commodity programs with green payment programs. In traditional commodity programs, farmers are paid to produce particular products. In green payment programs, farmers are paid for the environmental services associated with their land, such as nonpoint pollution control (Lynch & Smith, 1994; Smith, 1995). Green payments are being explored in the United States, through programs such as the Natural Resource Conservation Service's Conservation Reserve Program, authorized under the Farm Security and Rural Investment Act, 2002. In Europe, reform of the European Union Common Agricultural Policy opened the door for farmers to be paid for their provision of environmental "goods," including reductions in nonpoint pollution. Although some countries, such as those in Scandinavia, rely on the "polluter pays" principle, the United Kingdom represents one example of the "polluter is paid" principle. In particular, the United Kingdom's Nitrogen Vulnerable Zones scheme and the Nitrate Sensitive Areas scheme are two mechanisms that allow farmers to earn income by reducing nonpoint externalities associated with organic and inorganic nitrogen-based fertilizers (Hanley, 2001).

11.4.3 Land Policies

In recent years, urban and exurban spread (sprawl) have been recognized as a major environmental problem. Urban areas cover a small percentage of the total land surface area, but their environmental consequences can be acute. As the landscape becomes increasingly urbanized, the natural system becomes simplified, and ecologically and hydrologically disruptive elements (impervious surface areas, invasive species) are introduced. Recent studies show that exceeding a threshold of 10–15% impervious cover in a watershed disrupts the habitat for aquatic ecosystems and natural species (Booth, 1991; Jones & Clark, 1987; Klein, 1979; May et al., 1997; Steedman, 1988; Weaver, 1991). Hicks & Larson (1997) report that as little as 8% impervious surface area dramatically reduces the number of aquatic macroinvertebrates present in associated streams. U.S. EPA's National Water Quality Inventory (U.S. EPA, 2002a) identifies urban runoff as one of the leading causes of impairments in estuaries and lakes.

Population and economic growth are obvious drivers of urban and exurban spread. However, development patterns and their ecological consequences are shaped by a host of public policies, including zoning, architectural standards, transportation, and pricing of public services (e.g., water and sewer). It has long been recognized that land use and related policies often encourage urban and exurban spread. Concerns over the economic, social, and environmental consequences of sprawl have become acute and have led to widespread interest in "Smart Growth" policies. Several states (e.g., Arizona, Pennsylvania, and Maryland) and many communities (e.g., Portland, Ore., and Boulder, Colo.) have Smart Growth initiatives. A potential benefit of the reforms and initiatives that come under this broad umbrella would be elimination or reduction of the adverse water quality consequences of urban and exurban development. For this benefit to be fully realized, it will be essential for water quantity and quality impacts to be explicitly considered in the development of community growth strategies. Furthermore, because water quality is influenced by the collective actions of communities, coordination and cooperation among municipalities will be crucial to achieving desired outcomes.

Although there is growing understanding of the relationships between land use and land cover and water quality, there is still limited understanding of how urban and suburban growth policy reforms affect land use and land cover, and water quality. Research to support multiobjective land use planning is essential.

Land use policy reforms are important for water quantity as well as quality. For example, impervious surface area affects not only water quality but also groundwater recharge and stream flow regimes. Leapfrog development affects water supply infrastructure costs. As with water quality, the relationship between land policy reforms and water quantity is an area in which research is needed to help guide community decisions.

11.5 Local versus Federal Authority

A basic principle of the economic theory of federalism is that economic efficiency in the provision of public goods is generally best served by delegating responsibility for the provision of the good to the lowest level of government that encompasses all of the associated costs and benefits. The assumption underlying this principle is that policy choices consistent with the collective preferences of the affected group are more likely when made by decision-makers who represent their direct interests. National regulatory policies may not be sensitive to local conditions. For example, the Clean Air Act and Safe Drinking Water Act in the United States call for U.S. EPA to set uniform maximum allowable levels of pollutants for the entire country. Similarly, the federal Water Pollution Control Act sets uniform minimum surface water quality standards and technology-based maximum effluent standards for industrial and municipal point sources for the entire country. However, several states (e.g., Arkansas, Ohio, and Oregon) use ecoregion characteristics for establishing water quality standards because national criteria are either over- or underprotective of water resources in the ecoregions.

On the other hand, national standards reduce options for local authorities to develop policies that are sensitive to local preferences. Current national point source controls like the

National Pollution Discharge Elimination System (NPDES) constrain the freedom of subnational authorities to devise policies (for both goals and means) that correspond to local costs and benefits. Economic criticism of uniform "command and control" policies is centered on the costs imposed by uniformity and the limited use of information about local conditions possible when devising local solutions. The uniform command and control approach leaves little latitude for allocating pollution abatement among individual sources within the point source category, or between the point and nonpoint categories, to minimize costs. Both the uniform command and control policy instruments and national water quality goals can lead to pollution control levels in which the incremental costs for control exceed the incremental benefits at some sites. An air pollutant trading program was developed with the 1990 Clean Air Act Amendments to include market incentives for attaining air quality standards. Water pollutant trading among point and nonpoint sources has been used in a few instances to achieve water quality standards (Idaho Department of Environmental Quality, 2000; Michigan Department of Environmental Quality, 2004), but the practice is not widespread.

However, although much can be said in favor of decentralized planning so that state and local authorities are responsible for managing problems at the watershed level, some dimensions of water pollution control call for national involvement, because uniformity of outcomes and centralized decision-making are efficient for some problems. National pesticide regulation is an example. National registration and cancellation policies are a crude instrument for addressing the range of societal issues associated with the use of pesticides. However, nationwide cancellation of pesticide use is an optimal decision when the expected marginal damage to human health and the environment from the use of a pesticide is so large that it always exceeds its marginal benefit. EPA pesticide decisions to date appear to fit this characterization fairly well (Lichtenberg, 1992; Cropper et al., 1992). Centralized responsibility for decisions of this type reduce decision-making costs and thereby improve the cost-effectiveness of environmental protection.

In cases where pollutants are transported or move from one jurisdiction (i.e., state) to another, optimal policies for upstream jurisdictions should take into account the control benefits that are received in downstream jurisdictions. This accounting is unlikely with decentralized approaches for the simple reason that the political fortunes of upstream decision-makers depend on the preferences of upstream voters but not downstream voters. Moreover, even when the physical impacts of pollution occur entirely within a given political boundary, costs may still spill over into other jurisdictions. There is ample evidence in the economic literature that people are concerned about resources even if they do not use them (Fisher & Raucher, 1984). Thus well-being in one jurisdiction is diminished when its residents are concerned about another jurisdiction's poor water quality and its aquatic ecosystem impacts. These nonuse values are especially important in the case of unique natural resources. Indeed, some evidence suggests that the high level of concern for water quality that is found in public opinion surveys is often a concern for the quality of water where others live and work rather than one's own water (Bord et al., 1993).

11.6 Regulatory versus Economic Control Mechanisms

Regulatory command and control instruments have been the dominant approach to environmental policy in the United States and other developed countries (Opshoor et al., 1994). Command and control instruments generally involve mandated use of specific pollution control technologies; or adherence to input restrictions, product standards, emissions quotas or other regulations. A number of elements of the current U.S. water quality control policy, which utilizes command and control instruments, are considered to be problematic: the focus on regulating pollution control technology rather than actual water quality improvements, the use of uniform national technology-based effluent standards rather than watershed-based strategies, too little attention to nonpoint sources, and a regulatory structure that is very inefficient and results in private expenditures on water quality protection that are well in excess of those that would be required to achieve the same gains under more cost-effective regulations (Davies & Mazurek,1998). The current structure of water quality protection appears to be grossly inefficient in that the costs of water pollution control are well in excess of the benefits (Freeman, 1982, 2002). Substantial research demonstrates that the uniform technology-based effluent standards of the NPDES system are not cost-effective for point sources (Freeman, 1994; U.S. EPA, 2001). This is because the same effluent standards are applied to all discharges within classes and categories of industries without considering costs or the impacts of their discharges on water quality, which may vary among plants in a watershed and across watersheds.

There is now great interest in more comprehensive, results-oriented, watershed-based approaches that address the diverse set of stressors on water quality, that are cost effective, and that pursue water quality goals in the context of a reasonable balance between costs and benefits (USDA & U.S. EPA, 1998; NRC, 2000a). The use of economic incentives and market-based approaches has the potential to achieve water quality goals at lower costs than command and control instruments (Anderson et al., 1997; Opshoor et al., 1994; U.S. EPA, 2001). Economics research generally suggests that economic incentives produce more efficient outcomes and better long run incentives than command and control (e.g., U.S. EPA, 2001). Among the main reasons

- Economic incentives allow polluters flexibility to choose least-cost options for achieving discharge reductions.
- Economic incentives can, under certain circumstances, induce least-cost allocations with less information on pollution control costs than command and control approaches.
- Economic incentives can, under certain circumstances, provide greater motivation to innovate with clean technology than command and control policies.

Research also demonstrates that there can be large reductions in pollution control costs when using pollution control instruments that allocate load reductions cost-effectively among the subset of polluters targeted for control (e.g., Horan & Shortle, 2001). Table 11-1 summarizes studies compiled by U.S. EPA (2001) that estimate the difference between the estimated cost of achieving water quality goals cost-effectively, and the cost of achieving them by existing regulations. The table shows that substantial cost savings could result from policy reforms that improve the cost-effectiveness of water pollution controls.

However, no one instrument can be the answer in all situations. Design of the instrument and the context of the problem need to be considered. A poorly designed economic incentive may not perform as well as a well designed command and control instrument. And, there will be cases when the cost of designing and enforcing economic incentives with desirable economic and ecological features may be too great. In such instances, incentive mechanisms may do no better, or worse, than a command and control approach.

11.7 SWRM and the Law

Sustainable development has become a global watchword and goal for both developed and developing countries. In addition, there has been considerable discussion of sustainable development in a legal context, as being distinct from a political context, for human, social, economic, and environmental progress (Brundtland, 1993). The legal rules, or rights and obligations, of states and nations arising from sustainable development, however are still unclear (Wouters, 2000). Malanczuk (1995) has suggested a distinction be made between the concept of sustainable development and the legal principles and rules (and their normative clarification and advancement) for sustainable development. Clearly, water law must be part of sustainable water resources, but its specific role in integrated water resources management is still uncertain. In addition, a number of legal scholars have noted some of the legal barriers to restoration of aquatic ecosystems and the use of adaptive management for water resources (Gilliam & Brown, 1997; NRC, 2000b).

Table 11-1. Potential Savings from Using Economic Incentives to Control Water Pollution.

Substance Controlled	Source Year, Source	Geographic Area	Traditional Regulatory Approach	Ratio of Costs: Traditional Approach vs. Incentive Approach
Biochemical oxygen demand (BOD)	Johnson (1967) T	Delaware Estuary	Equal proportional treatment	3.13 at 2 mg/L 1.62 at 3 mg/L 1.43 at 4 mg/L
BOD	O'Neil (1980) T	Lower Fox River, Wisc.	Equal proportional treatment	2.29 at 2 mg/L 1.71 at 4 mg/L 1.45 at 6.2 mg/L
BOD	Eheart et al. (1983) T	Willamette River, Ore.	Equal proportional treatment	1.12 at 3.8 mg/L 1.19 at 7.5 mg/L
BOD	Eheart et al. (1983) T	Delaware Estuary	Equal proportional treatment	3.0 at 3 mg/L 2.92 at 3.6 mg/L
BOD	Eheart et al. (1983) T	Upper Hudson River, N.Y.	Equal proportional treatment	1.54 at 5.1 mg/L 1.62 at 5.9 mg/L
BOD	Eheart et al. (1983) T	Mohawk River, N.Y.	Equal proportional treatment	1.22 at 6.8 mg/L
Heavy metals	Opaluch & Kashmanian (1985) O	Rhode Island jewelry industry	Technology-based standards	1.8
Selenium	Young and Congdon (1994) O	Central Valley, Calif.	Best management practices	1.2
Nitrogen	Moore (2000) O	Long Island Sound	Equal treatment	1.46 at 3.5 mg/L
Nitrogen	Shabman and Stephenson (1998) O	Long Island Sound	Equal treatment	1.56 at 3.5 mg/L
Phosphorus	Faeth (2000) O	Minnesota River Valley	Equal treatment	2.7 at 1 ppm/L
Phosphorus	Faeth (2000) O	Rock River, Wisc.	Equal treatment	1.74 at 1 mg/L
Phosphorus	Faeth (2000) O	Saginaw Bay, Mich.	Equal treatment	5.9 at 1 mg/L

This compilation is taken from U.S. EPA (2001). It draws from previous compilations by Tietenberg (1990) and Anderson et al. (1990). Note: T refers to original citation in Tietenberg (1990), A to original citation in Anderson et al. (1990), and O to original publication of paper.

Given the uncertainty associated with ecological responses to natural and anthropogenic disturbances, adaptive management has become the recommended approach of ecosystem management (Holling, 1978; NRC, 1992, 2000b; Walters, 1986). Water law, however, is not a flexible system, so an important question is whether the rigid aspects of existing water law will permit the use of adaptive management (Neuman, 2001). This is particularly relevant in the arid West with the Prior Appropriation doctrine. With adaptive management, the management rules and strategies change frequently in response to a continuous flow of new monitoring and scientific information. The laws affecting water management in the West are not just the laws of the past 30–40 years, but 150-year -old property rights. In many areas of the West, the water has been fully allocated and nature is not on the list of water rights holders (Neuman, 2001). Water transfers are limited both by law and in practice.

Legal issues are not just associated with 150-year-old water doctrine, but also with the philosophy underlying the more recent National Environmental Policy Act, Clean Water Act and Endangered Species Act legislation, and associated regulations. These Acts were passed with an assumption about the "balance of nature." Ecological theory now acknowledges that there is no "equilibrium or balance of nature," but rather a paradigm of nonequilibrium theory (Botkin, 1990; Meyer, 1993). The idea underlying the Acts passed in the 1960s and 1970s was that there was a relatively stable dynamic equilibrium, and that command and control approaches could regulate human impacts on ecological systems by establishing a fixed set of management goals (Coleman, 1998; Profeta, 1996; Tarlock, 1994). Because ecological systems are not in equilibrium, but rather dynamic systems that are far from equilibrium, a fixed set of "command and control" management rules can be incompatible with dynamic sustainable systems. There is an inherent conflict among laws and regulations that are passed and promulgated to ensure the risks to investment can be understood, and the need to continuously adapt management practices to accommodate changing, dynamic ecosystems. This is an area in need of open and active discussion and research among ecologists, economists, sociologists, attorneys, investment and financial analysts, and policy analysts.

11.8 Water Governance

Rogers and Hall (2003) explain the principles of effective water governance developed by the Global Water Partnership. Water governance is the range of political, social, economic, and administrative systems that are in place to develop and manage water resources, and deliver water services, at various levels of society (Rogers & Hall, 2003). The Global Water Partnership finds that to be effective, water governance must be open and transparent, inclusive and communicative, coherent and integrative, and equitable and ethical (Rogers & Hall, 2003). These qualities are described below:

- Open and Transparent – Being open means being accessible, including using language that the general public can understand. Being transparent means that anyone should be able to understand the steps taken in formulating policy. More open competition and transparent processes may reduce problems of corruption in water governance.
- Inclusive and Communicative – Broad participation (inclusiveness) in policy development improves the quality, relevance, effectiveness, and confidence in the resulting policies. Governments at all levels need to use inclusive approaches when developing and implementing policies. Direct communication between institutions of governance, actors, and stakeholders helps them to work together for their common good.
- Coherent and Integrative – Water governance policies and actions need to be consistent among agencies and institutions, spatial scales, and government levels. Policies and actions need, to the extent possible, to consider all water uses and users and their interconnections with, and impacts on, each other.
- Equitable and Ethical – Water governance policies should ensure that all users and stakeholders have opportunities to maintain or improve their well-being. Water governance must be based on the ethical principles and laws of the society in which it

functions. Water governance policies must be enforced impartially and fairly (Rogers & Hall, 2003).

Recent emphasis on community-based programs and watershed management (U.S. EPA, 1995, 1996, 1997a,b, 1999, 2001; NRC, 1999) moves authority for water resources management into the hands of stakeholders rather than governments. Stakeholder associations have the freedom to be more flexible in management than governments, which must follow often slow and cumbersome protocols to develop and implement policies. The potential role and contributions of civil society organizations should be considered for SWRM.

CHAPTER 12.0

ELEMENTS OF SWRM: INSTITUTIONAL CONSTRUCTS

Dubline Principle 2 – Water development and management should be participatory
Enlibra Principle 1 – Solutions transcend political boundaries
Enlibra Principle 3 – Collaboration, not polarization
Enlibra Principle 8 – Change a heart, change a nation
Panarchy Principle – Ecological, economic, social, and institutional systems are interlinked in a hierarchy of adaptive cycles

12.1 Civil Society and Institutions

What is civil society? There is no single definition on which everyone agrees. The London School of Economics Centre for Civil Society definition is:

> Civil society refers to the arena of uncoerced collective action around shared interests, purposes, and values. In theory, its institutional forms are distinct from those of the state, family, and market, though in practice, the boundaries between state, civil societies, family, and market are often complex, blurred, and negotiated. Civil society commonly embraces a diversity of spaces, actors, and institutional forms, varying in their degree of formality, autonomy, and power. Civil societies are often populated by organizations such as registered charities, development nongovernmental organizations, community groups, women's organizations, faith-based organizations, professional associations, trades unions, self-help groups, social movements, business associations, coalitions, and advocacy groups. (www.lse.ac.uk/collections/CCS/introduction.htm).

Given this definition, there are hundreds of civil society organizations and institutions in many communities, and thousands of organizations and institutions in the United States (see Chapter 2.0).

Civil society organizations and institutions are critical in moving toward SWRM. One of the most important Enlibra principles is "Collaboration, Not Polarization" (www.epa.gov/adminweb/leavitt/enlibra.htm). Collaboration is enhanced when partnerships are forged. Engaging stakeholders not only in the process, but also in partnerships, is critical for identifying, building, and maintaining linkages among environmental, economic, and social sectors. Developing partnerships and implementing these integrated programs is an area where civil society institutions and organizations can make significant contributions. In fact, Peter Drucker (1999) has commented that:

> The 21st century will be the century of the social sector organization. The more economy, money, and information become **global**, the more **community** will matter. And only the social sector nonprofit organization performs in the community, exploits its opportunities, mobilizes its local resources, solves its problems. The leadership, competence, and management of the social sector nonprofit organization will thus largely determine the values, the vision, the cohesion, and the performance of the 21st century society.

As has been stated repeatedly throughout this report, the environmental, economic, and social sectors are interdependent and when an integrated, systems approach is used for management, the interactions can lead to complementary programs, leveraging of funds, partnerships, and greater cost effectiveness of all projects, programs, and agencies. A key part of this process is forming partnerships and engaging stakeholders in the process. Regardless of the organization or institution, there are three areas in which nearly all civil society institutions contribute to the process:

1. communication
2. education and outreach, and
3. liaison with government agencies.

12.2 Communication

Every organization or institution has some way of exchanging knowledge or information with its members, whether it is through printed or electronic media or orally. By definition, communication is the process of exchanging information or knowledge between or among individuals. Effective organizations ensure that the message received by their membership represents the information or knowledge that was meant to be sent. Civil society institutions have specific visions and missions, which entice individuals with similar beliefs and values to become part of these institutions. These civil society organizations, therefore, can provide insight into the beliefs and values related to water held by their membership, as well as the mental models of the organization or institution related to water. These organizations and institutions can serve as the initial check point for whether the information underlying individual and community beliefs is based on scientific principles and current knowledge. They can also act as interpreters between different groups—translating discipline- or group-specific jargon into language their membership understands. These organizations often have platforms, such as conferences, where results and achievements can be showcased.

12.3 Public Education and Outreach

Nearly all civil society organizations and institutions also have an education and outreach function. Public involvement is necessary for the development of sustainable water resources, and public education and outreach is necessary for public involvement. People need to understand the trade-offs that are involved in choices regarding water management to be able to make decisions that have a good chance of achieving their goals. There are a number of groups that include sustainability education and outreach in their programs. Table 12-1 is a sampling of education and outreach programs and activities used by U.S. agencies and organizations to inform people of issues related to sustainability.

Table 12-1. Sampling of Education and Outreach Programs/Activities Used by U.S. Agencies and Organizations to Inform People of Issues Related to Sustainability.

Agency/Group	Programs/methods
U.S. EPA Office Transportation and Air Quality Planning www.epa.gov/otaq/transp/traqpapo.html	Partnerships with communities and governments to develop programs to educate public and policy makers about transportation related threats to air quality, and encourage activities that will reduce the impacts, such as car-pools, use of public transportation, and making public transport systems more usable.
U.S. EPA Office of Water, Office of Wastewater Management http://cfpub.epa.gov/npdes/stormwater/menuofbmps/pub.ed.cfm	Fact sheets about the impacts of stormwater discharges on water bodies and the steps that can be taken to reduce pollutants in stormwater runoff. Requirements for public education included in Regulation language.
U.S. EPA Office of Policy Economics and Innovation www.epa.gov/ecocommunity/about.html	Community-based Environmental Protection: place-based policies to change behavior through partnerships, institutes, research projects, and protection programs.
University of Connecticut http://nemo.uconn.edu/about.html	Nonpoint Education for Municipal Officials: educational program for local land use officials addressing relationship of land use to natural resource protection.
NOAA, U.S. EPA Office of Wetlands, Oceans, and Watersheds www.epa.gov/nep	National Estuaries Program: to improve quality of estuaries of national importance—workshops, National Estuaries Day, pamphlets, classroom activities.
U.S.D.A. Cooperative State Research, Education, and Extension Service www.reeusda.gov/nre/sustain	Incorporate concepts and principles of sustainability into policies, practices and programs through partnerships, serving on committees, documenting active programs, providing educational materials, and supporting projects demonstrating sustainability.
U.S. EPA Office of Groundwater and Drinking Water www.epa.gov/safewater/protect.html	Source Water Protection program: informational materials, case studies of protection programs, planning tools.
NOAA Coastal Services Center www.esc.noaa.gov	Informational materials on smart coastal growth, hazards, and habitat; and National Spatial Data Infrastructure
NOAA Office Ocean and Coastal Resource Management www.ocrm.nos.noaa.gov	Resources and documents on coastal issues written for the general public.
NASA Earth Science Enterprise www.earth.nasa.gov/eseapps	Web site describing how satellite images are being used to study environmental interactions.
Center for Watershed Protection www.cwp.org/outreach.html	Workshops, advertising/education campaigns, resource center, templates of ordinances to use to protect local watersheds.
River Network www.rivernetwork.org	Web site to link activists with river information, resources, and services.
Smart Growth Network www.smartgrowth.org	Web site with resources for planners, decision-makers, and activists; conferences, publications, grants, and success stories.
American Farmland Trust www.farmland.org	Web site & magazine, consulting, Steward of the Land award, participation at conferences, publications.

12.4 Liaison with Governments at All Levels

Nongovernmental organizations often work with government institutions, and so are well placed to act as intermediaries between government and the public, representing the interests of each group to the other, and helping them see where they can work together. This liaison function with governments at all levels is critically important. A recent WERF study documented (again) that governments at all levels are not trusted by the public in general, nor by many interest groups (WERF, 2004). People join specific civil society organizations and institutions because they believe in their mission and their vision. Civil society institutions can provide the element of trust that is missing in many interactions with government agencies, to affect changes not only in their membership, but also public opinions and behaviors.

Civil society organizations can also facilitate communication between government institutions and/or levels that would not occur through regular government channels. Coordination among institutions and entities at different scales is a necessity for SWRM. The Western Water Commission recommended that Regional Coordinating Councils be developed to coordinate and manage western water resources. These Regional Coordinating Councils were specifically not to be a governmental entity, but rather a civil society organization or institution, consisting of governmental agencies, tribes, private sector businesses, watershed associations, and NGOs. The key is in identifying the appropriate groups or organizations to facilitate this interaction and coordination.

CHAPTER 13.0

ELEMENTS OF SWRM: PLANNING

Dublin Principle 2 – Water development and management should be participatory
Enlibra Principle 2 – Science for facts, process for priorities
Enlibra Principle 4 – Recognition of benefits and costs
Panarchy Principle – The complexity of systems emerges from a small set of controlling processes
Panarchy Principle – Water resources and related elements are nonlinear, uncertain, and unpredictable

13.1 Planning

There are more than a thousand books (i.e., Amazon.com; 1163 books) and millions of references on the Internet (i.e., Google.com – 17,500,000 citations) on strategic planning for private sector, civil society organizations, and government entities. The purpose of this chapter is not to expand on these references or books but, rather, to reinforce the necessity of strategic planning if sustainable water resources are to represent a management outcome. The basic elements of strategic planning have not really changed since the 1950s, and the W*ater Resources Council Economic and Environmental Principles and Guidelines for Water and Related Land Resources Implementation Studies* (WRC, 1983) are still applicable for integrating economic and environmental considerations into traditional engineering water resources plans.

13.1.1 Planning Elements

Although there are numerous ideas, approaches, practices, and procedures for strategic planning, common elements among most plans are listed below:

- Vision – the desired future state;
- Mission – purpose of the organization, agency, institution developing the plan;
- Core Beliefs – what is considered truth;
- Core Values – what actions derive from core beliefs;
- Issues – what problems, issues, are there in the community, watershed, organization, institution, etc.;
- Risks and Rewards (Benefits) – what are the risks of changing the status quo; what are the benefits of changing the status quo;
- Goals – broad statements of how the vision will be achieved;
- Strategies – actions needed to achieve the goals;
- Tactics – specific details to implement the strategies; and
- Performance Measures – quantitative indicators or metrics that indicate progress toward achieving the goals and vision.

These elements are generally found in one form or another in most strategic plans. The order of the elements listed above—vision, mission, etc.—however, does not reflect the order typically followed in developing a strategic plan. The usual order and process of developing strategic plan elements is shown below:

- issues,
- risks and rewards,
- core beliefs,
- core values,
- vision,
- mission,
- goals,
- strategies,
- tactics, then
- performance measures.

13.1.2 Issues

A common reason given for not developing strategic plans is that the process is overwhelming. Because water is essential for life, it interfaces with nearly every environmental, economic, and social issue in the community, watershed, or region. However, not all these issues are of equal priority nor of equal visibility in the community or watershed. The key is to pick one or two high priority or highly visible issues and initiate the process. The issues will evolve as the plan is refined, additional stakeholders become engaged, and progress is made in planning, implementation, and forming linkages with other sectors. The first step is choosing an entry point into the planning process (Global Water Partnership, 2004). As discussed in Chapter 11.0 on policy development ("muddling through"), and in the Global Water Partnership Handbook, major reforms in any system rarely occur, nor are they essential to catalyze change (Global Water Partnership, 2004). It is the incremental progress toward attaining goals that creates an atmosphere of success. Initiating the process and taking the first steps are the most critical in moving toward SWRM.

There are usually multiple water issues within any community or watershed—flooding, stormwater runoff, point and nonpoint source pollutants and discharges, impaired 303(d)-listed stream reaches, diminishing water supply, increased taste and odor problems, fish consumption advisories, mosquito control—that can provide an entry point for sustainable water resources planning. Once entry points or issues have been identified, it is important to clearly define the issues and establish priorities on these issues through stakeholder input. The next step is to determine the interactions of these issues with other sectors or resources, such as agriculture, transportation, energy production, human health, education, industry and businesses, as well as watershed land use, wetlands, riparian areas, and similar resources (Global Water Partnership, 2004). With this information, one or two issues can be targeted for planning and subsequent resolution.

Selecting the highest priority issue is not always the most desired entry point. If this issue has highly polarized the community, if it represents an entrenched issue that is nearly intractable, it is not a good candidate for demonstrating success through the planning and adaptive management process.

13.1.3 Risks and Rewards

In determining which issue(s) might be selected as the entry point into the planning process, the risks (cost, downside, failure, etc.) and rewards (benefits, upside, low-hanging fruit, success, etc.) need to be identified and evaluated for each issue. Stakeholder input is critical for ensuring all facets of the issue(s) are considered and addressed. Getting stakeholder buy-in at the beginning of the planning process is essential for their involvement during implementation and issue resolution. However, this input must not fall into the trap of endless consultation at the expense of action (Global Water Partnership, 2004). Periodic stakeholder involvement must occur, not only throughout the planning process, but, importantly, throughout the implementation and management process. The role, process, and benefits of stakeholder involvement are described by Barbour et al. (2004).

13.1.4 Core Beliefs and Values

Regardless of what we say we value, our true beliefs are expressed in our actions, where we spend our time, and how we spend our money. Identifying the core beliefs and values of the community (i.e., social mindscapes, Chapter 9.0) must be an integral part of the planning process. If the core beliefs are not based on facts and accepted research results, social marketing approaches might be required to change the fallacious core beliefs. Planning is an opportunity to engage sociologists, cognitive psychologists, and behavioral organization specialists as part of the planning team. These disciplines have the expertise and tools to evaluate and document the core beliefs and associated mental models and social mindscapes of communities, watershed associations, various organizations, and government agencies.

13.1.5 Other Planning Elements

Vision and mission statements need to incorporate the core values of the community, watershed, basin or region for which the plan is being developed. Stakeholder involvement during this phase is critical. This is an area where civil society institutions and organizations can provide information and insight into the vision of their respective communities and also aid in conducting the visioning sessions. Formulating the draft vision and mission statements is not a group activity; one or two individuals need to draft the initial statements and then iterate with agencies, organizations, institutions, and the private sector to refine the statements. Unpacking the final vision and mission statements (i.e., documenting what meanings are intended by key words or phrases in the vision and mission) should accompany these statements. Goal statements might be developed around the three primary sectors—ecological, economic, and social/cultural. Strategies then might reflect the institutional, legislation and policy, governmental and various management tools and instruments (see Chapters 16.0–18.0) available to achieve goals.

The Water Environment Federation, American Water Works Association, American Society of Civil Engineers, and most professional organizations have a number of useful books, articles, or reports on strategic planning that can be consulted for guidance on how to develop vision and mission statements, goals, strategies, tactics, and performance measures. These topics have been extensively discussed for industry, business, nonprofit organizations, and government agencies, so they will not be developed further here.

In 2002, the World Summit on Sustainable Development, held in Johannesburg, South Africa, called for all countries to develop integrated water resources management and water efficiency plans by 2005 (see Chapter 2.0). The "Why, What, and How" of developing integrated water resources management and water efficiency plans has been addressed by Jonch-Clausen (2004) in the Global Water Partnership Technical Advisory Committee 10 Report. The substance of this report, along with references to other Global Water Partnership Technical Advisory Committee Reports and related literature, can be found in the Global Water Partnership

Handbook (Global Water Partnership, 2004). This handbook focuses on developing strategies for integrated water resources management and water efficiency plans.

The Jonch-Clausen (2004) report and Global Water Partnership Handbook (Global Water Partnership, 2004) can be found at www.gwpforum.org. These two reports, in conjunction with the WERF framework and guidelines (Thornton et al., 2005) and a general book on strategic planning, are suggested resources for preparing sustainable water resources plans.

One of the objectives of planning is to provide a path for moving from the current position to a desired future position. In many cases, identifying the desired future condition and its attributes can be difficult. Alternative futures is a tool that has been used by some communities to identify the desired future condition and its attributes.

13.2 Alternative Futures

Alternative futures analysis is the consideration of potential future conditions or scenarios. It is well suited as an approach to help communities make decisions about water and land use (U.S. EPA, 2002b). Alternative futures can be applied in one of two ways. One approach for incorporating alternative futures is to propose a set of decisions and/or policies as a scenario and analyze the effect of that scenario on indicators of interest (e.g., economics, environment, society) (U.S. EPA, 2002b; Steinitz, 1999). In this application stakeholders are informed and educated about the effects of their actions on the indicators. A second approach for incorporating alternative futures analysis is to propose an endpoint (e.g., increase the supply of adequate clean water for a community by 20%) and consider the actions that would be necessary to achieve that end (Georgia Basin Futures Project, 2004; Dator, 1993). In this application stakeholders are informed and educated about the trade-offs required to achieve their desired end. In both approaches, alternative futures analysis works to improve informed decision-making through education of decision-makers and stakeholders about the results of their decisions expected based on scientific knowledge.

Unfortunately, the potential number of alternate futures is almost infinite, and consideration of all of them is virtually impossible. However, for the purpose of planning, this large number of alternative futures can be reduced to a fairly small number of policy alternatives that have the greatest potential for influencing chosen indicators. This is the philosophy of "muddling through" policy (Chapter 11.0). For example, in the Willamette Basin alternative futures analysis (U.S. EPA, 2002b) three scenarios were evaluated: continuation of current policies towards development and conservation, adoption of more conservation-oriented policies, and adoption of more development-oriented policies. As another example, in the SpaceCast program (Air University, 1994), four alternative futures relevant to the future of the space program were developed based on the interaction of five global elements:

1. the number and distribution of people,
2. types of geopolitical organizations and their interactions,
3. economics,
4. effects of new technologies, and
5. environmental constraints.

In this way, the essential elements of complex systems are captured in a fairly small number of alternative futures (U.S. EPA, 2002b). This reduction of a complex system to the consideration of a smaller number of controlling processes is consistent with the idea in

Panarchy Theory that complex living systems are primarily controlled by a few critical processes (Holling, 2000, 2001; Gunderson & Holling, 2002).

Alternative futures analyses frequently utilize the input of stakeholders to develop future scenarios (U.S. EPA, 2002; Steinitz,1999). The primary method of stakeholder involvement is through workshops (Dator, 1993). In the Georgia Basin Futures Project, a computer interface is being developed to collect information on the characteristics of desirable future scenarios from stakeholders and educate them about the tradeoffs necessary to achieve desirable futures (www.basinfutures.net).

The Georgia Basin Futures Project (www.basinfutures.net) is an example of a sustainability project that addresses ecological, economic, and social issues using alternative futures analyses. The project is being implemented in the Georgia Basin, in British Columbia and the State of Washington, and is led by the Sustainable Development Research Institute in Canada. The definition of sustainability for the project is, "living within natural limits while maintaining the current standard of living and enhancing the well-being of the community and the quality of life." The project involves determining likely development scenarios for the basin and researching the environmental, social, and economic consequences of those scenarios. This information will be incorporated into an integrated assessment model that will use a "backcasting"method. In backcasting, a desired outcome is selected and the model determines the approaches and tradeoffs necessary to achieve the outcome. The project team intends the model to be used to inform the public and decision-makers, as well as to poll these groups to determine the outcomes they desire and the tradeoffs they are willing to accept in the basin. Although this project focuses on development in the basin, the questions being asked are the same as those that are necessary for the development of sustainable water resources.

In the alternative futures process the indicators chosen for evaluation are important. In order for the alternative futures analysis to be relevant and useful to the stakeholders, the indicators chosen must be relevant to the stakeholders. The indicator responses then become the basis for the alternative futures analysis. Indicators act as a rallying point for stakeholders; groups may desire the same endpoint or response for an indicator (e.g., increase in upland forests) even if they disagree as to the details of how it should be achieved. The selection of appropriate indicators is important both for assessing the current resource condition, as well as for alternative futures analyses. These indicators must be selected judiciously from the environmental, economic, and social sectors for monitoring and assessment.

CHAPTER 14.0

ELEMENTS OF SWRM: MONITORING AND ASSESSMENT

Dublin Principle 2 – Water development should be participatory
Enlibra Principle 2 – Science for facts, process for priorities
Enlibra Principle 7 – Reward results, not programs
Panarchy Principle – Water resources and related elements are nonlinear, uncertain, and unpredictable

14.1 Monitoring

14.1.1 Indicators of Sustainability

The U.S. EPA Science Advisory Board (Young & Sanzone, 2002) developed a list of reporting categories that are useful for describing the condition of six ecological attributes essential for assessing and reporting ecological condition (see Table 2-1). The essential ecological attributes, and the reporting categories associated with them, suggest indicators that could be monitored to assess water resources sustainability from an ecological perspective (see Chapter 2.0 for additional discussion).

An Environmental Sustainability Index (ESI) was developed collaboratively by the World Economic Forum's Global Leaders for Tomorrow Environment Task Force, the Yale Center for Environmental Law and Policy, and the Columbia University Center for International Earth Science Information Network (World Economic Forum et al., 2002a). This collaborative work on developing national indices of sustainability and environmental management could also be useful for characterizing water management units and monitoring changes and trends in these units over time.

The 2002 ESI quantifies, at a national level, progress toward environmental sustainability by tracking changes in the scores of 22 indicators (World Economic Forum et al., 2002a). These indicators are constructed from 67 variables. Progress is tracked in five areas: environmental systems, reducing environmental stresses, reducing human vulnerability to environmental stresses, social and institutional capacity to respond to environmental trends, and global stewardship. Monitoring and tracking the 67 variables and 22 indicators for these five elements of sustainability makes it possible to assess a country's current environmental quality and track its progress over time.

The 2005 ESI integrates 76 variables and data sets into 21 indicators of environmental sustainability in the same five broad categories listed above (Esty et al., 2005). The selection of the indicators and variables is based on the pressure-state-response environmental model. The 2005 ESI results are available for 146 nations, with Finland, Norway, Uruguay, Sweden, and Iceland receiving the five highest ranks. The United States ranked 45 out of 146 nations.

Although the ESI is generally correlated with per capita income, development *per se* does not determine environmental condition. For some indicators there are negative relationships with per capita income. Environmental sustainability, therefore, will not necessarily emerge from economic development, but requires the concerted focus of local, state, and federal governments, the private sector, communities, and NGOs (World Economic Forum et al., 2002a).

An Environmental Performance Index has also been developed for use in comparing efforts of nations to manage a set of common policy goals in the areas of air and water quality, ecosystem protection, and climate change (World Economic Forum et al., 2002b). These four indicators are based on 13 variables. This index is more focused and results oriented than the ESI.

Many of the variables used to develop these indices are available for the United States at the national, state, or county level. Thus, it should be possible to evaluate these data sets at the larger water management scales, or to calculate similar indices to classify water management units. Components of the ESI that would be applicable at the regional and watershed water management levels include environmental systems, reducing environmental stresses, reducing human vulnerability, and selected societal and institutional capacity indicators. Variability in the elements of Global Stewardship is expected to be minimal among water management units, so these elements are not included in our analysis. Table 14-1 lists a modified set of ESI elements and the associated indicators and variables that might be considered as part of a core set of indicators for monitoring sustainable water resources.

The indicators and variables included in Table 14-1 represent a preliminary set of indicators or variables that might be considered for monitoring, assessing, and ultimately managing for sustainable water resources. The purpose was not to generate an exhaustive list, but rather to highlight the need to include indicators and variables, not just from one area such as hydrology, ecology, economics, or sociology, but from all of these sectors, if sustainable water resources are to become a reality. The four major categories of indicators and variables in Table 14-1 are important for sustainable water resources because they provide insight into the economic and social sectors as well as ecology and hydrology.

14.1.2 Additional Monitoring Indicators

Every local, state, and federal agency that deals with water, every water management or water research organization, and every institution addressing water has a set of indicators and variables that are monitored. Many of these indicators can be considered as additions to Table 14-1. Monitoring information is the only way of assessing the effectiveness and performance of management practices. Although monitoring provides the ultimate benchmark for measuring success, it is expensive, so additional indicators or variables should be selected only if the resulting information adds value to an assessment or evaluation of the performance of management practices.

Table 14-1. Indicators for Tracking SWRM Based on the Environmental Sustainability Index and Its Elements.

Component	Indicator	Variable
Environmental Systems	Water Quantity	Annual change groundwater level (USGS, Soil and Water Conservation Commissions, other state agencies, water management districts)
		Consumptive use as % total water withdrawal (state agencies, USGS). Per capita interbasin water transfers (water management districts, USACE)
	Water Quality	Dissolved oxygen (state agencies, USGS, U.S. EPA, university studies, other studies)
		Phosphorus (state agencies, USGS, U.S. EPA, university studies, other studies)
		Suspended solids (state agencies, USGS, U.S. EPA, university studies, other studies)
		Conductivity (state agencies, USGS, U.S. EPA, university studies, other studies)
		Areas/miles of 303(d)-listed waters (states, U.S. EPA).
		% surface waters under fish consumption advisories (state agencies)
		% wetland loss (state agencies)
		% impervious surface area (state agencies)
	Biodiversity	Fish Index of Biotic Integrity (IBI, Steedman, 1988) scores (state agencies)
		Benthic IBI scores (state agencies)
		Breeding bird surveys (USGS)
	Land	% land with low anthropogenic impact (Last of the Wild Project, analysis of current GIS data)
		% area protected (federal and state agencies, NGOs)
		% land with high anthropogenic impact (Last of the Wild Project, analysis of current GIS data)
	Air Quality	Urban population weighted NO_2 concentration (states, U.S. EPA)
		Urban population weighted SO_2 concentration (states, U.S. EPA)
		O_3 noncompliance areas (states, EPA)
Reducing Environmental Stresses	Water Stress	Fertilizer consumption/hectare arable land (USDA National Agricultural Statistics Service)
		Pesticide use/hectare cropland (USDA National Agricultural Statistics Service)
		Industrial organic pollutants/available freshwater (U.S. EPA)
		% area under severe water stress (USDA, NOAA)
	Air Pollution	Fossil fuel consumption per populated land area (states, U.S. EPA)
		Anthropogenic NO_x emissions per populated land area (states, U.S. EPA)
		Anthropogenic SO_2 emissions per populated land area (states, U.S. EPA)
		Anthropogenic VOC emissions per populated land area (states, U.S. EPA)
		Vehicles in use per populated land area [Department of Transportation (DOT), U.S. EPA, states]
	Natural Resource Management	Productivity overfishing (FWS, NMFS, state fisheries)
		Percentage of total forest area that is certified for sustainable management (USDA Forest Service, state forestry)
		World economic forum survey on subsidies (World Economic Forum)
		Salinized area caused by irrigation as percentage of total arable land (USDA, states Ag)
		Agricultural subsidies (USDA, AES, state ag)
	Ecosystem Stress	% change forest cover (USDA FS) from 1990–2000
		Acres/miles of water resources with ANC < 50 mg/L (states, U.S. EPA)
	Waste and Consumption	% of product imported per capita (economic census, state and local economic agencies)

Component	Indicator	Variable
	Pressure	Waste disposal at landfills per populated land area (state and local agencies)
		Paper recycling rate (state and local agencies)
		Glass recycling rate (state and local agencies)
		Total radioactive waste [U.S. Dept. of Energy (DOE), state agencies]
		Ecological footprint per capita (www.footprintNetwork.org)
	Population Pressure	Net population migration (U.S. Census)
		% change projected (U.S. Census) from 2000–2025
		Total fertility rate (U.S. Census)
Reducing Human Vulnerability	Basic Human Sustenance	Proportion food insecure (U.S. Census Current Population Survey, USDA Economic Research Service
	Environmental Health	Child asthmatic % [Centers for Disease Control (CDC), state health departments]
		Childhood obesity % (CDC, state health departments)
		Child death rate from respiratory disease (CDC, state health departments)
		Death rate from intestinal infectious disease (CDC, state health departments)
		Under-5 mortality rate (U.S. Census)
	Economics	% below poverty level (U.S. Census)
		% homeless (states)
		% in subsidized housing (U.S. Census)
		% unemployment (U.S. Census)
Social and Institutional Capacity	Environmental Governance	Ratio of gasoline price to world average [DOT, state DOT]
		% land under protected status (The Nature Conservancy, U.S. Fish & Wildlife Service, U.S. EPA, states)
		% variables missing from the Consultative Group on Sustainable Development Indicators "Rio to Joburg Dashboard"
		Knowledge creation in environmental science, technology, and policy
	Eco-Efficiency	Energy efficiency (DOE, states)
		Hydropower and renewable energy production/total energy consumption (DOE, states)
	Private Sector Responsiveness	Dow Jones Sustainability Group Index (DJSGI)
		Number of ISO 14001 certified companies/billion dollars GDP (World Bank)
		World Economic Forum Survey on private sector environmental innovation (World Economic Forum)
		Participation in the responsible care program of the Chemical Manufacturer's Association
		Science literacy scores 8th, 12th grades (Department of Education, states)
		Secondary education completion rate (Department of Education, states)
		Gross tertiary enrollment rate
		Number of researchers/million inhabitants

Additional information can be found at http://ciesin.columbia.edu/indicators/ESI/index.html.

Although procedures have been developed to formally evaluate the value of information, these procedures typically relate to economic indicators or variables (Lawrence, 1999), statistical decision theory (Morgan & Henrion, 1990), or the role and value of information technology in decision-making. A recent study investigated the value of information for water quality management by considering the value of understanding the relationship between agricultural practices, as well as economic drivers of these practices, and nonpoint source loading to the Chesapeake Bay (Borisova et al., 2005). Similar procedures and approaches for evaluating the value of information for water resources management should be reviewed for possible inclusion in water resources management.

14.2 Assessment

14.2.1 Frontier Analysis

Data envelopment analysis and value efficiency analysis are two methods for integrating environmental, economic, and social indicators such as those listed in Table 14-1 (Bowlin, 1998; Charnes et al., 1994; Halme et al., 1999; Korhonen et al., 2001). These, and similar methods, have been referred to as frontier analysis. Frontier analyses might be used for assessing not only the current quality of life in a community or county, but also what is needed to move to a desired future quality of life. These techniques explore the "production" of quality of life within assessment units (i.e., community, county, watershed) (Marshall & Shortle, 2005). Quality of life indicators are broadly categorized in three dimensions—social, environmental, and economic; the same three dimensions being explored and integrated for SWRM. The concept underlying frontier analysis is that quality of life has ecological, economic, and social constraints that define the maximum production of quality of life. This maximum is unique for each community and county, so each has a unique upper boundary, or frontier, in the quality of life that can be attained (Figure 14-1) (Marshall & Shortle, 2005). Frontier analyses are nonparametric analyses that investigate how undesirable outcomes (e.g., cancer risk, percent of developed land, miles of impaired streams, percent of population below the poverty level, percent of illiterate population) can be minimized and how desirable outcomes (e.g., high teacher/pupil ratio, percent of population over 25 years that are high school graduates, high amenity index score, miles/acres of wetlands, miles of streams in good condition) can be maximized (Bloomquist et al., 1988; Deller et al., 2001; Marshall & Shortle, 2005).

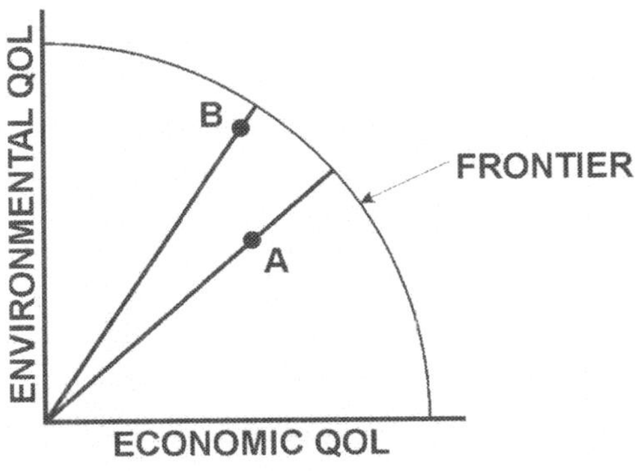

Figure 14-1. Frontier Analysis Uses Indicators of Economic and Environmental Quality of Life (QOL) to Determine How Far Communities (A,B) Are From the Maximum in desired Outcomes (the Frontier or Envelope of All Possible Outcomes). Community B is more efficient in its use of inputs than Community A because it is closer to the frontier (after Marshall & Shortle, 2005).

Frontier analysis provides an estimate of where a community, county, or watershed is with respect to a theoretical maximum, or to other communities, counties, or watersheds in the region, in producing a desired quality of life (Figure 14-1). The analysis provides a benchmark for "how far" a community is from the best that could be attained or in comparison to the best communities. This information might be used as part of alternative futures planning, for evaluating management tools or instruments to increase the efficiency of the community in moving toward a higher quality of life, or in selecting additional indicators to monitor that reflect community values that currently are not being tracked. Frontier analysis is one of many social science analytical tools and procedures that should be considered for assessing the current status and condition of the integrated socioeconomic and ecological systems.

14.2.2 Adaptive Environmental Assessment

A primary assessment objective is to inject knowledge into the decision-making process so that knowledgeable decisions can be made on issues of SWRM. Adaptive environmental assessment is the process of synthesizing and integrating monitoring, modeling, and empirical information for use in decision-making. Effective water management decisions require successful integration of ecological assessment results, analyses of economic benefits and costs, and stakeholder social and cultural concerns. Through the synergy of interactions among local stakeholders, scientists, engineers, and economists, information can be converted into knowledge and used to make sound policy and management decisions (Norton et al., 2004).

A Pellston workshop was held in September 2000 at the University of Michigan Biological Station to address ecological assessments, and specifically to focus on linking science with the decision-making process. The information generated at this workshop provided the basis for a SETAC publication, *Ecological Assessment of Aquatic Resources: Linking Science to Decision-Making* (Barbour et al., 2004). The ecological assessment process that emerged from the workshop incorporates many of the elements that form the framework and guidelines for sustainable water resources (Figure 14-2). The ecological assessment process includes the following elements:

- formulating the right assessment questions,
- engaging the local community in the stakeholder network,
- designing data collection networks if the information is not available at the spatial/temporal scale needed for the assessment,
- analyzing and interpreting the results of the ecological assessment,
- incorporating socioeconomics in the assessment to add value,
- translating information into knowledge about the ecological resources, and
- injecting this ecological knowledge into the decision-making process so that it is relevant to decision-makers, stakeholders, and the public (Chapters 9.0, 11.0, and 13.0) (Barbour et al., 2004).

Interdisciplinary assessments require interaction and involvement of all individuals, not simply a group of individuals from different academic fields working on their individual parts. Two of the lessons learned from a programmatic review of U.S. EPA's interdisciplinary Water and Watersheds Research Program were that

1. Interdisciplinary science was more applied and relevant to management and policy than disciplinary research; and

2. Stakeholder input and communication was invaluable to good science (Levinson & Thornton, 2003).

The following subsections summarize the ecological assessment process described by Barbour et al. (2004).

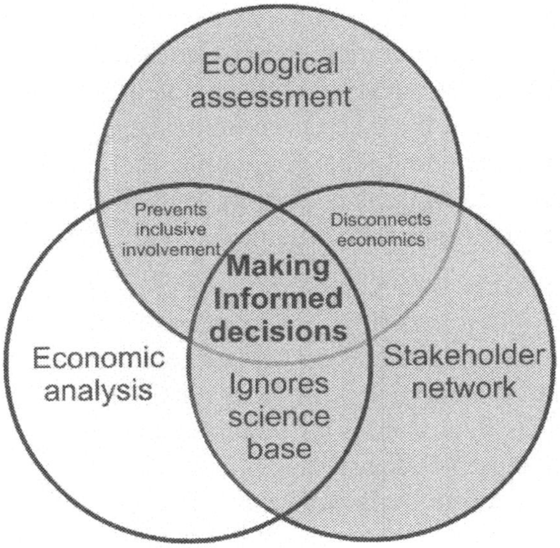

Figure 14-2. Making Informed Decisions Requires the Integration of Ecological, Economic, and Social (Stakeholder) Information (after Barbour et al., 2004).

14.2.2.1 Question Formulation

Posner (1973) stated that the initial representation of a question was the single most critical factor governing the likelihood of a satisfactory answer to the problem. The questions formulated organize and focus the assessments and guide the development of the structured technical questions that direct assessment activities. They articulate the purpose and scope of the assessment, the valued ecological attributes that are the assessment focus, and the level and types of integration that will be required for implementation (Norton et al., 2004). Questions that initiate an assessment can be broad in scope and unanswerable with existing information. However, subsets of the broader question can be addressed, providing at least a partial answer for managers and local stakeholders. This can help explain why the broad general question, as phrased, cannot be answered, what parts can be answered, and what would be needed to provide a complete answer. Questions can be formulated at many different stages of the assessment process, and may be refined iteratively as knowledge increases. Involvement of all disciplines and stakeholder interests is needed for success. Effective questions ensure that

- important stressors, effects, and processes are addressed by the assessment,
- the design and scale of the data collection support the needs of the assessment,
- the ecological endpoints of interest are identified and clearly linked to the measurements,
- stakeholder concerns are reflected and potential conflicts are identified early in the process, and
- appropriate answers are provided at the right stage of the management process, enabling decision-makers to adequately evaluate their alternatives (Norton et al., 2004).

14.2.2.2 Stakeholder Input

Stakeholders are defined as any person or organization that has a vested interest or "stake" in the waterbody, its use, protection, restoration, or future activities that will influence the overall condition of the waterbody. The most obvious stakeholders in an assessment are the people who have to live with the result of a decision, but they also include the ecologists, engineers, economists, communication specialists, and managers who implement the assessment process (Bastian et al., 2004). Bringing the full complement of stakeholders into the assessment process ensures that ecological information is gathered and evaluated in light of socioeconomic concerns. The process of engagement should be led by communication professionals and community facilitators. These experts ensure that the right people are included, that important view points are represented, that roles, expectations and the process for involvement are defined, and that a constructive and meaningful dialogue is maintained.

Engaging local stakeholders ensures that the assessment is relevant to management questions and local concerns, that legal requirements for involvement are met, and that the assessment is scientifically sound. Local knowledge can further contribute to the assessment by quantifying values, contributing to site selection, and providing insights into possible management alternatives. Successful engagement builds understanding in the community and provides the foundation for long-term support of assessment activities. The process of gaining the trust and involvement of the local community is not a quick process, however, and it is a very fragile relationship that can easily be broken. A time frame of 3 to 5 years to develop trust and a good working relationship is not uncommon (Levinson & Thornton, 2003). This investment in building stakeholder trust and involvement will pay dividends in subsequent assessments and management decisions (Bastian et al., 2004).

14.2.2.3 Designing and Implementing Data Collection for the Ecological Assessment

The ecological assessment provides the scientific basis for decisions on protection, restoration, or maintenance of the water resource. In the process of designing data collection for ecological assessment, ecologists transform the broad general questions developed previously into more structured technical questions (Stevenson et al., 2004). These technical questions provide the basis for designing the collection of new observations if the needed information is not available, and for combining existing information. Depending on the purpose for the assessment, the design might include the identification of a benchmark from which an assessment of condition can be made, or quantification of the relationship between the ecological indicators and the extent of human disturbance (Stevenson et al., 2004).

Effective ecological assessments identify the appropriate indicators to represent the ecological attributes of interest. They consider the scale of the assessment (i.e., spatial extent of the aquatic resource under investigation and the temporal extent of the magnitude, duration, and frequency of the exposure and associated ecological response). They ensure that the quality of measurements is known, that sampling designs are appropriate to the question, that any statistical or modeling procedures are valid, and that the confidence in the results can be described (Stevenson et al., 2004).

14.2.2.4 Analyzing Data for Ecological Assessments

The selection and implementation of the appropriate analysis techniques is critical to the underpinnings of the scientific results. Characterization of the aquatic resource by partitioning the natural variability and establishing a framework for the ecological assessment is central to the process (Stevenson et al., 2004). Linking the stressor/response relationships to the assessment condition provides the basis for diagnosing the problems and considering the management alternatives. The scientific integrity of water quality management decisions and the ecological monitoring data on which they are based are fundamental to the value of these programs and to their public acceptance.

14.2.2.5 Quantifying Economic Value

Knowledge of the economic value of the ecological resource is an increasingly vital component of decisions regarding water resources management. Economists select and implement appropriate valuation methods, take into account nonmonetary values, and consider how impacts on future generations should be weighed in today's decisions (Fisher et al., 2004).

Economic valuation should be a part of the environmental assessment. Considering ecological benchmarks and attainment goals can enhance valuation methods. The value of aquatic resources often varies depending on socioeconomic pressures in addition to the quality of those resources. Most importantly, economic assessment tools can be used to communicate the values provided to humans by ecosystems so that the consequences of different decisions and trade-offs can be evaluated (Fisher et al., 2004).

14.2.2.6 Communicating Results

Translating complicated scientific principles and results for the public and nonscientific managers is perhaps the greatest challenge in linking science to decision-making. Essentially, volumes of raw scientific data need to be consolidated, sorted, and transformed into meaningful recommendations for influencing management decisions, without losing the accuracy or substance of the findings. The reporting of the ecological assessment and economic valuing needs to be structured for clarity, conciseness, and accuracy so that there is little chance of misconstruing the intent of the report (Preston et al., 2004).

Substantive involvement of all disciplines and stakeholder groups is needed for successful communication of results, but the involvement of communications experts is key. They provide the expertise needed to facilitate a two-way dialogue that hones scientific findings into useful information. Effective communication provides results to managers and decision-makers in a way that conveys

- what is known,
- what can reasonably be speculated from what is known,
- what is unknown, and
- possible consequences of what is unknown on various management decisions (Preston et al., 2004).

When successful, effective communication provides the mechanism that integrates information, providing a sound foundation for decisions on protecting and managing aquatic resources.

14.2.2.7 Integrating the Process

Ecological, economic, and stakeholder social issues must be integrated to effectively inform decisions on management and protection of aquatic resources. For ecological assessments to make a difference, however, they must integrate ecological, social, and economic perspectives into a convincing story. How the message is conveyed is an influential factor in making appropriate decisions (Norton et al., 2004). Many business executives are being trained in telling stories because of the importance of getting clear and unambiguous corporate messages on their business to shareholders and the public (McCreary, 2003). Likewise, ecologists, economists, and engineers need to be trained to translate science into formative messages.

In addition to stories, understanding how different individuals reach decisions can aid in preparing and presenting the ecological assessment messages. Psychologists and organizational behaviorists have considerable insight into how different types of individuals process and use information in making decisions (Cialdini, 1998; Drucker et al., 2001; Hammond et al., 1998) (see Chapter 12.0). Understanding these styles, and using this insight in preparing, presenting, and transmitting the results of an ecological assessment can increase the use of this information in making decisions.

Injecting ecological knowledge into the decision-making process involves more than sound, objective science. It also requires considering and integrating economic, social, and behavioral perspectives and approaches into the assessment process. Using interdisciplinary teams with trained facilitators or brokers is not a luxury, it is a necessity for conducting ecological assessments that will make a difference in the decision-making process (Barbour et al., 2004).

We believe this ecological assessment process is consistent with, and builds on, the original concept of adaptive environmental assessment and management developed by Holling (1978). In addition, we think this approach will contribute the knowledge needed to adaptively assess and manage water resources through time so that sustainable water resources will be the outcome.

CHAPTER 15.0

ELEMENTS OF SWRM: ADAPTIVE MANAGEMENT– PRINCIPLES

Dublin Principle 2 – Water development and management should be participatory
Enlibra Principle 2 – Science for facts, process for priorities
Enlibra Principle 3 – Collaboration, not polarization
Enlibra Principle 7 – Reward results, not programs
Panarchy Principle – Ecological, economic, social, and institutional systems are interlinked in a hierarchy of adaptive cycles
Panarchy Principle – Adaptive management is required for dealing with nonlinear, uncertain, and unpredictable systems such as water resources and related elements

15.1 Adaptive Environmental Assessment and Management

Adaptive environmental management or "learning by doing" (Lee & Lawrence, 1986) has become the recommended approach for ecosystem and natural resources management, including water resources (Christensen et al., 1996; Jackson et al., 2001). Although adaptive environmental management represents the preferred approach for water resources management, the original context proposed by Holling (1978)—adaptive environmental *assessment* and management—reflects a critical part of the management process.

Adaptive environmental assessment and management, as originally proposed by Holling (1978) incorporated elements of modeling, systems analysis, and experimentation in the context of natural resources management. The approach was an outgrowth of the systems ecology and systems analysis movements in the late 1960s and 1970s. The understanding of the system and its linkages was embedded within the model constructs. Systems analysis provided a mechanism for analyzing the linear and nonlinear dynamics and feedback loops in the system and its interactions with its environment. These tools permitted different simulation experiments to be performed to evaluate how the system might respond to different natural resource management practices. Because ecological systems are dynamic and nonlinear, and modeling is an imperfect art, monitoring provided the information necessary both to refine the model structures and adapt or modify the management practices to achieve the desired system response.

This movement has contributed to the shift from "the balance of nature" concept to the concept that "ecosystems are dynamic, nonequilibrium systems." With the additional insight from chaos and complexity theory (Levin, 1990), adaptive management has become the only feasible approach for moving toward sustainable water resources (Coleman, 1998).

15.2 From Concept to Practice

Neuman (2001) describes five key steps for adaptive management:

1. Establish the desired management objectives for the resource.
2. Propose a management scheme to achieve these desired objectives.
3. Develop hypotheses and experiments to test if the management regime will achieve the objectives.
4. Design and implement the experiments and monitoring programs to conduct the experiments and test the hypotheses.
5. Adapt the management program, if needed, to achieve the desired results.

Moving from concept to practice will require at least four steps or actions (Christensen et al. 1996):

1. defining goals and objectives based on sustainability, not desired goods and services,
2. reconciling spatial scales (the spatial scale of the processes important in sustainable systems) and political boundaries (this will require cooperation among multiple stakeholders),
3. reconciling temporal scales (important temporal scales may be intergenerational, rather than fiscal years), and
4. making the system adaptable and accountable (humans are part of, not apart from, ecological systems and are accountable for the sustainability of these systems).

Given the uncertainty associated with ecological responses to natural and anthropogenic disturbances, adaptive management has become the recommended approach for ecosystem management (Holling, 1978; NRC, 1992; 2000b, Walters, 1986). In this approach, management policies are set based on imperfect knowledge about the system to be managed (Holling, 1978). As we continue to learn more about environmental, economic, and social systems and how they interact, more effective management policies will come to light. Furthermore, the system being managed is changing (Holling, 1978), such that policies that were useful at one time can become less useful, or even detrimental, as the system changes over time. In the case of sustainable water resources, changes in population and land use are inevitable and are likely to change the effectiveness of existing water resources management policies, or the water resources goals and priorities of the community.

In Panarchy Theory, policies are treated as hypotheses and management practices are treated as experiments. Monitoring provides the information needed to test these hypotheses and refine or develop new experiments to maintain desired system goods and services, and accommodate uncertainty and surprises (Holling, 2001). Policies are reviewed regularly to determine if they are achieving the desired result, and if they reflect the current state of scientific knowledge and society's desires.

CHAPTER 16.0

ELEMENTS OF SWRM: ADAPTIVE MANAGEMENT TOOLS – QUANTITY

Dublin Principle 1 – Freshwater is a finite and vulnerable resource
Dublin Principle 2 – Water development and management should be participatory
Enlibra Principle 4 – Recognition of benefits and costs
Enlibra Principle 5 – Markets before mandates
Enlibra Principle 7 – Reward results, not programs
Panarchy Principle – Ecological, economic, social, and institutional systems are interlinked in a hierarchy of adaptive cycles
Panarchy Principle – Adaptive management is required for dealing with nonlinear, uncertain, and unpredictable systems such as water resources and related elements

16.1 Desirable Attributes of Water Allocation Processes

Desirable attributes of water allocation processes include

- *Flexibility*: Water rights should be transferable so that water can be shifted in location, season, and use in response to changing social and economic conditions. Flexibility in the short term and in the long term are required for an allocation process to be efficient and foster economic growth and change.
- *Security of tenure*: There should be secure expectations of water availability by established rights so that users have a basis for long-term investment and planning decisions. This is essential both for the efficient allocation of water and for the efficient allocation of resources used in combination with water, such as land.
- *Recognition of opportunity costs*: Opportunity costs of water use and transfer should be implicitly or explicitly accounted for in the decisions of water rights holders. Explicit recognition could be provided by water prices, which may in turn be set administratively or determined in markets. Liability rules are another mechanism that can provide for explicit signals. Administrative allocations can reflect implicit recognition of opportunity costs.
- *Fairness*: No uncompensated costs should be imposed on third parties, nor should water transfers be compulsory.
- *Administrative ease*: The process should be understandable and easy to administer. In other words, transaction costs should not be exorbitant (Ciracy-Wantrup, 1967; Howe et al., 1986; Trelease, 1965).

16.2 Allocation Methods

16.2.1 Administrative Water Allocation

Many developing countries use highly centralized administrative water allocation mechanisms. Ideally, the administrative authority (e.g., a river basin commission) would identify demands for alternative instream and offstream uses, along with the cost of meeting them, and compute the efficient allocation. It might then achieve this allocation either by marginal cost pricing, which would give users an economic incentive to choose their optimal quantities, or simply by administrative decree. In the long run, the former would be preferable because it would provide users with incentives to discover new ways to conserve water and discourage new users from connecting to systems with higher scarcity rents. Whether the results are fair depends on whether the administrative authority chooses to compensate those who lose from reallocations.

There are many problems with this scenario. First, in an economy of any complexity, the transaction costs associated with acquiring and processing information on water demands for various instream and offstream uses would be exorbitant. In practice, agencies often have limited resources and abilities to manage water systems. Second, being political creations, government agencies are highly susceptible to rent-seeking activities that distort decisions to favor the interests of politically powerful groups. This kind of activity has been a major obstacle in efforts to reform water allocations in California and other places. The poor performance of administrative allocation mechanisms is a major reason for current interest in water policy reform (World Bank, 1993).

16.2.2 Water Markets

Although market allocation of water still is relatively rare, water markets have emerged as a leading candidate for improving the efficiency of water allocation. Market transfers have a long history in the western United States, and, as noted above, some states are introducing reforms to improve the performance of these markets.

A fundamental requirement of water markets is well-defined water rights that are transferable and exclusive. Transferability and exclusivity meet the first two "desirable" attributes listed in Section 16.1. Moreover, markets have desirable features with respect to fairness. Market transactions are not compulsory and at least the immediate parties involved in a transfer are compensated. Because rights holders can gain from a sale of rights, they also have an incentive to recognize the opportunity costs of water. Finally, another key attribute of markets is that they have low transaction costs compared to administrative mechanisms.

To illustrate, suppose that there are legally defined, transferable, exclusive, and enforceable water rights. Industrial users would be willing to pay \$10 for the right to use a given amount of additional water, while agricultural rights holders would be willing to sell rights for \$6. As a result, there is a potential gain from trade of \$4. Clearly, both could gain from a voluntary exchange of rights as long as the transaction costs are less than \$4. Generalizing, if there are no transaction costs, then trades would be expected to eliminate all differences in incremental values between uses, thus satisfying the equimarginal rule for efficient water use (see Chapter 10.0).

In practice, water markets cannot be expected to meet this ideal because of high transaction costs and the difficulties in defining water rights. Well-defined water rights would have the following attributes (Saliba & Bush, 1987):

- *Completely specified and enforced,* so that all individuals know the privileges and restrictions associated with holding water rights, and the penalties for violation. A complete description of rights for diversions must cover quantity diverted and consumed, uses, timing, quality, and places of diversion and return. Similarly, instream rights descriptions must cover uses, quantity, quality, timing and location.
- *Comprehensive,* so that all attributes and uses of water that generate value are represented by water rights.
- *Exclusive,* so that the benefits of water rights and their transfer accrue only to buyers and sellers, and not to third parties.
- *Transferable,* so that rights can move into their highest-valued uses.

These characteristics are very similar to those referenced above as desirable attributes of a water allocation process. Yet, several characteristics of water resources make it difficult to define rights with all of these attributes (Saliba & Bush, 1987; Young, 1986). The definition and measurement of water rights is complicated by limited ability of water rights holders to observe and control water quantity and quality between locations, seasons, and years. Interdependencies between water users are inherent in water management because water is mobile and can be used and reused. Property rights transfers can change the quality or quantity of water for other uses. There are also a variety of instream uses for which demands cannot be effectively represented in markets because the benefits are nonrival (i.e., shared more or less equally by all users with access) or nonexclusive (i.e., available to users whether or not they have obtained rights). These include instream recreation, wildlife habitat, and dilution of pollutants. The characteristics of water can also make for high transaction costs in water markets compared to other commodities. These costs include searching for willing buyers and sellers, conveying water, evaporation and filtration losses, negotiating conditions and prices, enforcement, and acquiring permission from regulatory authorities when required.

These imperfections suggest limits to what water markets can accomplish. However, water markets in combination with regulatory safeguards to protect against certain problems have considerable merit. Compared to other allocation mechanisms, markets offer a more responsive and lower cost mechanism for transferring water from low-valued uses to high-valued uses when economic conditions change. Economic flexibility is increased as water and the allocation of related resources are freed from bureaucratic constraints. Because water rights holders can benefit from selling water rights when they adopt less water-intensive goods and technologies, markets can provide a continuing incentive to consider the opportunity costs of water use and thus conserve.

Analysis of water transfers in the United States provides empirical support for the use of water markets. For instance, Vaux and Howitt (1984) estimate potential economic gains from water trade in California to be as much as $219 million annually by 2020. Saliba (1987) finds that water transfers in the western United States move water to higher valued uses within and between sectors, and that prices correspond to changing market conditions.

16.2.2.1 Issues in Implementing Tradable Water Rights

Several issues must be addressed when shifting from administrative water allocation to market allocation. These include the definition of water rights and their initial allocations, and protection of third party interests and investment in and ownership of water management infrastructure (Howe et al., 1986; Rosegrant & Gazuri, 1994). Some of the major issues are illustrated below.

Prior versus Proportional Rights Rights to water for diversion can take a variety of forms. Two common forms are priority and proportional. Under the Prior Appropriations Doctrine, which evolved in California and spread to other western states, senior rights holders, who established rights before junior rights holders, have first priority. In contrast, proportional rights entitle holders to shares in available water.

A key difference between these approaches is how they handle variations in supply. The prior rights approach allows varying degrees of supply reliability to be purchased through a mix of junior and senior rights. The desired mix will reflect the risk preferences of the purchaser and the relative cost of the rights. However, the heterogeneity of the market increases transaction costs. Under the proportional rights approach, additional reliability can be achieved by purchasing a greater share of the flow. The homogeneity of rights promises lower transaction costs, but this must be balanced against the need to hold additional shares. Howe et al. (1986) conclude that neither system is ideal; the advantages and disadvantages depend on the variability of water supplies and the heterogeneity of water users.

Withdrawals versus Consumption A crucial issue in defining water rights is whether the rights are defined with respect to withdrawals or consumption (withdrawals less return flows). Upstream and downstream users are connected by return flows. Rights in consumption give users an incentive to return water as well as to conserve withdrawals, thus providing some protection for the interests of downstream users. Rights in withdrawals do not afford that protection. On the other hand, reasonably accurate determination of return flows can be costly because of uncertainty over the fate of water. Accordingly, withdrawal rights are more conducive to an active market that can exploit potential gains from trade. Consumptive rights are preferable for avoiding costs for downstream users because of reduced return flows.

Water rights in the western United States are mixed in their definition of rights and in the complexity of the procedures used to determine flows. Transaction costs aside, the consumption rights approach is the more appealing in principle.

Instream Uses Significant benefits accrue from instream uses of water. These include instream and near-stream recreation, hydropower, dilution of pollutants, and fish and wildlife habitat. Some protection of these uses can be provided by the creation of rights to instream flows. Such rights have long existed in some countries, such as the United Kingdom, and are now evolving in others, including the United States (Anderson, 1982; Howe & Salles, 1996).

There are, however, limits to what can be accomplished by instream rights. Instream use values of water are often nonrival and nonexclusive, both of which pose problems for efficient markets. This is because the provider of instream rights cannot capture the value that accrues to other uses. Thus, the market tends to provide less than the efficient level of instream rights. Uncertainty about cause-and-effect relationships also can complicate matters. Adequate protection of instream water use values will require public involvement in water resource management. The challenge is to choose procedures that provide efficient protection of these values.

Water rights transfers in the western United States are subject to review processes to protect downstream users. Colby (1995) argues that the review procedures have worked reasonably well in Colorado, New Mexico, and Utah. Transaction costs in these states, on average, are not a large proportion (6%) of the economic value of water. Colorado, which uses a judicial review process only, has the highest costs. New Mexico and Utah use administrative processes as the first forums for review. Transaction costs also were found to correspond in reasonable ways to the potential size of the externalities and the value of the transaction.

Initial Rights Allocation The creation of water markets entails the creation of a new set of tradable rights in water. A fundamental issue is the initial allocation of rights. Rights may be allocated to existing users more or less in proportion to their historical use. But there are many other possibilities. The initial allocation of rights is of huge importance to the political acceptability and fairness of a shift to water markets.

16.3 Water Pricing and Allocation

The large and increasing financial and environmental costs of water resources development mean that the needs of growing human populations generally must be met with available supplies rather than through developing new supplies. A relatively new twist in this problem is managing flow regimes to protect aquatic and related ecosystems. This context requires addressing several fundamental economic questions: how to price water to encourage patterns of economic development and water use that promote conservation, how to allocate scarce water resources between competing uses, and how to account for ecosystem values in water allocation.

The desirability of pricing water to recover the costs of water supply systems and encourage conservation of scarce supplies is relatively uncontroversial. However, there are important implementation issues.

16.3.1 Basic Principles and Issues in Water Pricing

Where allocative efficiency is the goal, water prices should, in principle, be based on the volume consumed (*volumetric pricing*) and cover the opportunity cost of the water consumed. Users will increase consumption up to the point at which the incremental economic benefit just equals the price, thus satisfying the equimarginal principle of efficient use. This pricing principle is referred to as *marginal cost pricing*. In the case of groundwater, the optimal price would include the marginal user cost of groundwater as well as pumping and delivery costs.

In regions where water is plentiful, the opportunity cost of water would be relatively small, so that the major component of price would be to cover delivery costs. Where water is scarce, a *scarcity rent* may be a large component of the price. Because the costs of supplying water vary between peak and offpeak use periods, seasonally, and between types of users (e.g., by elevation or distance), optimal rate structures in principle should vary correspondingly.

In practice, marginal cost water pricing is neither easy nor the norm. One barrier to marginal cost pricing is high transaction costs. Complex rate structures are costly to construct and administer. Further, implementing marginal cost pricing in pure form requires information on the social opportunity cost of water. Scarcity rents are not easy to determine if there are no freely functioning water markets to provide data on the value of water, or if there are significant nonmarket values (Randall, 1981). Where water is abundant or the transaction costs are high, the economic benefits of marginal cost pricing may not be worth the costs (e.g., Young & Havemann, 1985).

Marginal cost pricing also faces political and financial barriers. If water costs are high, charging prices approximating them may be politically difficult and harmful to the poor (e.g., Martin et al., 1984). Increasing block rate structures can be used to diminish the impacts on the poor and other groups, while still retaining desirable incentive effects (Harrington, 1987). In systems where average costs exceed marginal costs, marginal cost pricing alone will not recover the total costs of operation. This problem can be addressed by multipart pricing schemes that combine volumetric and nonvolumetric fees.

In practice water prices typically are set to recover some proportion of the historical costs, rather than the opportunity costs, of water systems. In the case of self-supplied households, farms, and industries, this usually means no price even though the water may have a scarcity rent. *Flat rate* pricing is common; users are charged for access but not according to the volume of water use. Flat rate pricing saves metering costs, but does not offer incentives to conserve water.

Volumetric pricing and multipart pricing are also common and can be designed to give desirable, if not optimal, incentives. Estimates of the demand for water indicate that water consumption is responsive to prices (Gibbons, 1986), so average cost volumetric pricing will recover costs while providing incentives to conserve. However, average cost pricing can encourage "urban sprawl" and excess migration to water scarce regions if the marginal costs for new connections are higher than the system's average cost. This problem can be addressed by adjusting connection fees and volumetric prices according to cost variations, although at an increase in transactions costs.

Metering farm water use for irrigation is generally more expensive than metering domestic and industrial water use. Per acre charges offer a lower cost alternative, but create only indirect incentives to save water.

Although water prices below historical costs have been common, water authorities in regions with water scarcity problems are revising pricing strategies both to improve cost recovery and to ration water supplies. The evidence supports the effectiveness of these reforms (e.g., Brookshire et al., 2001).

In summary, a practical pricing strategy is to use flat rates to cover system costs when water is relatively abundant and service costs are relatively constant, and these conditions are expected to persist. However, where water scarcity is a current or potential future problem, volumetric or multipart pricing schemes are more appropriate.

16.3.2 Other Demand Management Strategies

Price incentives have proven effective in managing water demands and offer desirable long-run incentives. However, other instruments are also available. These include educational and technical support programs, cost-sharing subsidies, and tax credits to facilitate adoption of water conservation practices and technologies. Quotas and restrictions on certain water uses are also sometimes used, especially in times of drought emergency.

16.4 Privatization

Privatization involves transferring some or all of the production, distribution, operations, and/or assets of public water supply systems to the private sector (Gleick et al., 2002a). Over the past decade, there has been an increasing global trend, and increasing controversy, over privatization of water resources. At the end of 2000, 93 countries, including the United States, had partially privatized water or wastewater services (Brubaker 2001). Currently, two French companies—Vivendi and Suez Lyonnaise des Eaux (now Ondeo)—dominate the privatization sector, serving about 200,000 million people world-wide (Gleick et al., 2002a). The primary assumption for privatization is that private sector businesses can provide water more efficiently than government or public sector agencies.

Privatization can take many forms. There is no single model, arrangement, or agreement that is used in privatizing water supply. Some of the functions that might be privatized are listed in Table 16-1.

Table 16-1. Water Supply Functions That Could be Privatized (from Gleick et al., 2002b).

Capital improvement planning and budgeting	Pricing decisions
Capital improvement financing	Billing/revenue collection management
Capital improvement design	Payroll management
Capital improvement construction	Financial or risk management
Facility operation	Establishment, monitoring, and enforcement of water
Facility maintenance	quantity/quality and other service

Although there are a number of functions that can be, and have been, privatized, there are also a number of concerns about privatization. Some of these are listed in Table 16-2.

Table 16-2. Concerns about Privatized Waters (from Gleick et al., 2002b)

Unsurping basic government responsibility	Neglecting the potential for water-use efficiency and conservation
Bypassing underrepresented and underserved communities	Lessening protection of water quality
Worsening economic inequities and water affordability	Lack of dispute resolution procedures
Failing to protect public ownership of water and water rights	Irreversible decision
Lack of public participation and contract monitoring	Transferring assets out of the community
Ignoring impacts on ecosystems and downstream users	

To begin addressing concerns about water supply management and privatization, Gleick et al. (2002b) proposed a set of principles and standards to guide water management, regardless of whether it was in the public or private sector. These principles and standards were subsequently updated and illustrated using examples from around the world (Palaniappan et al., 2004). These principles and standards are briefly discussed below. Within each principle, there are a number of associated standards. Additional information can be obtained from Gleick et al. (2002b) and Palaniappan et al. (2004).

16.4.1 Principle 1: Manage Water as a Social Good

This principle states that water is both a social and an economic good. This is a theme that has been echoed throughout this report. As a social good, water should be available to meet basic human and ecological needs.

Standard 1.1 Meet basic human needs for water. All residents in a service area should be guaranteed a basic water quantity under any privatization agreement.

Standard 1.2 Meet basic ecosystem needs for water. Natural ecosystems should be guaranteed a basic water requirement under any privatization agreement.

Standard 1.3 The basic water requirement for users should be provided at subsidized rates when necessary for reasons of poverty.

This principle and these standards, although discussed within the context of privatization, have been discussed throughout this report as integral to SWRM. This principle and these standards reinforce the distinction between needs and desires for water. Needs are nonnegotiable, whereas desires can be addressed through a variety of management options.

16.4.2 Principle 2: Use Sound Economics in Water Management

The Dublin Principles state that water is an economic good and market practices should be used to ensure it is efficiently and cost-effectively allocated. As noted in Principle 1 above, water is also a social good, so it must be affordable to enable human needs to be met.

Standard 2.1 Water and water services should be provided at fair and reasonable rates.

Standard 2.2 Whenever possible, link proposed rate increases with agreed-upon improvements in service.

Standard 2.3 Subsidies, if necessary, should be economically and socially sound.

Standard 2.4 Private companies should be required to demonstrate that new water supply projects are less expensive than projects to improve water conservation and water use efficiency before they are permitted to invest in such projects and raise water rates to repay the investment.

Economic approaches, practices, and tools for implementing these standards are discussed in Chapters 10.0, 11.0, and 17.0.

16.4.3 Principle 3: Maintain Strong Government Regulation and Public Oversight

Local communities should have the responsibility and the authority to provide water and water services, whether through privatization or public service. Water as a social good is best protected through strong governments with public oversight.

Standard 3.1 Governments should retain or establish public ownership or control of water sources.

Standard 3.2 Public agencies and water service providers should monitor water quality. Governments should define and enforce water quality laws.

Standard 3.3 Contracts that specify the responsibilities of each partner are a prerequisite for the success of any privatization.

Standard 3.4 Clear dispute resolution procedures should be developed before privatization.

Standard 3.5 Independent technical assistance and contract review should be standard.

Standard 3.6 Decision-making in the water sector should be open, transparent, and include all affected stakeholders. Negotiations over privatization contracts should be open, transparent, and include all affected stakeholders.

This principle and these standards highlight the purpose, responsibility, and accountability of governments to their constituents. Existing water laws, regulations, and policies were discussed in Chapter 12.0. This principle is an area where civil society institutions can also contribute by ensuring both the public and private sector responsibilities and accountability are clearly defined, being followed, and revised, when necessary.

CHAPTER 17.0

ELEMENTS OF SWRM: ADAPTIVE MANAGEMENT TOOLS – QUALITY

Dublin Principle 1 – Freshwater is a finite and vulnerable resource
Dublin Principle 2 – Water development and management should be participatory
Enlibra Principle 4 – Recognition of benefits and costs
Enlibra Principle 5 – Markets before mandates
Enlibra Principle 7 – Reward results, not programs
Panarchy Principle – Ecological, economic, social, and institutional systems are interlinked in a hierarchy of adaptive cycles
Panarchy Principle – Adaptive management is required for dealing with nonlinear, uncertain, and unpredictable systems such as water resources and related elements

Current water quality policy is dominated by command and control mechanisms. However, there is great interest in mechanisms that are cost effective and pursue water quality goals in the context of a reasonable balance between costs and benefits (USDA & U.S. EPA, 1998; NRC, 2000a). Improving water quality programs with regard to cost effectiveness and benefit–cost ratio would help answer of the following questions:

- How should responsibility for pollution load reductions be allocated among sources (i.e., who should be targeted by water quality policies)?
- At what level should the maximum pollutant load (i.e., TMDL) for a waterbody be set?
- What types of environmental policy instruments should be used to achieve the desired outcome?

17.1 Targeting and Pollution Load Allocation

In theory, optimal water quality management would entail controls on all who discharge pollutants into water resources. Concern for control costs suggests allocating discharge reductions among this set cost-effectively. However, if the transaction costs associated with such a comprehensive approach are extensive, it may be optimal to take the second-best approach, which limits those who are directly regulated. Considering monitoring and enforcement costs as well as pollution control costs, there are economic advantages to targeting those (1) whose discharges can be monitored at reasonable cost, (2) who are large contributors to ambient pollution concentrations, and (3) who can reduce pollution loads at relatively low cost. The second-best approach then allocates discharges cost-effectively among dischargers that meet these criteria. However, it will rarely be the case that these conditions are all satisfied. The first criterion applies to many point sources, but is generally violated by nonpoint sources. And although some point sources may be comparatively large contributors to pollution loads, their contributions could be dwarfed by the collective contributions of smaller point sources and nonpoint sources. Finally, marginal control costs might not correlate positively with metering costs or the magnitude of the load. Accordingly, it sometimes is not easy to answer the targeting question.

Nonetheless, targeting is important. Expanding the set of those who must comply with pollution controls increases public monitoring and enforcement costs. It also increases the chances of regulating agents who cause little or no problem, thus increasing compliance costs without increasing benefits. Yet failure to target broadly enough diminishes effectiveness and limits opportunities for cost-effective allocations. The most obvious example of this problem is the consequences of focusing pollution controls on point rather than nonpoint sources. Research demonstrates that careful targeting that considers the three factors mentioned above can greatly improve the cost-effectiveness of pollution controls (e.g., Horan & Shortle, 2001).

17.2 Total Maximum Daily Load

The U.S. EPA's total maximum daily load (TMDL) program has focused attention on the question of how to determine the level of pollution loads that can be allowed in individual waterbodies and have them still attain their designated uses. The answer to this question hinges on the designated uses of the water resource, and the relationships between pollution loads and the level of water quality required to sustain the designated uses (NRC, 2001). The latter involves biophysical issues, but the former raises the essential question of social choices about water quality objectives. Benefit–cost analysis offers one approach to the problem, but its applicability is limited when there is substantial uncertainty about costs and benefits, or limited acceptance of economic values being used as the basis for societal choice. Alternative methods for engaging the public in setting environmental goals have been developed and are being refined through practice (e.g., Bobertz, 1991; see Chapter 13.0).

17.3 Choice of Environmental Policy Instruments

Once water quality goals have been determined, their achievement requires choosing and implementing policy instruments that will lead dischargers to adopt pollution prevention and/or abatement practices to attain those goals. Much attention and many resources have been devoted to developing water quality plans. By comparison, little attention is given to the essential means for achieving water quality goals (i.e., the choice of policy instruments to induce the behavioral changes required to achieve targeted load reductions). Yet, this choice is essential to the economic and ecological performance of water quality programs.

Because of the complexities of water quality protection, the physical, economic, informational, and institutional context of environmental policy strongly influences appropriate choices for environmental policy instruments. There are two fundamental elements of the choice of policy instruments; choice of the basis for defining and measuring compliance with the policy, and choice of the mechanism for compliance (i.e., the lever used to induce compliance).

17.3.1 Choice of Compliance Basis

Economically and ecologically desirable candidates for compliance bases will have three characteristics, more or less:

1. correlated with water quality conditions,
2. enforceable, and
3. targetable in time and space (Braden & Segerson, 1993).

Economists generally agree that the optimal basis for compliance is metered discharges when discharges can be metered accurately and routinely at reasonable cost. Discharge-based instruments are commonly used for point sources of pollution. Instruments that base compliance on the use of specified technologies are also commonly used for point sources, although they do not meet criteria (1) and (3).

Discharge-based instruments are plausible for point sources, but are difficult to implement for nonpoint sources. The diffuse nature of nonpoint source loads makes routine observation and accurate metering prohibitively expensive (given current monitoring technology). One option for nonpoint source compliance basis is *emissions proxies*; site-specific environmental performance indicators that are constructed from observations of site-specific data. Such indicators are often available, and can vary from simple formulas that aggregate a few variables, to data intensive and computationally demanding computer models (Novotny & Olem, 1994). Examples of widely used environmental performance indicators for agriculture are annual average gross soil loss (for managing sediment pollution) and the excess of nutrient inputs to crop and/or livestock production over the nutrients contained in farm products (for managing nutrient pollution) (NRC, 1993). Issues in the use of a proxy include how well it serves as an indicator of environmental stressors, and how burdensome it is to compute and enforce (Braden et al., 1991; Braden & Segerson, 1993).

Another option for a nonpoint source compliance base is *inputs or practices* that are correlated with pollution flows and loads (see Plott, 1966; Holtermann, 1976; Shumway, 1995; Hayami & Ruttan, 1985). For example, in agriculture these would include polluting inputs such as fertilizers and pesticides, and farming practices that affect the movement of these chemicals into the environment. Issues in the use of inputs or practices as compliance base are described in detail in Griffin and Bromley (1982), Dinar et al. (1989), Shortle and Abler (1994), Shortle et al. (1998), and Shortle and Horan (2001). Another option that has received considerable interest from economists is the *ambient concentrations* of pollutants in environmental media (see Segerson, 1988; Miceli & Segerson, 1991; Shavell, 1987; Lichtenberg, 1992; Menell, 1990; Wetzstein & Centner, 1992). However, the economic literature is generally skeptical about the economic and environmental merits of this approach (Weersink et al., 1998; Shortle & Horan, 2001).

17.3.2 Choice of Mechanism for Compliance

Mechanisms are the levers used to induce compliance. It could be a command about the acceptable level of the compliance basis backed up by civil or criminal penalties for noncompliance. It could be a "carrot" in the form of a subsidy or payment to the polluter for choosing a desired course of action as measured by the compliance basis, or a "stick" in the form of charges for choosing an undesired course of action as measured by the compliance basis. Examples of mechanisms are listed in Table 17-1 and discussed below.

Table 17-1 Mechanisms for Water Quality Policy.

Information/ Education Programs	Research and Development Programs	Direct Regulations	Economic Incentives
Provision of information (i) To polluters (technical assistance) (ii) To investors, consumers, activists (e.g., U.S. Toxics Release Inventory) (iii) To consumers (green product or process) Challenge regulation and voluntary agreements	Technological basis for discharge standards[a]	Prohibition (of inputs, processes or products) Technology specification (for production, recycling or waste treatment) Performance specification (discharge permits)[b]	Pollution charges Subsidies (i) Lump sum for capital cost (ii) Marginal for desired results[c] Liability law provisions Tradable performance specification (tradable permits)

Adapted from Russell and Powell (1999).

Notes:

[a]In setting a technology-based standard, the amount of allowed pollution is determined via an engineering study in which a legally designated technology is applied on paper to a particular polluting operation with known uncontrolled pollution load (raw load). The result of this exercise is an achievable discharge amount.

[b]Performance specification can be based on any of several rules or methods that range from uniform percentage reduction by all sources to modeling that determines the cheapest way to attain a given ambient quality standard.

[c]The deposit-refund system, for example for drink containers, is a self-financed marginal subsidy study for container return.

Education programs in the form of moral persuasion (e.g., informing citizens that they too cause water pollution) and technical assistance (e.g., providing advice on choice and implementation of best management practices) have been used extensively for addressing nonpoint pollution problems. Examples include agricultural extension programs to educate farmers on the impacts of conventional farming practices on water quality, programs to help farmers adopt best management practices, and urban storm drain stenciling programs (OECD, 1998; Coburn, 1994). There is growing interest in using information programs for point source problems. One form is mandatory information disclosures to inform citizens about environmental performance of firms or their products or working condition. Examples are the Toxic Release Inventory provisions of the federal Community Right to Know Act, California's Proposition 65, product labeling schemes (e.g., energy efficiency ratings), and drinking water quality reports. Another is voluntary pollution initiatives such as U.S. EPA's 33/50 program, which encourages manufacturers to voluntarily reduce emissions of 17 target chemicals by 50%.

There are many examples of product and design standards. Pesticide registration is the principal method for protecting the environment, workers, and consumers from pesticide hazards in developed countries, and is used increasingly in developing countries (Dinham, 1993; OECD, 1986). In the European Union and United States, ammonia, phosphorous, and nitrogen pollution from agriculture is controlled by standards governing the amount and timing of manure applications, and restrictions on the numbers of farm animals allowed per unit area. (Hanley, 2001; Ribaudo, 2001). Architectural standards are used to control urban runoff.

Pollution trading, in which individual sources of pollution are provided with limited rights to pollute and allowed to trade these rights in markets, has gained popularity in a number of settings and is increasingly accepted as a cost-effective approach for achieving environmental quality goals (Horan, 2001; U.S. EPA, 2001; Montgomery, 1972; Hahn, 1989; Tietenberg,

1995). In most of these programs, municipal and industrial sources of pollution that are regulated by the NPDES are able to avoid costly discharge reductions at their own facilities by paying agricultural sources to reduce their emissions (Malik et al., 1993; Shortle, 1987; Hanley et al., 1997; Letson, 1993; Horan et al., 2002a, b).

There has been a movement at both the state and federal levels to hold agricultural producers liable for damages resulting from chemical use, but only if they failed to apply registered chemicals in accordance with the manufacturer's instructions and any related laws (Ribaudo, 2001).

17.3.3 Environmental Policy Instruments

The alternative control mechanisms can be combined with alternative compliance measures to define a range of instruments. Examples are presented in Table 17-2.

Table 17-2. Water Pollution Control Instruments.

	Compliance Measure		
Mechanism	**Products/Resources/ Practices**	**Emission or Emissions Proxies**	**Ambient Concentrations**
Taxes/Subsidies/ Fees	Charges on water use, fertilizer, pesticide purchases Cost-sharing or other subsidies for structural or nonstructural practices pollution control Tax benefits for use of structural or nonstructural pollution control practices Cropland retirement subsidies Purchase of development rights Conservation easements Charges on impervious surface area	Effluent charges Charges on modeled nutrient loadings Changes on nutrient applications in excess of crop needs Charges on estimated net soil loss	Ambient taxes
Standards	Pesticide registration Restrictions on fertilizer application rates Urban architectural design standards Mandatory use of technologies	Restrictions on modeled nutrient loadings Regulations on modeled nutrient applications in excess of crop needs	Monitoring
Markets	Point to point effluent trading Point effluent/nonpoint Input trading	Point effluent for estimated nonpoint discharge trading Land retirement contracts	Contracts/bonds Contracts involving the adoption of conservation or nutrient management practices
Liability rules	Negligence liability rules		Strict or negligence liability rules
Voluntary Programs	-------------------------------------Moral persuasion----------------------------------- -------------------------------------Technical assistance---------------------------------		

Water quality protection is a complex issue that cannot be addressed through a single instrument. The biogeochemical complexities, and economic, informational and institutional context all affect the appropriate choice of instruments. Combining various policy instruments in Table 17-2 provides greater flexibility, cost-effectiveness, and adaptability in protecting water quality. These mechanisms also were discussed in Chapters 10.0, 11.0, and 16.0. There are no single, universally appropriate instruments. In addition, the advantages and disadvantages of each of these instruments is relatively well known and documented in the literature. The information needed to develop the appropriate combinations of instruments for your community, watershed, basin, or region, therefore, should be available.

CHAPTER 18.0

ELEMENTS OF SWRM: ADAPTIVE MANAGEMENT TOOLS – CONSERVATION, REUSE, AND RECYCLING

Dublin Principle 1 – Freshwater is a finite and vulnerable resource
Dublin Principle 2 – Water development and management should be participatory
Enlibra Principle 2 – Science for facts, process for priorities
Enlibra Principle 8 – Change a heart, change a nation
Panarchy Principle – Ecological, economic, social, and institutional systems are interlinked in a hierarchy of adaptive cycles

The terms water conservation, reuse, recycling, and reclamation are perceived differently by water professionals, government officials, and the public. In many cases, these terms have negative connotations: water conservation can imply giving up some of the desired uses of water; recycling can imply putting less than pure water back in potable water supplies. The definitions we are using for these four terms are

1. Water conservation – reducing water use by improving the efficiency of various uses of water without decreasing services (Gleick et al., 2003).
2. Water reuse – beneficial use of treated water (NRC, 1998).
3. Water recycling – capture, use, and reuse of water by a single user (Asano & Levine, 1998).
4. Water reclamation – treatment of wastewater to make it reusable (Asano & Levine, 1998).

These terms all refer to the general area of demand management in contrast to supply management. Supply management has traditionally received attention in water resources management. With increased awareness of the social, economic, and environmental impacts of dams, increased water supply through impoundment is unlikely to be a major part of future water resources management (World Commission on Dams, 2000). This chapter focuses on water conservation, and water reuse and recycling. For our discussion, water reuse and recycling are used synonymously.

18.1 Water Conservation and Use Efficiency

Reducing water use by improving the efficiency of various uses of water without decreasing services, as noted above, is, by definition, water conservation. One of the more comprehensive water conservation and use efficiency studies has been conducted for the State of California (Gleick et al., 2003). This study, *Waste Not, Want Not: The Potential for Urban Water Conservation in California*, focuses only on the urban sector in California and does not address agricultural use or irrigation. The urban sector represents about 20% of water use in California. The study concluded that one-third of California's current urban water use—more than 2.3 million ac-ft—can be saved with existing technology. This potential for conservation and efficiency improvement is so large that no new water-supply dams or reservoirs would be needed for decades, even when accounting for projected population and economic increase! (Gleick et al., 2003). In addition, at least 85% of this water can be saved at costs less than would be required for new sources of supply (Gleick et al., 2003).

Current urban water uses and the best estimates for conservation and potential percent reduction of the current urban water use in California are shown in Table 18-1. Although a number of assumptions were made in conducting the conservation study, in part because the required water estimates were neither measured nor monitored, these assumptions were fully documented in the study. As noted in the table, an estimated range is given for various categories for residential outdoor use (Table 18-1). The estimated efficiency in water use can be attained with existing technologies such as low flow/volume toilets, showerheads, washing machines, and dishwashers for residential indoor uses; mulching, drip/bubbler irrigation, turf management, and landscape design for residential outdoor use; and correcting leaks, fixing faucets, using gray water, and similar procedures for commercial and industrial uses.

Table 18-1. California Urban Water Use in 2000 and the Potential to Improve Efficiency and Conservation. (Reprinted with permission from Gleick et al., 2003. Copyright 2003, Pacific Institute for Studies in Development, Environment and Security, Oakland, Calif.)

California Urban Water Use by Sector	Current (2000) Water Use (ac-ft/yr)	Best Estimate of Conservation (ac-ft/yr)	Potential to Reduce Use (%)	Minimum Cost-Effective Conservation (a) (ac-ft/yr)
Residential Indoor	2,300,000	893,000	39	893,000
Residential Outdoor	983,000 to 1,900,000 (b)	360,000 to 580,000 (c)	25 to 40	470,000
Commercial/Institutional	1,850,000	714,000	39	658,000 (e)
Industrial	665,000	260,000	39	(e)
Unaccounted-for Water	695,000	(d)	(d)	(d)
Total	6,960,000 (+/- 10%)	2,337,000	34	2,020,000

(a) Minimum cost-effective conservation is that for which economically relevant data were available and estimates of the cost of conserved water were less than $600/ac-ft. The figure for indoor uses in the residential sector assumes natural replacement of devices when accelerated replacement would cost more than $600/ac-ft.
(b) This is a range of estimated outdoor residential water use. The study's best estimate is 1,450,000 ac-ft/yr.
(c) This is the range of conservation potential for this sector, based on the best estimate for residential outdoor use.
(d) No independent estimate of unaccounted-for water was made. The study adopts here the 10% estimate from the California Department of Water Resources. No separate estimate of the potential to reduce unaccounted-for water was made in this analysis.
(e) Combined commercial, institutional, and industrial cost-effective savings estimated at around 660,000 ac-ft/yr.

Similar water conservation studies at the state level have not, to our knowledge, been made for other states. The approach used for California, however, should be applicable not only for other states, but also for individual municipalities. Some of the recommendations made for California are also applicable for other states and municipalities (Table 18-2).

Private sector businesses also are focusing on water use efficiency and conservation. A water sustainability tool has been developed by the Global Environmental Management Initiative (GEMI) to help corporations and businesses assess their current water use and develop plans and strategies to not only increase the efficiency of their water use, but also assess the potential risks and impacts of water shortages to their suppliers. Some of the elements of this water sustainability tool include: current areas of key water use, including sources and potential vulnerability; risk prioritization for those risks linked to the corporation's water uses and sources; opportunities to proactively address costs and potential risks, including conservation and increased water use efficiency; developing strategies for moving toward sustainable water; and roles of the business functions in developing and implementing these strategies (Figure 18-1) (GEMI, 2002).

The GEMI water use tool uses a life cycle perspective to view all phases of water use throughout the process (Figure 18-2). This perspective considers not only conservation and water use efficiency, but also the potential for reuse and recycling. Daimler Chrysler used a life cycle process similar to the GEMI model in building a zero liquid discharge wastewater facility in Toluca, Mexico (Zacerkowny, 2001). The Toluca facility was initiated in 1998 to reduce the demand on the local aquifer. This facility treats all the wastewater with no liquid discharge, and has halved its water costs, while doubling the amount of water available for production expansion (Zacerkowny, 2001).

Below is a review of various reclamation, reuse, and recycling projects, which begins with a review of international studies, (water reuse and recycling has been more extensively used in other countries than in the United States), and then reviews United States studies. The United States review documents any inherent differences between the west, northeast, and southeast regions.

Table 18-2. Recommendations for Water Conservation and Improved Water Efficiency (after Gleick et al., 2003).

Sector	Recommendations
Technology	♦ Existing technologies are available to greatly reduce urban water use without reducing the goods and services desired. ♦ New technologies are constantly evolving
Economics	♦ The power of proper pricing of water is underestimated. ♦ Economic innovation and financing mechanisms lead to cost-effective water conservation.
Regulation	♦ Smart regulation is more effective than no regulation ♦ Appliance standards are powerful conservation tools that also help educate consumers.
Information	♦ Ignorance is not bliss: Data and information are keys to successful conservation. ♦ Some specific data needs should be a top priority. ♦ Meter and measure all water uses. ♦ Appliance labeling is a powerful educational tool. ♦ Standardize water-use terms. ♦ Educate decision-makers about conservation opportunities. ♦ Give agencies and industries an opportunity to share success stories. ♦ State and local water agencies should work more closely with industry associations and national agencies on data collection. ♦ Reconcile data reported from individual water agencies, industry associations, and various other agencies. ♦ Be aware of the water implications of nonwater policies. ♦ Promote reclaimed and recycled water as a secure source for water supply. ♦ Integrated management practices should be encouraged at water districts or within specific industries.

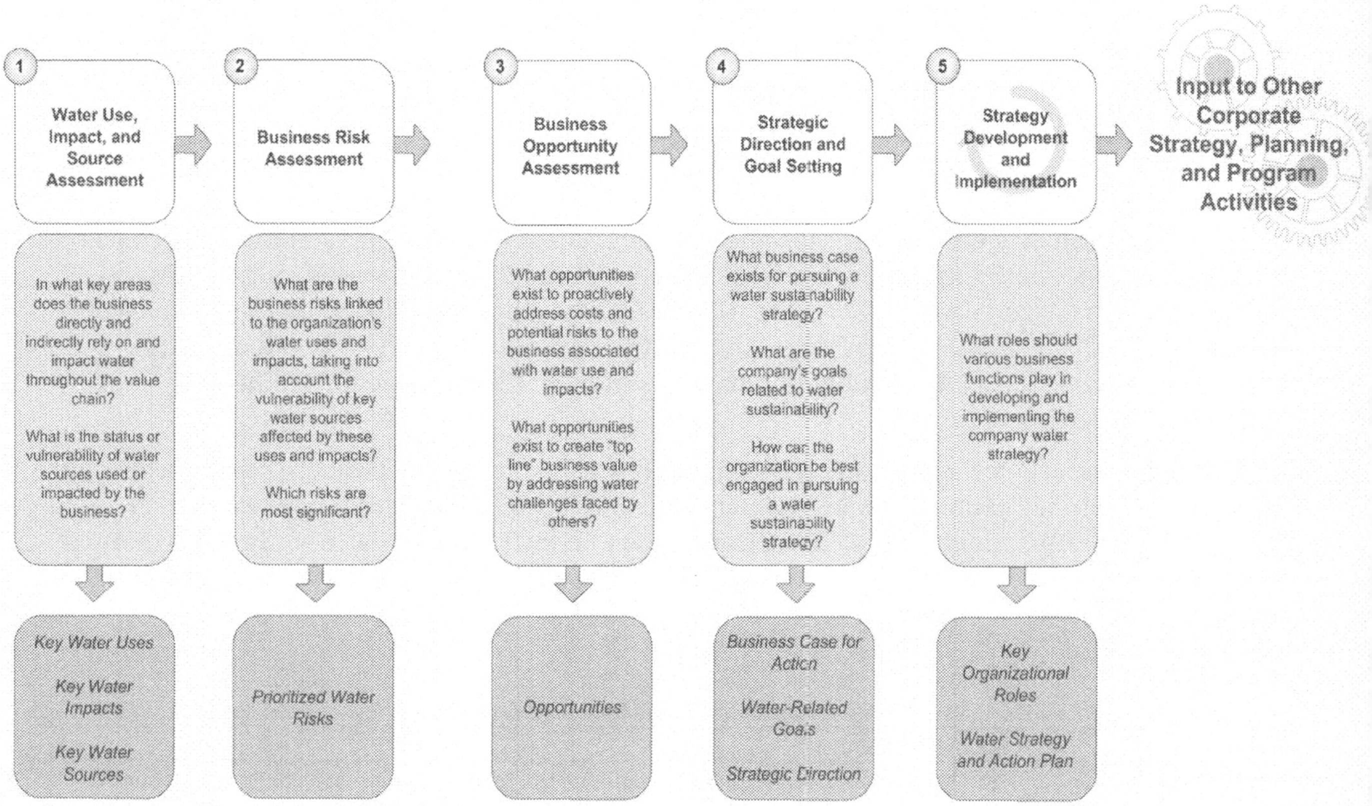

Figure 18-1. GEMI Water Sustainability Tool Roadmap. (Reprinted with permission from GEMI, 2002. Copyright 2002, Global Environmental Management Initiative, Washington, D.C.)

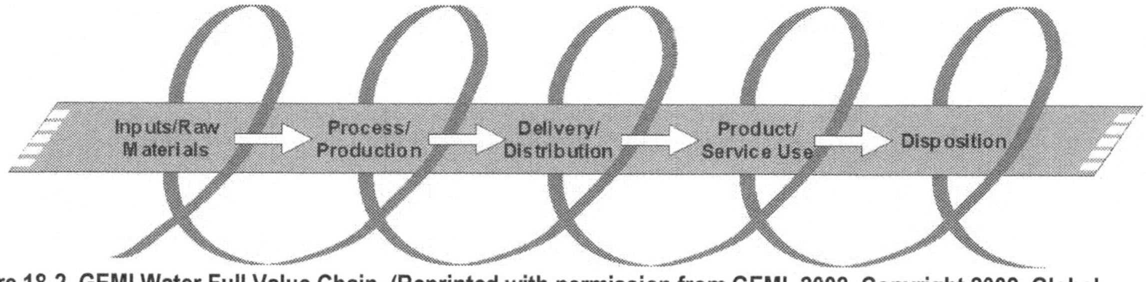

Figure 18-2. GEMI Water Full Value Chain. (Reprinted with permission from GEMI, 2002. Copyright 2002, Global Environmental Management Initiative, Washington, DC.)

18.1.1 International

Postel (1999) cites a number of countries that have stretched water supplies by reusing municipal wastewater for irrigation. Reuse of treated wastewater will be necessary in a number of developing countries, as urban water use is likely to double over the next three decades. Levine et al. (1997) observes that wastewater reuse standards are highly variable, ranging from high-quality effluent required in high-income countries, to open issue reuse regulations in low-income countries. Table 18-3 summarizes water reuse characteristics for selected countries.

Table 18-3. Overview of Water Reuse in Selected Countries (after Crook et al., 2005).

Country	Water Reuse Activity[a]	Reuse Guidelines/ Regulations	Percent Water Consumption by Sector			Water Reuse Applications[a]		
			Urban	Industry	Agriculture	Irrigation	Urban	Potable
Argentina	Medium	Yes	16	9	75	++		
Australia	High	Yes	12	6	70	++	+	
Belgium	Medium	No						+
Canada	Medium	Yes	11	68	7		+	
Chile	Low		5	11	34	+		
China[b]	Medium	No	5	18	77	++		
Egypt[b]	Medium	No	6	8	36	+		
France	Medium	Yes	15	73	12	++	+	
India[b]	High	No	5	3	92	+++		
Israel	Intensive	Yes	29	7	54	+++	++	
Italy	Medium	Yes	14	33	53	++		
Japan	High	Yes	19	17	64	++	+++	+
Jordan	High	Yes	22	3	75	++		
Kuwait	Intensive	Yes	37	2	60	+	++	
Namibia	Medium	No	28	3	68	+		+
Mexico[b]	Medium	No	17	5	78	+++		
Morocco[b]	Medium	No	5	3	92	++	+	
Oman	High	Yes	5	2	93	++	+	
South Africa	High	Yes	17	11	72	++	+	
Saudi Arabia	Intensive	Yes	9	1	90	+++	++	
Spain	High	Yes	13	18	68	++	+	
Sweden	Low		35	30	4	+		
Tunisia	Medium	Yes	14	3	83	++		
United Arab Emirates	Intensive	Yes	24	10	67	+	++	
UK	Medium	Yes	65	8	2	+	++	

[a]Subjective evaluation based on literature: +++ numerous, ++ occasional, and + isolated projects

[b]Countries using raw sewage for irrigation

Some international areas that have already instituted successful water recycling initiatives include Australia, the Mediterranean and Middle East, Latin America, and Africa.

18.1.1.1 Australia

Australia, one of the driest continents, has historically experienced problems with water resources. Since Australia suffers extreme droughts and floods, which creates an uneven distribution of water resources, some river basins are approaching the limits of their allocation capacity. Traditionally, storage dams have been constructed to help with this issue, but the expense and the lack of acceptable sites has caused this to become an unacceptable approach in most areas of the country (Simpson, 1999). In the last few decades, Australia has become one of the leading countries in the world in water reuse.

Planned water reuse in Australia is for nonpotable uses. Unplanned water reuse does occur, particularly in the Murray Basin (Radcliffe, 2004). The primary uses of reclaimed water include

- landscape irrigation—both public (parks, golf courses) and private (residential property);
- toilet flushing;
- environmental uses (dune stabilization, streamflow augmentation);

- crop irrigation;
- silviculture;
- commercial uses (vehicle washing, turf farm irrigation);
- construction uses (dust control, soil compaction); and
- industrial cooling, boiler feed, and process water (Radcliffe, 2004).

Water use reductions of 20–30% have also been achieved through pay-for-use pricing initiatives. Anderson (1996) estimates that another 10–30% are saved through consumer outreach materials and the installation of water-efficient appliances. Anderson (1996) also forecasts recycling scenarios for the 21st century. The four scenarios include a traditional water supply approach (no reuse), water conservation and recycling, residential nonpotable recycling, and potable recycling.

Simpson (1999) documented that the Queensland community, one of the wetter areas of Australia, favored reuse even though the community, in general, opposed the disposal of sewage effluent into sensitive waterways. An education program was developed to focus on water quality of reclaimed water to overcome the negative image generated by the terminology used by the water industry. This concept of potable reuse was largely community-driven in this area, as a result of enabling the community to have an informed opinion based on sound knowledge provided by the professional members of the industry. This led to the formation of a Water Reuse Strategy and a policy on water reuse.

Specifically, Queensland Government's Department of Natural Resources developed a draft Queensland Water Recycling Strategy (The State of Queensland, Department of Natural Resources, 2000). The strategy stated three major benefits of water recycling: it conserves supplies (e.g., total available water supply); it is environmentally responsible as it preserves the health of waterways, wetlands, flora and fauna; and it makes economic sense by saving resources, which can in turn reduce pollution in waterways and assist tourism and fishing industries. Eight action plans were developed by the Queensland government and others to provide sustainable water recycling in Queensland:

1. change laws to support recycling,
2. provide guidance for using recycled water,
3. provide technical training and information for the workplace,
4. increase community awareness and participation in water recycling,
5. encourage further research into key aspects of water recycling,
6. use demonstration projects to raise awareness of water recycling,
7. support water recycling projects, and
8. implement the Queensland Water Recycling Strategy (QWRS) (also referred to as the Water Re-Use Strategy due to protest) through collaboration.

Thomas et al. (1993) noted that more stringent regulations on wastewater land application systems in Australia have necessitated a multidisciplinary planning approach to land application. Long-term impacts of nutrients, salt, and other contaminants on downstream water quality need further research.

18.1.1.2 Mediterranean and Middle East

Angelakis et al. (1999) discussed three major barriers contributing to water issues in the Mediterranean:

1. uneven distribution of precipitation and runoff spatially and temporally,

2. uneven occupation of territory, and
3. cost of making the necessary quantity of water available at the right time and place (often this is more of a problem than the actual scarcity of water).

Many projects have been instituted in the Mediterranean basin that follow local and national reclamation and reuse guidelines from Israel, a pioneer in the development of wastewater reuse practices, as well as France, Tunisia, and Cyprus. Currently there are several research and pilot projects dealing with wastewater reclamation and reuse in Greece, Italy, Malta, Morocco, Portugal, and Spain.

Appropriate water quality requirements for treated wastewater are essential to protect public health, increase water availability, prevent coastal pollution, enhance water resources, and enforce nature conservation policies. The establishment of guidelines is critical for the management and safe implementation of wastewater reuse for irrigation. Reuse water quality criteria for irrigation may relate to chemical constituents to control potentially toxic substances or pollutant accumulation in crops and soil; or biological constituents such as the presence of coliform bacteria and pathogenic bacteria and viruses. Guidelines for use of reclaimed water should promote the development of best management practices by considering local conditions including

- quality of reclaimed wastewater,
- soil type,
- climate,
- relevant crops, and
- local agricultural practices.

Mediterranean countries with established national guidelines for the use of reclaimed wastewater are Cyprus, France, Israel, Italy, Jordan, and Tunisia (Angelakis et al., 1999). In addition, guidelines have been developed for a variety of reuse applications of municipal wastewater in Andalusia, Spain (Blancas et al., 1997).

Jordan has recognized the need to maximize the safe reuse of wastewater effluent and has formed a plan to achieve this goal. The agricultural sector, which consumes 70% of Jordan's water, has been determined to be the major user of treated wastewater effluent. Domestic and industrial wastewater effluent standards specifically for agricultural reuse have been incorporated into law. In addition, new wastewater treatment plants are being designed to incorporate reuse programs, and existing facilities are being renovated to meet the new standards and promote reuse.

Juanico and Friedler (1999) discuss the feasibility of reclaimed wastewater in semi-dry countries such as Israel. Wastewater effluent is the main source of water in the rivers of Israel. The two major topics discussed by Juanica and Friedler are water quality management in rivers, and allocation of reclaimed wastewater flow in rivers and streams. Water quality management officials in this area must consider that the main water source to the river has a pollution loading. This may reduce the capability of the river to absorb other pollution inputs.

The impact of the use of reclaimed water for irrigation on soil and groundwater was researched in a Muszkat et al. (1993) investigation of the effects of 20 years of effluent irrigation of citrus crops in Israel. Organic pollutants such as toluene and phthalates were found in the water table even after 66 ft (20 m) of travel through the unsaturated zone. Caution was advised regarding the potential long-term effects of irrigation with reclaimed wastewater.

18.1.1.3 Latin America

Lack of treatment and recycling of domestic wastewater is the most critical issue in Latin America. Currently less than 14% of wastewater receives any treatment and only 6% receives acceptable treatment before discharge (Cavallina & Young, 2002). Untreated wastewater is also used for agricultural purposes, which may pose health risks in many localities. Cities Feeding People is funding a project to integrate systems for the treatment and recycling of wastewater in Latin America. The project is managed by the Centro Panamericano de Ingenieria Sanitaria Ciencias del Ambiente, which is based in Peru. The objectives of the project are to document and inventory Latin American and Caribbean experiences in domestic wastewater treatment and recycling.

Agricultural water reuse has been the practice in Mexico for more than 100 years. Almost 860,000 acres are irrigated with wastewater, generally with little or no treatment (Crook et al., 2005). Recent efforts have begun to evaluate priorities for reuse of municipal wastewater discharges based on consideration of geography, treatment technology, and crops (Jimenez & Ramos, 1997). However, these reuse practices have brought with them an increased prevalence of parasitic infections among agricultural workers exposed to raw wastewater irrigation, and metal accumulation in soils and crops as a result of long-term application (Cifuentes & Siebe, 1997).

Significant efforts at environmental cleanup are being combined with new strategies to use reclaimed water to meet the needs of South America. Reclamation technologies and reuse applications used in Chile are being monitored and evaluated by Peru, Argentina, and Brazil (Del Piero, 1998). Recommendations include developing policies to permit toilet flushing with reclaimed water and establishing pricing fees for volume-based water use.

18.1.1.4 Africa

Namibia is the driest African country south of the Sahara Desert. Despite many years of access to high-quality recycled water, the residents of Windhoek still oppose its use for personal consumption. For this reason, most of this purified wastewater irrigates gardens. However, in times of drought, local freshwater reservoirs are strained in Windhoek, Namibia's capital city, which relies heavily on treated effluent to boost water supply. During the drought of 1995, reclaimed water accounted for more than 30% of the clean water piped into homes.

The effects on groundwater quality of long-term irrigation with wastewater were researched by Farid et al. (1993). A 3000-acre site near Cairo, Egypt, was also investigated following 75 years of irrigation with wastewater ranging from raw sewage to primary-treated effluent. The groundwater was determined to have elevated levels of nitrogen, phosphate, heavy metals, and fecal coliforms. Reuse of municipal wastewater and agricultural drainage water was a key factor in the Egyptian plan to develop new lands for agriculture (Mukhamer & Ibrahim, 1993).

South Africa experiences highly variable and uneven distribution of rainfall, high evaporation rates, low yielding aquifers, and a growing urban population (Odendaal et al., 1998). By reusing industrial-process water, they have reduced overall water consumption. However, less than 3% of available treated wastewater is directly reused in the country (Grobicki & Cohen, 1998).

18.1.2 United States

Water reclamation and reuse has a long history in the United States. Historical uses were primarily those that did not require high-quality water (Crook et al., 2005). Reclaimed water, however, is slowly being valued as a resource with the trend shifting toward urban uses such as irrigation, toilet and urinal flushing, indirect potable reuse, and industrial applications and uses (Crook et al., 2005). The predominant uses of reclaimed water in the United States are shown in Table 18-4.

Table 18-4. Uses of Reclaimed Water in the United States (after Crook et al., 2005).

Category of Use	Specific Types of Use
Potable water supply augmentation (indirect potable reuse)	Groundwater recharge, surface water augmentation
Nonpotable urban uses (other than irrigation)	Toilet and urinal flushing, fire protection, air conditioner chiller water, commercial laundries, vehicle washing, street cleaning, decorative fountains and other water features
Landscape irrigation	Parks, playgrounds, cemeteries, golf courses, roadway rights-of-way, school grounds, greenbelts, residential and other lawns
Groundwater recharge	Aquifer storage and recover, seawater intrusion control, ground subsidence control
Agricultural irrigation	Food crops, fodder crops, fiber crops, seed crops, nurseries, sod farms, silviculture, frost protection
Industrial uses	Cooling, boiler feed, stack scrubbing, process water
Impoundments	Ornamental, recreational (including full-body contact)
Environmental uses	Stream augmentation, marshes, wetlands, fisheries
Miscellaneous	Aquaculture, snowmaking, soil compaction, dust control, equipment washdown, livestock watering

Water recycling has become increasingly important in the United States because of the rapid growth in the southern and western states, which lead to increased demand for potable and nonpotable water. The two states in the United States that extensively use water recycling are California and Florida. Uses of recycled water in these states include irrigation for landscaping and agriculture, nontraditional uses such as cooling towers and high-rise buildings, and toilet flushing. In a 1992 study performed by U.S. EPA, 18 states had some reuse and reclamation guidelines, 18 had complete guidelines, and 14 states had no provisions for reuse (Angelakis et al., 1999).

Water recovered from treated sewage has already become an integral part of most state water supplies. Despite high cost and worries over public perception, the use of recycled water has increased. It is being used to make snow for ski areas, grow hay, make newsprint and concrete, dye carpets, wet landfills, and for cooling water in oil refineries. Surtevant (1993) describes plans for the first U.S. project to use reclaimed water for artificial snowmaking.

U.S. EPA published revised Guidelines for Water Reuse (U.S. EPA, 2004a) for utilities and regulatory agencies in the United States because some states do not currently have standards for water reuse. The guidelines address various aspects of water reuse, including treatment recommendations, water quality limits, monitoring frequencies, and other factors related to water reuse. The treatment and water quality limit guidelines for various reclaimed water applications are listed in Table 18-5.

18.1.2.1 California

California has been a pioneer in the area of water recycling, with many successful projects documented for a variety of types of reuse. Activities that benefit from reuse include landscape irrigation, agricultural irrigation, industry, and recreation. The use of recycled water also supports and enhances aquatic habitat, fish, and wildlife (Watereuse Association http://www.wateruse.org/). California uses more than 450,000 ac-ft of reclaimed water annually, which is equivalent to the water consumed by two-thirds of Los Angeles in a year. Throughout the state, independent sewer and water utilities are working together on reuse projects.

Table 18-5. U.S. EPA Guidelines for the Water Reuse (from Crook et al., 2005).

Type of Use	Reclaimed Water Quality	Treatment
Urban uses, crops eaten raw, recreational impoundments	♦ pH = 6-9 ♦ ≤10 mg/L BOD ♦ ≤2 NTU turbidity[a] ♦ No detectable fecal coliform colonies/100 mL[b] ♦ ≥1 mg/L Cl_2 residual[c]	♦ Secondary ♦ Filtration ♦ Disinfection
Restricted access area irrigation, processed food crops, nonfood crops, aesthetic impoundments, construction uses, industrial cooling[d], environmental reuse	♦ pH = 6–9 ♦ ≤30 mg/L BOD ♦ ≤30 mg/L total suspended solids ♦ ≤200 fecal coliform colonies /100 mL[e] ♦ ≥1 mg/L Cl_2 residual	♦ Secondary ♦ Filtration
Groundwater recharge of nonpotable aquifers by spreading	♦ Site specific and use dependent	♦ Site specific and use dependent ♦ Primary (minimum)
Groundwater recharge of nonpotable aquifers by injection	♦ Site specific and use dependent	♦ Site specific and use dependent ♦ Secondary (minimum)
Groundwater recharge of potable aquifers by spreading	♦ Site specific ♦ Meet drinking water standards after percolation through vadose zone	♦ Secondary ♦ Disinfection ♦ May also need filtration and advanced wastewater treatment
Groundwater recharge of potable aquifers by injection, augmentation of surface supplies	Includes the following: ♦ pH = 6-9 ♦ ≤2 NTU[a] ♦ No detectable fecal coliform colonies /100 mL[b] ♦ ≥1 mg/L Cl_2 residual ♦ ≤3 mg/L TOC ♦ ≤0.2 mg/L TOX ♦ Meet drinking water standards	♦ Secondary ♦ Filtration ♦ Disinfection ♦ Advanced wastewater treatment

[a] Should be met before disinfection. Average based on a 24-h period. Turbidity should not exceed 5 NTU (nephelometric turbidity units) at any time
[b] Based on 7-day median value. Should not exceed 14 fecal coliform colonies/100 mL in any sample.
[c] After a minimum contact time of 30 minutes.
[d] Recruiting cooling towers
[e] Based on 7-day median value. Should not exceed 800 fecal coliform colonies /100 mL in any sample.

In 1993, the California Department of Water Resources, in cooperation with the U.S. Bureau of Reclamation and several southern California water agencies, began to evaluate the feasibility of a regional water recycling plan. Among the potential reuses were irrigation, industrial processes, groundwater recharge, and groundwater injection to prevent seawater intrusion. The study defined the regional water needs and described an effort to match these needs with resources, including consideration of methods to convey the water. The overall goal was to identify opportunities for maximizing water reuse in southern California.

The Irvine Ranch Water Recycling Program is one of the longest and most successful reuse projects in California. The Irvine Ranch Water District imports about 50% of its water to meet the district irrigation and domestic requirements. To reduce dependency on these imports, the Irvine Water District started recycling water for agricultural use in about 1967. Since then, treated wastewater has been reused for crop, golf course, park, school ground, greenbelt, and highway median irrigation, commercial toilet flushing, and industrial uses. Water reuse currently supplies about 15% of the annual water needs for the District (Young et al., 1998).

The Irvine Water District success occurred because of a consistent commitment to inform and educate the local community on the importance of water reuse in reducing the water shortage issue. The entire local community was told about the value of water for decades by the District (Young et al., 1998). Water reuse was promoted as a means of protecting the environment, saving money and energy, and providing water during droughts (D'Angelo, 1998).

The City of San Diego undertook a water reuse project following the 1991–1992 drought (D'Angelo, 1998). The project proposed mixing reclaimed water with imported freshwater in reservoirs. After a year's detention, conventional potable water treatment would be used to treat the water before distribution as public water supply. The San Diego City Council and San Diego Water Authority all understood the importance of public information, education, and acceptance in implementing this project. Public information and education campaigns, telephone surveys, focus groups, and similar procedures were used in the community. All survey results indicated there was support of reclaimed water for drinking, washing, and other domestic uses (Katz & Tennyson, 1997). However, despite the strong public support, water reuse became a political campaign issue. Communities were informed that the recycled water would be distributed to the less affluent areas in the community, and potential health dangers were highlighted (Recycled Water Task Force, 2003). The slogan, "Toilet to Tap," characterized how recycled water was conveyed during the political campaign (Recycled Water Task Force, 2003). The project was finally put on indefinite hold.

Hultquist (1995) drafted an overview of the current regulatory status of augmentation of ground and surface drinking water sources with reclaimed water in California. Extensive involvement of the community was considered critical to obtaining acceptance of the planned indirect potable reuse project in San Diego, California (Katz & Tennyson, 1997). The results of public surveys indicated that residents of San Diego perceived their region as an area of water scarcity and accepted the concept of the potable reuse project.

Neal et al. (1996) documented the planning of a water reuse system for the San Francisco Zoo. The reclaimed water system project involved a number of complex design issues, such as identifying suitable uses for reclaimed water at the zoo, retrofitting existing zoo facilities for reclaimed water use, and implementing public health safeguards mandated for reclaimed water use. The use of reclaimed water was also examined for its applicability at the Los Angeles Zoo (Stillwell et al., 1996). This reclamation project involved general irrigation and select animal exhibits. The zoo was able to conserve potable water resources without jeopardizing the health of the animals and with little risk to most of the plant collection.

Public education is critical to the success of large-scale water reuse efforts, such as the one currently under way in the South Bay region of San Francisco Bay Area, California. Cooperation between public agencies is also important, as success of the project requires complex interagency relationships (Garcia et al., 1997). In San Diego, local community involvement and acceptance has proved to be critical to the success of the planned indirect potable reuse project.

The Water Replenishment District of Southern California researched the economics of their indirect potable reuse projects (Alamitos Barrier Project and West Basin Project) and concluded that, in addition to current economics, the future cost of water and the value of drought-proofing should be considered (Leserman & Ferry, 1997).

Regulations in California regarding groundwater recharge have caused wastewater agencies to reconsider existing facilities. These guidelines require that total organic carbon concentrations in reclaimed water be below specified limits tied to the percentage of reclaimed water to be recharged. Regulating the presence of trace amounts of organics in reclaimed water is usually accomplished by dilution using surface water or groundwater from a less-contaminated source (Asano, 1993). The Clean Water Revival Project in Northern California, however, produces high-quality reclaimed water for direct injection into the groundwater. The project is designed specifically for disposal of peak wet-weather flows that cannot be carried through the existing outfall (Geselbracht et al., 1998).

18.1.2.2 Florida

Florida promotes reuse of reclaimed water and water conservation as major state objectives. Florida's ever-increasing population continues to put increasing demands on the state's water resources, leading to prolonged, severe water shortages. Reuse is necessary to help the state meet its water supply and wastewater management needs.

The Florida Department of Environmental Protection allows for the use of reclaimed water for irrigation of golf courses, parks, residential properties, highway medians, and other landscaped areas. They also suggest urban uses such as toilet flushing, car washing, dust control, and aesthetic purposes, including decorative lakes, ponds, and fountains. Reclaimed water is allowed for irrigation of edible food crops, as well as pasture lands, grasslands, and other fodder crops. Reuse projects can also include wetlands creation, restoration, and enhancement, as well as recharging groundwater using rapid-infiltration basins, absorption fields, and direct injection to aquifers that augment surface waters used for drinking water supplies. Industrial uses for reclaimed water can include plant wash down, processing water, and cooling water.

Several authors have reviewed reuse projects in Florida. Curran and Kiss (1995) analyzed data from the world's largest residential dual-water system in Cape Coral, Florida. Information about level of participation, nonpotable water demands, and potable water resources were included. Thabaraj (1998) presented a literature review of reclaimed water projects in Florida, indicating the beneficial results of water reuse in the state. Ten years of scientific study show that Tampa Bay's proposed indirect potable reuse project is technically feasible. The work continued to address five key implementation issues: public acceptance, permitability, cost, size of project, and legal/intergovernmental issues (Baird & Waller, 1997). McEwen and Richardson (1996) reviewed the key issues of treatment, cost factors, and implementation guidelines for evaluation and development of indirect potable reuse as an alternative water supply for municipalities.

The National Research Council (1988) released a study of Orange County, Florida: *Issues in Potable Reuse: the Viability of Augmenting Drinking Water Supplies with Reclaimed Water*. The report was to assess the public health implications of using recycled water as a component of the potable water supply. The purpose of the report was to identify considerations for ensuring reliability in evaluation of the suitability of water sources augmented with recycled water. Research needs regarding potable reuse safety testing and health effects were also indicated. The report indicated that indirect potable reuse should be an option of last resort. It should be adopted only if other measures, including other water sources, nonpotable reuse, and water conservation, have been rejected as technically or economically infeasible. This report conveyed a mixed message to the water supply industry.

Paret and Eisner (1993) interviewed users of reclaimed water in Florida and found satisfaction with reuse programs, especially when sharing of project cost was provided. Paret and Eisner (1993) also documented the importance of public education programs for successful reclamation projects. The reclaimed water public education program used in Largo, Florida, was beneficial because it included accurate definitions of terminology, demonstrations of reclaimed water, promotion of advantages over other sources, information about the environmental benefits, and safety education (Andrade, 1996).

The Florida Department of Environmental Protection documented some successful water reuse projects including

- *St. Petersburg Dual Distribution System* – More than 9,000 residential lawns, 51 schools, 86 parks, and six golf courses benefit from the highly treated reclaimed water provided by a separate piping system for landscape irrigation.
- *Conserv II* – Highly treated reclaimed water provides irrigation for more than 8,000 acres of citrus groves, foliage and landscape nurseries, tree farms, ferneries, and the Orange County National Golf Center. Tapping into Water Conservation II is a cooperative water reclamation venture between the City of Orlando, Orange County, Florida and the area's agricultural community. Most participants have had substantial financial and environmental benefits in the 13 years of using the program. The project, a combination of agricultural irrigation and rapid-infiltration basins that divert water into the ground is the largest of its kind in the world. The benefits are numerous. The water is free and abundant, so farmers are maintaining higher soil moisture levels, which protect their entire crop area. Some citrus growers increased crop yields by 19% and tree growth by up to 400%. For citrus growers this means a savings of about $128 per acre ($316 per hectare) per year (Woodard & Curran, 1999).
- *Project APRICOT* – Residential lawns and greenspace irrigation in Altamonte Springs benefit from reclaimed water provided by this project. Reclaimed water is also used by firefighters, a local car wash, and for street-sweeping equipment.
- *Orlando Wetlands* – A 1,220-acre wetlands system that provides a public park and nature preserve for all types of recreational activity including hiking, biking, and nature observation was created using reclaimed water from the Iron Bridge advanced wastewater treatment facility.
- *Tallahassee Spray Irrigation System* – Reclaimed water is used to irrigate feed and fodder crops. In addition, cattle also graze in this area during certain times of the year.

- *Gainesville, Southwest Reuse Project* – Reclaimed water is used for irrigation of residential lawns, golf courses, parks, and other landscaped areas.
- *Reedy Creek Utilities* – This utility provides the Walt Disney World Resort complex with reclaimed water for irrigation of five golf courses, landscaped areas at five hotels, highway medians, a water park, and a 100-acre tree farm that produces horticultural materials for use in the complex.

Florida also has many successful industrial uses of reclaimed water, mostly for cooling purposes. Examples are the Curtis Station Energy Center, Vero Beach Municipal Power Plant, McIntosh Power Plant, Falkenberg Road Resource Recovery Facility, the McKay Bay Refuse to Energy Facility, and University of Florida Co-Generation Facility.

18.1.2.3 South Carolina

Hilton Head, South Carolina, has instituted a number of urban water reuse applications such as parks and golf courses, over the past three decades. Wetlands water reclamation has also begun assisting the revival of wildlife. Previously, the wetlands had been drying because of changes in water flow patterns caused by increased development in the area (www.wateruse.org/pages/otherstates.htm1#wetlands).

18.1.2.4 Utah

The surge in water recycling in Utah has been propelled by improvements in technology, regulatory changes, and government subsidies for reclamation systems. In the Salt Valley of Utah, a cooperative effort between water districts, sewer districts, and municipalities has helped to develop a reuse project to provide treated wastewater effluent to local canal companies for agricultural and residential irrigation (Sorenson, 1994).

18.1.2.5 Colorado

A Denver, Colorado, direct potable water reuse demonstration project was designed to examine the feasibility of converting unchlorinated secondary treated wastewater to potable water quality (Lauer & Rogers, 1996). The findings demonstrated the ability of advanced water treatment processes to remove a broad spectrum of pollutants and render a product that satisfies current measures of safety, enabling direct potable reuse to be considered along with more conventional water supply development alternatives (Bernados, 1996).

18.1.2.6 Arizona

Doba and Pellatz (1996) documented a small city's experience with its new reclaimed water plant and dual distribution system. Areas researched included the interaction between the city of Flagstaff and the Arizona Department of Environmental Quality before system start up, rate establishment, policies on customer rebates for tying into the system, effect on peak water consumption, investment analysis, and public acceptance. WateReuse.org documents that most of the water used in Phoenix, Arizona, is recycled in some way, including irrigation of crops such as vegetables and melons, and in cooling systems. Because of the variability in the supply and demand of the wastewater, Phoenix is developing groundwater recharge systems and constructed wetlands.

18.1.3 Public Perception of Water Reuse

Although the public has generally been supportive of the idea of water reuse and recycling, there has also been serious opposition to some projects, as noted above in the San Diego example. Hartley (2003) compiled and analyzed information from several public perception surveys and noted positive responses to water reuse generally were higher when

- human contact with recycled water is minimal,
- public health protection is clear and obvious,
- environmental protection occurs through reuse,
- water conservation is promoted through reuse,
- treatment and distribution costs are reasonable,
- perception of wastewater as the source of the reclaimed water is low,
- water supply problems are known and appreciated by the community,
- role of reclaimed water as part of the overall water supply is clear and understood,
- perception of the treated high quality of the reclaimed water is high, and
- public confidence in the management of the public utilities is high.

Hartley (2001) identified five categories of factors that influence public perception and participation.

1. **Information and Context**: Information, knowledge, local context, and education all played an important role in shaping perception. Furthermore, the uncertainty or incompleteness of information influenced perception about water reuse.
2. **Communication and Dialogue**: Communication is the means of exchanging information. Although the effectiveness of the communication was important, the quality of the dialog was a major factor contributing to public perception. The quality factors were based on the sincerity, honesty, legitimacy, and credibility of the speaker; and the comprehensiveness, clarity, understandability, and factual accuracy of the communication.
3. **Trust and Trust-Building**: Perception of water reuse, and behavior in participation was influenced by trust, but it was a cyclical relationship. Trust was built through communication, especially when
   a. dialogue was ongoing,
   b. information was available from entities not sponsoring the project,
   c. questions could be asked,
   d. the public was involved early,
   e. information was available to everyone,
   f. behavior was noncoercive,
   g. there was willingness to listen to all views and expand discussion if necessary, and
   h. the public had some level of control in the process (e.g., contributing to agenda or ground rules).
4. **Fairness**: This perception was critical in public participation and perceptions of water reuse. Fairness applied in both the decision-making process, and the outcome. Was the process reasoned, did it involve everyone, was everyone genuinely listened to, and were all concerns addressed appropriately. Fairness in the outcome was based on whether burdens and/or benefits were being shared fairly (not necessarily equally).
5. **Motivation and Commitment**: Individuals needed to be motivated enough to participate in water reuse decision-making. Rarely was a single motive enough; usually people needed multiple motives to participate. Commitment represented an organizational-level

factor in public dialog. The organizations involved needed to show genuine commitment to public participation in the process.

These five themes were not independent, but tended to influence and overlap each other in multiple ways.

18.1.4 Water Reuse Criteria

A number of factors need to be considered before initiating water reuse or recycling projects to ensure that they are successful. These include

- community education to foster acceptance of the proposed programs,
- thorough planning,
- specifying reuse criteria, and
- special concern for potential health effects.

A major factor directing reuse is community-driven acceptance. Reuse projects need safe technology and assurances that public health will not be compromised. Some programs speak of "DID"—decide, inform, and defend. First, the utility must decide to go forward with the project, then inform the public, and defend the decision. This has resulted in both negative and positive outcomes.

Mills and Asano (1996) found that many agencies undertake water reclamation projects with inadequate planning. This may result from a sense of urgency because of problems such as drought conditions, or from a lack of experience in acknowledging which factors are vital to the project's success. Problems encountered by reclamation projects include

- permit approvals,
- lack of reclaimed water market,
- lack of reliable data, and
- institutional issues.

Permit approvals should be obtained before planning, as they can affect the overall feasibility of the project, the location of the facility, or specific design criteria. There is a need to secure reclaimed water users for a project. Not doing so may result in time delays, disappointing freshwater savings, or significantly higher costs per unit of water delivered than originally estimated. Reliable data should be obtained regarding the amount of reclaimed water that will be available for reuse. In addition, the potential reclaimed water demands, and the variation that occurs hourly, daily and monthly, needs to be considered. Institutional issues may come into play, especially during cooperative efforts. For instance, each agency may have its own interests to protect, which may stem from political reasons or previous negative encounters (Mills & Asano, 1996).

Reuse criteria should be established. However, it should be noted that several factors affect the establishment of reuse criteria including public policy, health protection, past reuse experience, technical feasibility, and economics.

Crook and Surampali (1996) indicate that although there haven't been any documented disease incidents related to reuse projects in the United States, adverse health effects in other countries have been well documented (Sepp, 1971; Lund, 1980; Feachem et al., 1983; Shuval et al., 1986). Factors that affect the quality of reclaimed water must be considered, including source water quality, wastewater treatment processes, treatment effectiveness, treatment reliability, and distribution system design and operation.

The presence of toxic chemicals, pathogenic organisms, and the effects of certain physical parameters also must be addressed (Crook & Surampali, 1996). The effects of pH, color, temperature, and particulate matter, as well as chemical constituents (e.g., chlorides, sodium, and heavy metals), are well documented, and permit limits have been established for most of these constituents. The most common parameters for water quality limits include biochemical oxygen demand, turbidity, total suspended solids, total and/or fecal coliform, nitrogen, and chlorine.

18.2 Urban Runoff Capture and Use

Urban runoff is typically not captured for reuse. As part of the general NPDES program, best management practices are required for treatment of stormwater runoff from industrial, commercial, and urban sites, but treated stormwater is typically discharged to receiving systems after treatment. The City of Santa Monica, California, not only built a facility to treat urban runoff, but is also reusing this water following treatment.

Santa Monica built a dry-weather runoff treatment facility that can treat up to 500,000 gal/day. Dry-weather runoff arises from inefficient use of potable water, over-irrigation, broken irrigation systems, washing pavement and business equipment, car washing, pool drainage, leaking water pipes, mains, and hydrants, and illegal dumping (Antich et al., 2001). The average daily flow is about 300,000 gal/day, representing a little more than 2% of Santa Monica's total water demand of about 13 mgd (Antich et al., 2001). This treated dry-weather runoff is reused for landscape irrigation and toilet flushing in dual-plumbed buildings.

The Santa Monica urban runoff recycling facility was located near the Santa Monica pier, a popular and prominent tourist area. The facility was designed by a team that included not only wastewater engineers, but also an architect, artist, and educators, and public relations specialists. The facility has colorful tile mosaics, water features, innovative architecture, and dramatic lighting (Antich et al., 2004). This comprehensive approach to the wastewater facility resulted in a positive response by the community.

Stormwater treatment and reuse represents another water source to help satisfy increasing demands and might help satisfy agricultural demands for irrigation water, industrial once-through cooling water or similar uses. Additional uses and treatment options for stormwater are expected as both stormwater treatment and water demands increase.

CHAPTER 19.0

PUTTING SWRM INTO PRACTICE

A separate framework and guidance document, intended to guide water managers in developing SWRM, has been developed. This chapter is based on that document. As a result, the format of this chapter is different from the remainder of this report.

Putting the Framework into Action: Steps in the Process

Unfortunately, most communities, agencies, and organizations do not consider moving toward sustainable water resources until they are in a crisis situation. The need for a water crisis to initiate change in water management practices was confirmed at a stakeholder workshop held in September 2004. When asked what would be required to implement a framework for SWRM, stakeholders answered that one or more of the following would be required (in priority order):

1. a crisis or defining event,
2. leadership, and
3. money.

Therefore, the first part of the process flow chart is applicable to communities or watershed associations that are facing a water crisis (Figure 19-1). In a water crisis, management options are restricted to those that can be implemented quickly. These management options are typically developed and implemented by government entities. The following text illustrates the information needed in each numbered step of the process flow chart to address a water crisis (water quantity and/or quality) at the watershed scale and eventually move toward SWRM. This document addresses the watershed scale because watershed management is actively encouraged by many agencies and makes sense for moving toward sustainable water resources. However, the guidance presented can and must be applied to communities, as well as basins and regions. Examples are provided from the Denver Drought Response Plan, which is considered a good model for managing a drought crisis (Rogers, 2003).

Leadership: Making It Happen

In 2002, New York State initiated a future-oriented planning effort (Project 2015) to consider the impact of an aging population on services, products, constituency make-up, and day-to-day work of multiple state agencies. Getting state agencies to work together can be a challenge. In New York, 36 state cabinet agencies had to work together for Project 2015 to succeed. In February 2002, Governor George Pataki called a cabinet meeting at his home on a Saturday to inform all 36 state agency Commissioners they had nine months to develop a coordinated plan to 1) assess the impact of demographic change on their agency, 2) prioritize the major impact areas, and 3) specify action steps or strategies they would implement over the next five years to address their top three impact areas. Governor Pataki made it clear that all agencies would coordinate and participate, there would be a plan on his desk in nine months, and there would be no exceptions. The work of the 36 agencies was presented to the Governor at a symposium in October 2002 (Rosenbaum 2005). Leadership can make things happen, including sustainable water resources management.

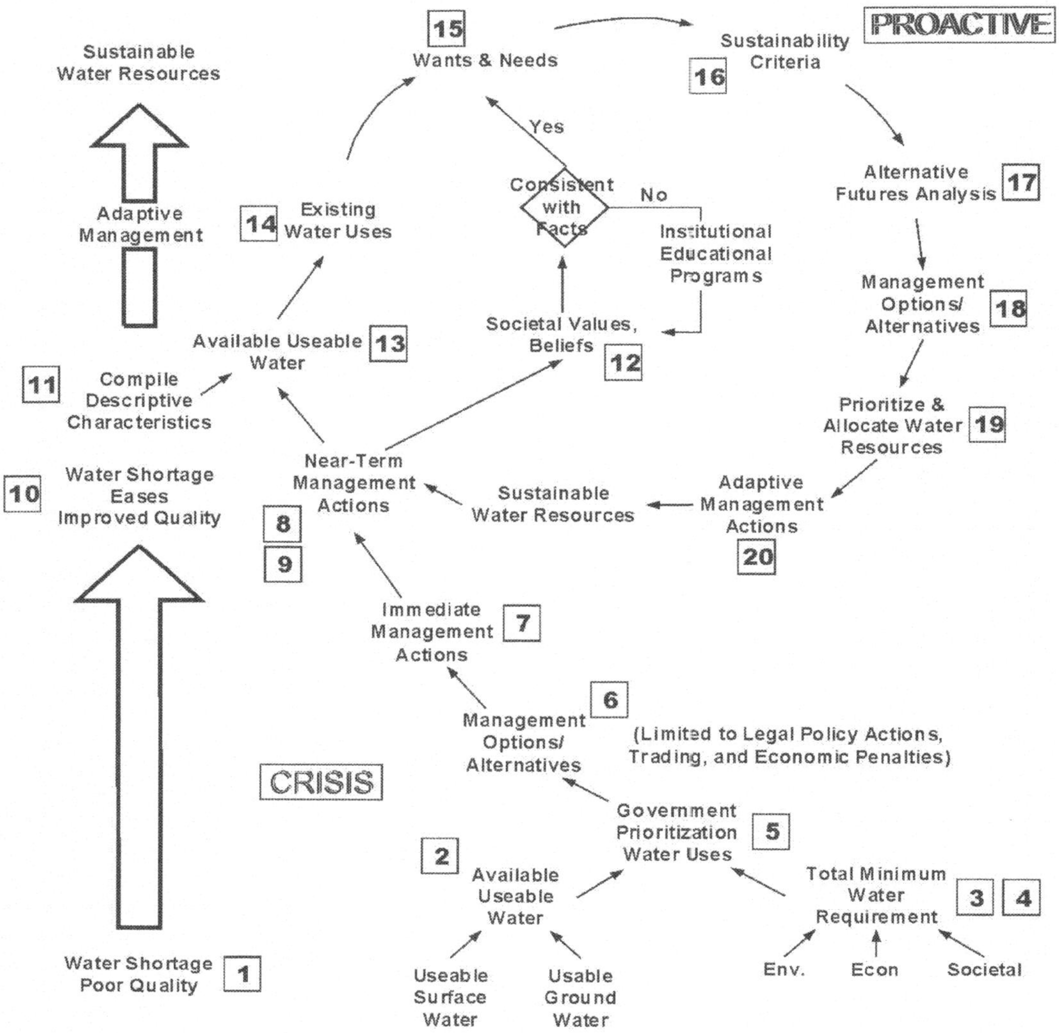

Figure 19-1. SWRM Process Flow Diagram with Two Paths – a Crisis Chart and a Proactive Chart.

Path One — Crisis Chart

1. Recognizing the Crisis

There can be multiple causes of a water crisis. A water shortage that has reached crisis proportions could be the result of drought, community growth and increased water use, aging infrastructure, increased upstream withdrawals, reduced upstream discharges, all of the above, and more. A water quality crisis may occur for exactly the same reasons as a water shortage (growth, aging infrastructure, etc.). In many cases, identifying the causes is secondary to treating the symptoms. You're running out of usable water—do something about it!

2. Estimating Available Usable Water

Estimating available usable water includes both water quantity and quality. Both surface and groundwater sources need to be considered, as well as any constraints on supply from outside the watershed.

3. Minimum Water Requirements

During a water crisis, minimum water quantity and quality requirements are typically determined only for socioeconomic needs and uses. Ecological requirements are assumed to be of lower priority than human health and economic sector needs. However, there might be legal, regulatory, policy, economic, or other reasons why the minimum ecological water requirements also need to be estimated at this step.

Unfortunately, estimating minimum ecological water requirements is a difficult process even under the best of circumstances. However, in crisis situations, best estimates of minimum ecological requirements suffice. Estimates of minimum water requirements for aquatic ecosystems might be based on minimum regulated low-flow releases below structures (e.g., dams), minimum regulated freshwater inflows to estuaries, guidelines or regulations to maintain the minimum flows required for NPDES discharges, or a percentage of the water required to maintain the minimum socioeconomic needs. While estimates of minimum low flow requirements are available for a few fish species, minimum water requirements for aquatic ecosystems, in general, are unknown.

4. Comparison of Available Usable Water and Minimum Water Requirements

A comparison can now be made between the estimates of available usable water versus the minimum amount of water required to maintain socioeconomic and ecological systems. Obviously, the key issue is whether there is sufficient usable water to satisfy the total minimum requirements. Although this comparison is relatively straightforward, confidence in the estimates of available and needed water can affect management decisions.

5. Prioritization

If the quantity of available usable water is sufficient to satisfy the minimum water requirement for socioeconomic and ecological uses, then any remaining water might be allocated using market-based economic principles. If the quantity of available usable water is less than the estimated minimum water required to sustain social (i.e., human health and

Prioritization: Guidelines for Water Restrictions During Drought

In 2002, the Denver Water Board adopted policy guidelines for developing a drought restriction program. Denver Water will follow these principles in restricting water use during a drought.

1. Avoid irretrievable loss of natural resources
   - Allow watering of irreplaceable trees.
   - Avoid killing perennial landscaping if possible.
   - Tailor water restrictions to landscape needs.
2. Restrict less essential uses before essential uses.
   - Restrict water use for misters, fountains, and other aesthetic water features first.
   - Avoid using water for cleaning driveways, sidewalks, or washing personal vehicles.
   - Curtail outdoor water use, along with restrictions on commercial use, before restricting domestic indoor use.
3. Affect individuals or small groups before affecting large groups or the public as a whole, allowing as much public activity as possible to be unaffected.
   - Preserve community pools rather than residential pools.
   - Restrict golf courses before public parks.
   - Restrict water use on less heavily used areas of parks where grass can go dormant before formal and informal playing fields, where recreational activity would either kill the grass or have to be prohibited.
4. Minimize adverse financial effects.
   - Try not to close businesses, although businesses that depend on purely discretionary water use will be affected.
   - Restrict seasonal commercial use, which is likely to be outdoors.
   - Restrict nonessential uses of water in businesses before affecting fundamental business functions.

welfare), economic (i.e., industry, agriculture, commerce), and ecological (i.e., wildlife support, waste assimilation, maintenance of natural resources that contribute to economy) uses, the highest priority will likely be water for human health and welfare (see sidebar, previous page). Prioritization of other uses might be based on socioeconomic principles, including nonmarket valuation approaches for estimating the value of ecological services in conjunction with water requirements for market-based goods and services.

6. Options and Alternatives

The number of options and alternatives for allocating water among competing required uses is much more limited under crisis situations than under noncrisis situations because some options and alternatives require more time to develop and implement than is available in a crisis. In general, water use restrictions and required water conservation measures are enacted to ensure there is sufficient water to meet the minimum required uses. Beach closings and restrictions on fisheries use (e.g., consumption advisories, closing shellfish beds) are commonly used strategies for water quality crises. Fines are used in lieu of market-based economic incentives to control water distribution and allocation. Communities in both the West and the East have used fines to limit water use for washing vehicles, watering lawns, filling swimming pools, etc., during drought periods, and fines are used to enforce bans on commercial fishing in contaminated waters. Some communities, such as Denver, Colorado, have developed specific drought response practices for water shortages (see sidebar - above).

Denver Management Options for Drought Response

The Denver Water Drought Response Plan identifies several management options for meeting water needs when supply is reduced. These include options for increasing water supply, decreasing water demand, and monitoring and evaluating water supply and use, as listed below:

1. Increasing water supply
    - Call back water rights others have been allowed to use.
    - Pay a water user downstream of Denver Water diversion to not place a river "call".
    - Seek waivers of minimum bypass requirements.
    - Develop groundwater sources.
    - Pump water not available through gravity flow.
    - Implement legal clauses allowing reservoir storage to be drawn below currently mandated minimum levels.
2. Decreasing water demand
    - Impose water use restrictions.
    - Add drought surcharges to water use charges.
    - Implement programs to enforce water use restrictions.
    - Provide incentive programs to encourage and reward immediate reductions of water use.
3. Monitoring and evaluation of water use
    - Historical water demand.
    - Weather-adjusted water demand.
    - Reservoir levels.
    - Water supply forecasting.
    - Snow pack readings
    - Precipitation monitoring
    - Monthly updates

7. Immediate Implementation

The timing of various water management options and alternatives becomes a significant selection factor during crisis periods. Those alternatives and options that can be implemented immediately (e.g., fines for lawn watering and vehicle washing, ordinances requiring low-flow shower heads in new/remodeled buildings and facilities, beach closings, fish consumption advisories) have priority over those that require years for implementation (e.g., improved water distribution and collection infrastructure, improved water treatment for water recycling/reuse, installing low-flow toilets and washing machines, implementing improved irrigation systems throughout the basin, watershed restoration).

8. Near-term Options

Some options might not be implemented immediately, but could be implemented within 2 to 5 years. These options include economic incentives for installing low-flow toilets and washing machines, initiation of water and pollution trading programs, locating and repairing home/business leaks, low-water landscaping, rain gardens, studying pollution sources, etc. A greater number of options are available when there is adequate time for implementation.

9. Retaining Management Practices

As water options and alternatives reduce or ameliorate the water crisis, an assessment of the practices that were effective in reducing water use or improving water quality should be conducted. The tendency is to return to previous water use practices once the crisis has been resolved, but if management practices invoked during the crisis are effective in efficiently allocating water among users, and/or improving water quality, these management practices should be retained. Water is a renewable, but finite resource. Management practices that promote efficient water use, contribute to integrating socioeconomic and ecological systems, and support adaptive management should be retained as elements of SWRM.

10. Move to the Proactive Chart

If the water crisis has been resolved and longer-term options and alternatives can be assessed, the water resources management process should move from the Crisis Chart to the Proactive Chart so that future crises can be averted.

Path Two – Proactive Chart

The time to begin implementing SWRM practices is before water needs or water quality issues reach crisis proportions. When you are not in a crisis situation, management options increase, and stakeholders can actively participate in development of the management plan.

The key is to start the process! There are always water issues in a community, even if there is no water crisis. Start with a water issue of concern in your community, watershed, basin, or region. Water can be related directly or indirectly to almost every management practice in the community or watershed. Starting with community or watershed issues is preferred because you not only know the actors, you are one of them.

Steps in the implementation process need to be considered at each scale, from the local community to the region. Each of the scales, from local community to region, however, is interactive, so it doesn't matter where along the scale continuum you begin. Cumulative effects of decisions made by local communities contribute to regional scale effects, and regional constraints affect what can be attained in local communities. In addition to the regional constraints, it is useful to determine the hydrologic effects on your region from large scale weather cycles. These large-scale cycles, such as El Niño and La Niña, influence periods of flooding and drought at regional, basin, and watershed scales.

The information needs of the Proactive Chart process are similar, but more extensive, than for the Crisis Chart process. The Proactive Chart has greater emphasis on integration of, and interactions among, the social, economic, and ecological sectors, than the Crisis Chart.

11. Continuing the Process – Descriptive Characteristics

One of the steps in the Proactive Chart process should be to compile information on the general characteristics of communities, watersheds, basins, and the region. Continuing the watershed management theme, the discussion of the Proactive Chart relates to the watershed scale, but is also applicable for the community, basin, and regional scale.

Maps are useful for conveying much of this information, and generally already exist for many watershed features. At a minimum, a watershed map showing water features (e.g., stream networks, lakes) and a land use map should be prepared. Tables also are useful for presenting information such as summaries of various land use types, and population statistics by socioeconomic sector.

12. Stakeholder Values, Beliefs, Interests, and Concerns

Understanding the water-related values, beliefs, interests, and concerns of the various population segments in the watershed is critical. All of us develop mental models to simplify, understand, explain, and predict things in a complex world. Everyone carries unspoken mental models that have been formed through their culture, background, education, and personal experience. These mental models reflect what we believe and value, and guide our actions. Because they are not always based on fact (e.g., water is an infinite resource; turn on the tap and water always comes out), it may be necessary to change mental models before behavior can be altered to support sustainable water resources. Changing a false mental model into a factual mental model requires first finding out what mental models exist in the community and watershed. In general, you'll need help from social scientists to determine and understand the values and beliefs (mental models) that exist within your watershed. Most social scientists get excited when approached by nontraditional clients. Everyone wants to feel that what they do can make a difference in society. Find them and engage them in the process.

If the stakeholder beliefs about water resources in the watershed are factually incorrect or unrealistic, educational campaigns need to be developed to provide factual information and educate the public on the issues. Social marketing techniques can be, and have been, used by government agencies and civic organizations to change the understanding and behavior of stakeholders on a range of issues ranging from smoking and drunk driving (designated drivers) to littering and water conservation.

Determining Stakeholder Values, Beliefs, Interests, and Concerns in the Upper San Pedro River Watershed

The Udall Center for Studies in Public Policy, University of Arizona, worked with elected officials, environmental activists, and community leaders in the U.S. and Mexico to identify local issues and concerns related to water in the upper San Pedro River watershed. The primary findings were:

United States

- There is widespread appreciation for the river.
- Residents of the watershed want to provide substantive input to water management and planning for the river.
- Opinions are divided over the nature and extent to which the river is in jeopardy.
- There is virtually unanimous support for water conservation and aquifer recharge.
- Many people see a need for growth management.
- Multiple jurisdictions, both local and national, should share in the river's management.
- The burdens of water conservation should be equitably shared among water users.
- There is a need for more study of the Mexican portion of the watershed.
- Mexico should be included in cooperative planning and management processes for the watershed.

Mexico

- There is concern about potential negative effects on Mexican farmers and ranchers from conservation and restoration efforts. Alternate options for livelihoods should be provided.
- Mexico desires that they benefit from protecting the San Pedro River to a degree similar to the United States.
- The use of groundwater for the copper mine is resulting in lowering of groundwater levels and affecting water levels in San Pedro River.
- There is support of conservation and restoration in the watershed on both sides of the border.

13. Available Usable Water

Water quantity and quality information are needed for surface water *and* groundwater *and* atmospheric vapor (precipitation, evaporation). This is the area in which most engineers and scientists are comfortable and experienced. It is also an area where not all the needed information is readily available.

14. Existing Water Uses

In addition to the quantity (volume) and quality of water in the watershed, existing water uses need to be quantified. The USGS collects and publishes information on water usage by state and major U.S. Water Resources Regions, but this information is typically not available for watersheds.

15. Water Uses – Needs and Wants

The next step in the process is to determine which water uses are required for the survival of the human communities present in the watershed and the natural systems on which society depends. These are water use needs. Those water uses that do not satisfy this criterion are categorized as desired water uses, or wants. Minimum water requirements for the water use needs are also determined. In addition, the quality of water required for all of the water uses identified here also needs to be considered.

Water Use Needs

Gleick (1996, 1998b) has estimated that humans need 5 L/day per person for drinking water, 20 L/day per person for sanitation and hygiene, 15 L/day per person for bathing, and 10 L/day per person for cooking, or approximately 50 L/day for domestic use. This is about 13 gal/day per person or 4,750 gal/yr per person as a basic minimum. This estimate is considered to be a true minimum to sustain life in a moderate climate with moderate daily activity. This minimum estimate does not include water needed for food production. In the United States, we currently use about 54,000 gal/yr per person for domestic purposes. Canada and Mexico use about 41,000 and 12,000 gal/yr per person, respectively (Gleick, 2000). Minimum water quality requirements for human use can be obtained by using water standards for aquatic life (because drinking water standards are typically for treated, not raw water).

> **What Makes a Use a Need?**
>
> In the San Pedro River watershed of Arizona, protection of water dependent natural systems is one of the primary management goals, and has been federally mandated for that portion of the river contained in the San Pedro Riparian Conservation Area. Therefore, estimates of the water needs for natural systems are an important element of management in the watershed. Several attempts have been made to estimate water needs for natural systems in the San Pedro River watershed. A water budget prepared by Putnam et al. (1988) for the period 1968-1976 estimated that approximately 55,000 ac-ft of water was used by natural systems in upper San Pedro River watershed. The Arizona Department of Water Resources estimated in a 1991 report that in 1990 approximately 61,000 ac-ft of water was used to maintain natural ecosystems in the San Pedro River watershed. The Semi-Arid Land Surface Atmosphere program has been researching water needs of the San Pedro River riparian ecosystem since 1998, and is contributing valuable information to current water management and budgeting efforts.

Estimates of minimum water requirements are also needed for agricultural, commercial, industrial, power generation, cooling water, and similar human uses, as well as wetland, riparian, stream, and river ecosystem requirements. One approach for establishing minimum requirements for nondomestic human uses is to estimate the water use assuming water conservation, recycling/reuse, and similar practices are fully implemented. Chapter 18.0 includes estimates of water use for different use categories with conservation, recycling/reuse, efficient irrigation, and similar practices in place. These estimates, then, become the baseline for minimum water requirements. Minimum water quality requirements can be obtained by assuming that state water quality standards, and federal water quality criteria will be attained. Water quality standards incorporate the designated waterbody uses, and the water quality criteria to protect those designated uses.

There currently is little information available on flow regimes required to sustain aquatic ecosystems. For ecological systems, minimum water requirements are typically established for specific species of interest (e.g., brook trout, salmon) rather than for the ecosystem. If minimum flow requirements have been established for specific species, these can be used to determine the minimum water requirements for natural systems. The Nature Conservancy has a sustainable waters program that will eventually provide estimates of minimum and maximum flows and required flow durations, by season, for various streams throughout the United States (www.freshwaters.org). Minimum water quality requirements can be obtained by assuming that state water quality standards and federal water quality criteria will be attained.

Desired Water Uses

Initial desired water use quantities can be estimated by compiling existing water use quantities (volumes). Future desired water use can be estimated using demographic projections of population and land use change within the watershed, and prorating future use based on current use. Population and land use projections are available from municipal, county, state, and regional planning agencies, U.S. Census Bureau, and from many trade associations. Initial water quality requirements can also be based on existing water quality standards for the water bodies.

16. Sustainability Criteria Comparisons

Four criteria have been established to determine if current water management practices support sustainable water resources:

1. Has the groundwater table elevation remained relatively constant over a 10-year running average? If the groundwater table elevation has been declining over a 10-year period, the current management of groundwater use is not considered sustainable.
2. Is there sufficient surface water and groundwater quantity to satisfy existing and projected future uses? If there is not sufficient water to satisfy existing or projected future uses, existing water management practices are not considered adequate to ensure a sustainable water resource.
3. Is the timing of water delivery adequate to satisfy existing and projected future uses? If the timing is inadequate, such that, for example, minimum flows cannot be sustained in the stream during the summer or fall season, then existing water management practices are not considered sustainable.
4. Is the water quality adequate to achieve existing and projected future water uses? If the water quality is not adequate to support existing or projected future uses, the existing water management practices are not supporting sustainable water resources.

If any one of these criteria is not met, it will be necessary to make changes in current water management practices to achieve sustainability. If all of these criteria are met, current water management practices are sustainable. If any of these criteria cannot be evaluated, collecting the information necessary to evaluate those criteria should be incorporated into the management plan. Whether current water management practices are sustainable, unsustainable, or undetermined, the next step is to determine the stakeholder vision and goals for the watershed.

17. Alternative Futures Analyses

The temptation to move directly from the determination of sustainability to assessing options for achieving the required/desired water uses should be avoided. The first step in deciding what changes to make is to take a careful look at stakeholders' desires and goals for the watershed and its water resources. With the data and information compiled through the activities above, sufficient knowledge of the watershed exists to create a vision of what the stakeholders want for the future. This collective watershed vision provides an overall direction for the future and leads to the formulation of goals and objectives for various social entities within the watershed, and prioritization of these goals when choices among competing goals have to be made. Alternative futures analysis can assist in moving toward a consensus watershed vision (see Chapter 13.0).

Alternative futures analyses will require the use of a facilitator and team with experience in conducting alternative futures studies. The process is very similar to the design charettes used by architectural firms in planning new developments or urban renewal projects. It involves engaging as many stakeholders as possible in dialog and visioning activities, and then consolidating the resulting information about desired futures into a handful of scenarios for evaluation. Three scenarios is an optimal number for most communities or watershed associations. Few groups will be able to move toward consensus if there are more than five alternatives from which to choose. The sidebar describes an alternative futures study. Table 19-1 is a listing of several alternative futures studies completed in the United States.

Insights from Study of Alternative Futures for Upper San Pedro River Watershed

An alternative futures study of the upper San Pedro River watershed was performed by the Harvard University Graduate School of Design with the Desert Research Institute, University of Arizona, Instituto del Medio Ambiente y el Desarrollo Sustenable del Estado de Sonora, and United States Army. In this study, watershed conditions in 2000 were compared to potential conditions in 2020 resulting from scenarios in three categories. The three categories were Plans Scenarios, in which existing development and growth plans were followed; Constrained Scenarios, in which development and growth was constrained to those areas currently developed; and Open Scenarios, in which development and growth occurred with less control than currently exists. Impacts on watershed land use/development, hydrology, vegetation, wildlife habitat, habitat for threatened and endangered species, species richness, and aesthetics were evaluated. The study also identified areas of high potential for conflict between development pressure and conservation needs. Insights from this study included:

- All alternatives resulted in loss of groundwater storage and drying of the San Pedro River, even with reductions in per capita water use,
- Policy decisions affecting irrigated agriculture in Arizona have the greatest impact on watershed hydrology, and
- Development control policies have the second greatest impact on watershed hydrology.

18. Management Options and Alternatives

There are a number of options available to stakeholders to move toward their desired future, including water conservation, water reuse, recycling, economic incentives, market trading, social marketing, new technologies, laws, regulations, ordinances, policies, compacts, and treaties. Chapters 16.0 through 18.0 include examples of the many management options that can contribute to SWRM.

As stated in the beginning of this document, there are few sound management practices, social norms, or economic approaches that can not contribute to SWRM. For example, Smart Growth practices can decrease the percent impervious area and increase forested watershed area, thereby improving water quality (by decreasing runoff volume), restoring natural stream flow and watershed hydrology, increasing groundwater recharge, and decreasing stream sedimentation and water pollution. While protecting water quantity and quality, these practices can also stimulate economic development in desired areas.

Table 19-1. Example Alternative Future Studies

- Alternative Futures for Monroe County, Pennsylvania
- Modeling Effects of Alternative Landscape Design and Management on Water Quality and Biodiversity in Midwest Agricultural Watersheds (Iowa)
- Alternative Futures for Changing Landscapes: The Upper San Pedro River Basin in Arizona and Sonora, Mexico
- Willamette River Basin Planning Atlas: Trajectories of Environmental and Ecological Change
- Southern Rockies Landscape Planning Project
- The Development of Alternative Future Growth Scenarios for the California Mojave Desert
- Biodiversity and Landscape Planning: Alternative Futures for the Region of Camp Pendleton, California

19. Prioritization of Water Needs and Wants

The first step for prioritizing water uses is to compare available usable water with required and desired water usage. Minimum water requirements for social, economic, and ecological sectors were estimated previously (see Step 15). Existing and projected future water uses also were estimated previously (see Steps 14 and 15). A comparison of the required and desired water use estimates with available water quantity and quality estimates will indicate if

- available water is of sufficient quantity and quality with adequate timing to meet the minimum required existing and future needs for the three societal sectors—social (human), economic (e.g., commercial, agricultural, industrial), and ecological (e.g., aquatic ecosystems—rivers, streams, lakes, wetlands); and
- water is of sufficient quantity and quality with adequate timing to meet the existing and future desired water uses, for each sector.

The alternative futures analysis and other knowledge of stakeholder desires and goals are used as guides for deciding which water uses will have priority. Water uses are prioritized based on their level of support for stakeholder watershed goals.

20. Adaptive Management

Monitor, Assess, and Adapt. Adaptive management is the foundation of SWRM. Plan → Do → Check → React → Plan… represents the iterative approach needed to manage highly complex, nonlinear socioeconomic and environmental systems whose behavior cannot be predicted. Management programs need to be evaluated on a regular basis to ensure that they are achieving the desired societal goals, are identifying and addressing existing problems, are based on the best available information about the current state of the system, are using the best available technology and techniques, and are identifying emerging problems.

Indicators that provide information to assess progress toward sustainable water resources in each of the sectors of society [ecology (including hydrology), economy, and society/culture], are necessary for successful adaptive management. See Chapter 14.0 for a list of potential indicators. Many of these water resources indicators are currently monitored by local, state, and federal agencies. Monitoring information for the indicators will need to be accessible to multiple users (i.e., through linked information systems) to be useful for adaptive management at multiple scales.

A Regional Emphasis

Sustainable water resources are achievable, but within a regional context. Although useful strides toward SWRM can be made at the community and watershed scale, a regional perspective is required to achieve true sustainability. Cumulative effects of local community decisions that affect watersheds, basins, and aquifers can be dealt with effectively only at the regional scale. To achieve sustainable water resources, we must consider surface water, groundwater, and atmospheric vapor. A regional scale is needed because, although watersheds are suitable management units for surface water issues, groundwater aquifers often do not follow watershed boundaries, and precipitation and evaporation patterns are the result of large-scale climate patterns. In most cases, the USGS Water Resources Regions correspond reasonably well with regional aquifer boundaries (Figure 5-4) and weather patterns; as such, these Water Resources Regions might serve as the logical starting point for developing regional management units for sustainable water resources. Factors such as local variability in transmissivity and connectivity of regional aquifer units, and intrabasin water transfers will affect how regional management units are ultimately defined (see sidebar).

Developing Regional Management Units

The San Pedro River watershed in Arizona is part of the Lower Colorado USGS Water Resources Region. This surface water region contains a large portion of the Southwest Alluvial Basin regional aquifer, and part of the Colorado Plateau regional aquifers. The Southwest Alluvial Basin regional aquifer is a system of mostly local aquifers, with little connectivity between aquifers, therefore, we suspect that including only a portion of this regional aquifer would not impair our definition of this as a regional management unit, reflecting a closed land water system. The Colorado Plateau regional aquifers, however, are a system of connected aquifers. As a result, the Lower Colorado water resources region is not as close to a closed water system as we would like. A more appropriate regional management unit for this system would be the entire Colorado River basin – this brings all of the Colorado Plateau regional aquifers into the management unit, resulting in a more closed water system. Some aspects of the Colorado River basin are already managed at the basin scale, i.e., water quantity by the Bureau of Land Management, and the Colorado River Water Users Association.

The Spring Creek watershed in Pennsylvania is part of the Mid-Atlantic Water Resources Region. The Spring Creek watershed is also part of the Chesapeake Bay watershed, which is currently being managed by the Chesapeake Bay Program, primarily for water quality. Since this Chesapeake Bay management unit is already in place, it could also be a reasonable regional management unit for SWRM. An analysis of the aquifers underlying the Chesapeake Bay watershed indicates that the watershed is very nearly a closed hydrologic system, and as such, is suitable as a regional management unit for SWRM.

Managing at the regional scale will require changes in how agencies, organizations, civic institutions, communities, and the private sector address water resources management. The need for these changes has been recognized. The Western Governors Association (2002) is already on record as stating that regional coalitions will be needed to move toward SWRM in the West. This is equally true in the eastern states. In the western states, regional water quantity management already occurs through the Colorado River Compact. In the East, regional water quality management occurs through the Chesapeake Bay Compact. Over time, these regional management entities will address other water resources issues (i.e., water quality in the Colorado

River basin, and water quantity in the Chesapeake Bay basin), and other regional water resources management entities will come into being.

Start Now

SWRM might appear to be overwhelming, but all the information in this document was readily available in the literature. What is new is the synthesis and integration, the framework, and the guidelines. The key to SWRM is to start. Pick an issue, form a team, and start the process. Work through the guidelines and move toward a sustainable outcome.

CHAPTER 20.0

CASE STUDIES OF SWRM APPLICATION

Below are two examples of the application of the SWRM process at community and watershed scales. One example is based on the San Pedro River located in Arizona and Mexico. This river was chosen for an example because it is an international basin, and is located in the arid southwestern United States. The second example is based on the Spring River, located in central Pennsylvania. This river was chosen as an example because is it located in the fairly wet northeastern United States. Both rivers have well-established issues associated with them, which have been the subject of study for a number of years.

20.1 San Pedro River, Arizona

20.1.1 Background

The San Pedro River is located in southeast Arizona, in Cochise, Graham, and Pinal Counties. The headwaters of the river are in Sonora, Mexico, near the town of Cannanea (Figure 20-1). From this area the river flows north, crossing into the United States near the border town of Los Corrales, Mexico, and eventually joining the Gila River, a tributary of the Colorado River, at Winkleman, Arizona. It is the largest undammed river in the Southwest. The hydrology of this area is evaporation dominated, and water quantity is the primary issue. During dry months, portions of the San Pedro River are maintained by groundwater, so the river is partially perennial, making it an oasis in this area of the arid Chihuahua and Sonoran Desert. The San Pedro River is located in the Basin and Range physiographic province, characterized by long mountain chains separated by alluvium-filled valleys (Lacher, 1994). The San Pedro River basin is a relatively closed hydrogeologic system. The aquifer underlying the watershed is effectively bounded by the same bedrock mountains that form the surface watershed boundary [Arizona Department of Water Resources (ADWR), 1990]. The aquifer is recharged primarily by runoff from these mountains, with the most active recharge areas being the alluvial fans at the base on the mountains (Freethey, 1982).

The San Pedro River corridor is important to regional biodiversity (BLM, 2000; TNC, 2004; SWCBD, 2004). It supports a large number of plant and animal species, including 13 endangered species; and is an important stop-over for birds on the north-south migration route through this area. The riparian corridor habitat that supports the regional biodiversity is a function of the hydrology of the San Pedro River. Concerns that this important regional resource was being threatened by local cattle ranching and groundwater withdrawals led to the creation of the San Pedro Riparian National Conservation Area (SPRNCA) in 1988.

The ecological importance of the San Pedro River corridor and the regulations and policies associated with the SPRNCA have driven extensive water resources research and planning in the watershed. In this case study, ecological issues related to water quantity became the driver for SWRM. Because this is an arid region, conflicts among competing water demands tend to be pronounced. The creation of the SPRNCA has given ecological water demands more weight than they usually are accorded in water allocation in arid regions.

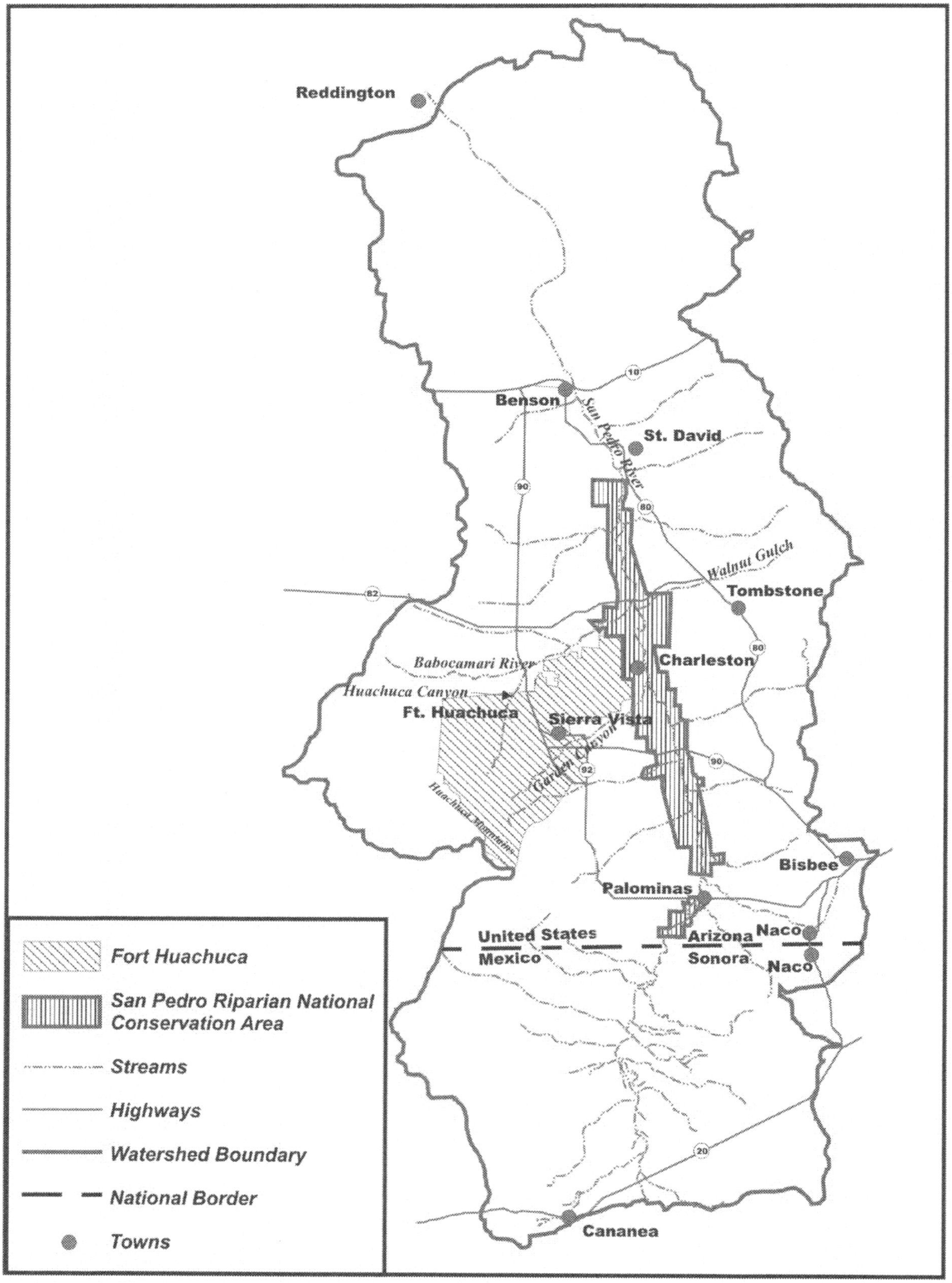

Figure 20-1. Upper San Pedro River Watershed (USGS, HUC 15050202).

This waterbody has been chosen as a case study because it is located in an arid region, is an international waterbody, and because there is considerable information available about it. The information presented in this case study was obtained through Internet searches. Additional information is available from avenues of research not pursued as part of this project.

20.1.2 Community Scale

The city of Sierra Vista, Arizona, is the largest Arizona community in the San Pedro River watershed. This is a growing community that includes the U.S. Army base, Fort Huachuca (Figure 20-1). Both Sierra Vista and Fort Huachuca rely on groundwater for water supply. A number of interest groups have suggested that groundwater withdrawal to supply water for this community has adversely affected the nearby San Pedro River within the SPRNCA. A recent hydrogeologic study of the Sierra Vista subbasin supports the theory that groundwater pumping in the area of Sierra Vista and Fort Huachuca is intercepting groundwater flow from the Huachuca Mountains to the San Pedro River and could be affecting the river hydrology (Pool & Coess, 1999). The Upper San Pedro Partnership (Partnership) is a coalition of community leaders, state agencies, and interest groups that has been appointed by the U.S. Congress to address the water issues in the Sierra Vista subbasin of the Upper San Pedro watershed (the watershed of the San Pedro River from the Arizona border to just north of the Babocamari River and Walnut Gulch) (www.usppartnership.com). The Partnership has created a large body of work about these issues that is being used to develop a plan for SWRM for the Sierra Vista community, and other water users in the subbasin.

Background information about this community that may be useful for water resources planning includes the following:

- Sierra Vista covers approximately 153.5 sq mi of land area (www.city-data.com/city/Sierra-Vista-Arizona.html).
- The midyear 2003 population estimate for this community was 40,430 [Cochise College Center for Economic Research (CCCER), 2004]. This number includes over 6000 active-duty soldiers stationed at Fort Huachuca with their families (www.militarydial.com/fort-huachuca-arizona.htm).
- The three most prevalent races recorded in Sierra Vista during the 2000 Census were "White non-Hispanic" at 65.4%, Hispanic at 16%, and Black at 11% (www.census.gov).
- Approximately 91% of the adult population have completed high school, and 26% have college degrees.
- Fort Huachuca is the largest employer in Sierra Vista and exerts a strong influence on the local economy. Sierra Vista is the economic and cultural center of Cochise County, which includes attracting shoppers from northern Mexico (CCCER, 2004).
- Temperatures in Sierra Vista usually range between 33 °F and 93 °F.
- The average annual rainfall is 14.7 inches (http://www.wrcc.dri.edu/cgi-bin/cliMAIN.pl?azsiev). The heaviest rainfall usually occurs in July and August, however there is also usually a rainy period during the winter (Steinitz et al. 2003).
- This area experiences periods of wetter or drier than normal weather as a result of the influences of cycles in Pacific Ocean surface temperatures (i.e., Pacific Decadal Oscillation, El Niño, and La Niña).
- This community is located in the Chihuahuan Seimi-Desert ecoregion (Bailey, 1995) in the Basin and Range landform (USGS, 2004).

20.1.2.1 Community Beliefs and Values

There has been some work to determine water-related values, beliefs, interests, and concerns in this community. A survey of Cochise County residents by The Nature Conservancy in the mid 1990s found that the San Pedro River was primarily valued as a water source for farming, ranching, and recreation, although the importance of the river as wildlife habitat, and for its scenic beauty was also acknowledged (Liverman et al., 1997). A 2004 survey conducted by KezziahWatkins consulting firm for the Partnership found that there was confusion among stakeholders about the status of water resources in the Sierra Vista subbasin and conflicting information being presented (KezziahWatkins, 2004). Both surveys found strong support for water conservation. The KezziahWatkins survey also documented a prevalent distrust of government and other public agencies, particularly with respect to promises of funding for water projects and the use of funds for water projects. This survey also documented concern about negative effects of water management options on the economy and growth of the community. This survey highlighted the need for better education of stakeholders about the water resources issues facing the community and the role of laws in dealing with these issues. The survey also highlighted community perspectives that could affect water management efforts, such as:

- the presence of an "us-versus-them" mentality between rural and urban area residents,
- the presence of very strongly held opinions about whether conservation measures should be voluntary or mandatory (regulated), and
- the strongly held belief that the burden of water conservation must be equitably shared (KezziahWatkins, 2004).

20.1.2.2 Water Quantity

The amount of water available to the Sierra Vista community is a subject that has been, and is still being, much studied. Water enters the Sierra Vista subbasin as precipitation, San Pedro River flow from Mexico, and groundwater flow from Mexico. The aquifer and surface watershed boundaries are practically the same. Except for the San Pedro River, surface water flow in the subbasin results from precipitation in the subbasin. As mentioned previously, the average annual rainfall in Sierra Vista is 14.7 inches (http://www.wrcc.dri.edu/cgi-bin/cliMAIN.pl?azsiev). The heaviest rainfall usually occurs in July and August, however there is also usually a rainy period during the winter (Steinitz et al., 2003). Table 20-1 summarizes surface flow information for the streams near Sierra Vista. The lowest flows reported at the San Pedro River stations occurred in May, while the lowest flows at the rest of the stations occurred in June. The maximum flows were typically recorded in August.

Table 20-1 Stream Flows Reported at USGS Gaging Stations.

Stream	Gaging Station ID	Drainage Area	Period of Record	Average Annual Flow
Upper Babocomari River	09471380	Unknown	Aug. 2000–Sept. 2003	4.54 cfs
Babocomari River	09471400	Unknown	Water years 2000–2002	1.5 cfs
Garden Canyon	09470800	8.8 sq mi	Water years 1959–1964, 1993–2000, 2002	1.37 cfs
Huachuca Canyon	09471310	Unknown	Oct. 2000–Sept. 2003	0.18 cfs
San Pedro River at Palominas	09470500	737 sq mi	1996–2002	24.4 cfs
San Pedro River at Charleston	09471000	1234 sq mi	1996–2002	31 cfs

The estimated groundwater storage in Sierra Vista subbasin is 31 million ac-ft, however, it is also estimated that less than half of this total is available for use (i.e., pumpable at cost that is not prohibitive) (http://www.svherald.com/articles/2004/08/12/local_news/features_and_opinions/columnist1.txt). Based on a groundwater model of the Sierra Vista subbasin, Freethey (1982) estimated groundwater recharge from the Huachuca Mountains, located just west of Sierra Vista, at 5500 ac-ft/yr, and an additional 4300 ac-ft/yr of groundwater recharge in the Babocamari River valley, located just north of Fort Huachuca (Pool & Coes, 1999). This gives an estimated local recharge rate of 9800 ac-ft/yr for the aquifer near Sierra Vista.

There are several aquifer recharge projects that have recently been put in place in the Sierra Vista area. Fort Huachuca and Sierra Vista both currently treat their wastewater and release it to recharge basins to recharge the local aquifer. In 2001, the Fort Huachuca facility recharged approximately 540 ac-ft [Upper San Pedro Partnership (USPP), 2002]. In 2002, the Sierra Vista facility recharged approximately 2600 ac-ft (USPP, 2004). In its 2004 Water Management and Conservation Plan (USPP, 2004), the Partnership reports that the aquifer in the Sierra Vista subbasin received approximately 3000 ac-ft of artificial recharge from municipal sources in 2002. Groundwater recharge projects in Arizona are permitted through the ADWR. Groundwater quality monitoring, to identify any adverse affects of recharge projects on groundwater quality, is required for permitted recharge projects (http://www.water.az.gov/recharge/).

20.1.2.3 Water Quality

With regard to water quality, the Arizona Department of Environmental Quality (ADEQ) reports some wells in the Sierra Vista area with nitrate concentrations of 5–10 mg/L. Concentrations of the other parameters reported for wells in the Sierra Vista vicinity indicate good water quality for the groundwater (ADEQ, 2004). Portions of the San Pedro River upstream and downstream of Sierra Vista are listed as impaired in the Arizona 303(d) list, however the reach adjacent to Sierra Vista is not impaired. The Babocamari River has not been evaluated for water quality condition.

20.1.2.4 Water Demand and Use

The ADWR has developed estimates of past water demand in the Sierra Vista subbasin, as well as projected future demand (USPP, 2004). The ADWR estimates for 2002 are listed in Table 20-2. An estimate of water demand for the Sierra Vista community can be made using this information. We can assume that irrigation and livestock water demand does not apply to the City of Sierra Vista. We can calculate a per capita use from the municipal use estimates (0.2 ac-ft/person) and use that to calculate a municipal use for Sierra Vista (8000 ac-ft) based on population estimates for the community (40,430 people; CCCER, 2004). Although we expect that there is industry located outside of Sierra Vista, we did not locate information that could guide us in estimating what portion of industry is located in Sierra Vista. Lacking this information, and knowing that Sierra Vista is an economic center, we will include all of the industrial demand (1300 ac-ft) in our estimate of Sierra Vista water demand. These assumptions and calculations result in an estimated water demand of 9300 ac-ft for the Sierra Vista community, approximately half of the demand estimated for the entire Sierra Vista subbasin.

One of the important issues in this community is competition for water between the community and the SPRNCA located just east of the community. The water requirements for the SPRNCA is the subject of much past and ongoing study. However, we did not find any reported

Table 20-2 Estimated Water Demand for Sierra Vista Subbasin in 2002 (from USPP 2004).

Sector	Groundwater Demand (ac-ft)	Surface Water Demand (ac-ft)	Groundwater Recharge (ac-ft)
Agriculture (800 ac irrigated)	2,500	-	-
Municipal (70,500 people)	13,300	600	4,900
Industrial	1,300	-	40
Livestock	160	-	-
Total	17,260	600	4,940

value for the SPRNCA water demand. In the past, ADWR (1991) estimated that 44% of the water demand in the entire San Pedro River watershed could be attributed to support of natural systems. If we assume the same proportion applies to the Sierra Vista subbasin, based on the 2002 estimate of anthropogenic water demand (Table 20-2) we would calculate an estimated water demand of approximately 14,000 ac-ft for support of natural systems in the Sierra Vista subbasin. Something less than half of this water demand would represent natural systems on the same side of the San Pedro River as Sierra Vista, which would be expected to be in fairly direct competition with the community for local groundwater resources.

If we assume a natural system water demand of 6000 ac-ft/yr and sum that with the estimated Sierra Vista water demand of 9300 ac-ft/yr, we get an estimated total local water demand of 15,300 ac-ft/yr. The assumption in this watershed would be that the majority of this demand would be met from groundwater resources. It is therefore interesting to note that this estimated local water demand is approximately one and a half times the estimated local groundwater recharge rate of 9800 ac-ft/yr. This indicates that there may indeed be competition between human and ecosystem water needs in this community.

20.1.2.5 Water Needs

We need to consider what portion of the water demand constitutes water needs, versus desired water uses. Water for support of human health and welfare would be considered a need. Multiplying the population estimate for Sierra Vista by Glieck's minimum per capita water requirement (13 gal/day) we calculate that approximately 600 ac-ft/yr of water would be required for the health and welfare of the population of Sierra Vista. We currently do not have the information necessary to estimate water needs for maintaining the important industrial sectors in Sierra Vista.

We did not locate any information on recommended San Pedro River flow levels for maintaining the SPRNCA, nor the endangered species present there, that might be considered natural system water needs. A request for water rights for protection of entire San Pedro ecosystem that was submitted to ADWR sometime in the past specified 37,000 ac-ft/yr (Liverman et al., 1997). We were not able to determine the result of that request. San Pedro River flows are under consideration as part of the ongoing adjudication of Gila River water rights (Lacher, 1994), however, we did not locate any information indicating that downstream flow requirements have yet been set for the San Pedro River under this process. Legislated flow requirements would be considered water needs. Water quality criteria are in place for the San Pedro River, based on its designated uses as assigned by the ADEQ. These are considered water quality needs.

20.1.2.6 Desired Uses

Desired water uses are those not necessary for the survival of the society of Sierra Vista and the natural systems on which it depends. Examples of desired water uses that are probably present in Sierra Vista include water for pools, watering landscaping, landscape water features, evaporative coolers, washing pavement, vehicles, and equipment, dust control, and leaks. Desired water use can also refer to the amount of water used over and above the minimum requirement for a necessary use. Current water demand could be considered the desired level of water use (see Section 20.1.2.4).

20.1.2.7 Sustainability Criteria

The next step in our community SWRM process is to determine if the four water resources sustainability criteria are being met. The first criterion is that the groundwater elevation must be relatively constant for the last 10 years. A groundwater table cone of depression centered on Sierra Vista was identified in 1973 (Roeske & Werrell, 1973). Through the mid 1980s the average rate of decline in this cone of depression was reported to be 1.4 ft/yr (Putnam et al., 1988). A number of groundwater level monitoring wells have been installed in the Sierra Vista subbasin since 2000. The Partnership reports (USPP, 2004) that annual rates of water level decline in the regional aquifer have ranged from 0.1 ft to 0.6 ft since the mid 1990s. This information indicates that groundwater elevation has been declining over the last 10 years and is not stable. However, it also suggests that the rate of decline may be decreasing and that groundwater levels may be becoming more stable.

The second criterion is that there be sufficient surface and groundwater to satisfy existing and projected water demand. The Partnership reports that current Sierra Vista subbasin water budgets estimate a 3000–6000 ac-ft/yr overdraft of the subbasin water resources (USPP 2004), indicating that there is not sufficient surface and groundwater to satisfy existing water demand. ADWR projects a total anthropogenic water demand of 22,600 ac-ft for 2030, and net groundwater use (total use minus incidental and artificial discharge) of 14,000 ac-ft (USPP, 2004). Both of these projections are greater than the 2002 estimates, suggesting that water resources will not be sufficient to satisfy projected water demand. A declining trend in San Pedro River base flow at Charleston gage (the gage closest to Sierra Vista), has been determined to be caused primarily to reduced wet-season runoff. The reduced wet-season runoff may be partially caused by reduced wet-season precipitation, fewer occurrences of high-intensity rainfall, increased capture of runoff as a result of increased vegetation, increased surface water diversions, and increased groundwater withdrawals by wells, and deep-rooted vegetation (Pool & Coess, 1999).

The third criterion is that the timing of water inputs be adequate to satisfy existing and projected water demands. There is information available on the seasonality of precipitation and flow inputs (BLM, 2000; NWS, 2003; http://www.wrcc.dri.edu/cgi-bin/cliMAIN.pl?azsiev; www.usgs.gov). However, we did not locate any information on the seasonality of water demand. The seasonality of riparian area water use is currently under study (D. Goodrich, ARS Southwest Watershed Research Center, http://www/ars.usda.gov/research/projects/ projects.htm?accn_no=404015). Therefore, this criterion cannot be evaluated at this time.

The fourth criterion is that water quality be adequate to support existing and projected water uses. Based on the most recent state water quality assessments (ADEQ, 2004), surface and groundwater quality in the Sierra Vista area are reportedly adequate to support their designated uses.

Since two of these criteria were not met, existing water resources management is determined to be unsustainable.

20.1.2.8 Alternative Futures

An alternative futures analysis of the upper San Pedro River watershed completed by Steinitz et al. (2003) included extensive evaluation of the effects both on, and caused by, Sierra Vista and Fort Huachuca. Indicators that were evaluated in this analysis included San Pedro River flows, groundwater levels, aesthetics, endangered species habitat, vegetation, and land use. The alternative futures scenarios that were evaluated included scenarios in which existing development plans and restrictions continued, scenarios in which development was only permitted in previously developed areas, and scenarios in which there were fewer restrictions on development than currently exist. Projections of future conditions based on these scenarios indicated that

- Standard water conservation measures alone would not be adequate to stop groundwater level declines in the Sierra Vista area,
- Conditions in Sierra Vista were considerably dependent on how Fort Huachuca was managed, and
- Areas along the base of the Huachuca Mountains and in arroyos within Sierra Vista have a high potential for conflict between conservation of critical animal habitat and development.

The Partnership is also evaluating water management scenarios using a preliminary decision support system developed by the University of Arizona (USPP, 2004). Although we did find information about public opinion surveys about water management options and issues (KezziahWatkins, 2004), we did not find information about surveys of stakeholders goals and desires for water resources in the Sierra Vista area. Currently, water resources goals appear to be directed primarily by influences from outside the community (e.g., The Nature Conservancy, BLM, U. S. EPA, ADWR, ADEQ, U.S. Fish and Wildlife Service, National Audubon Society).

20.1.2.9 Prioritizing Uses

The first step for prioritizing water uses is to compare available water with required and desired water uses. We have already determined that water resources in the Sierra Vista area appear to be overallocated. We did calculate an estimate of minimum water requirements for the support of the Sierra Vista population, however, we did not locate data on required water uses for the support of Sierra Vista economic sectors or the SPRNCA. Therefore, it is not possible to at this time to determine if there are adequate water resources to sustainably meet the minimum needs of Sierra Vista and its associated ecosystems, including the SPRNCA.

20.1.2.10 Management Options

There is a lot of work occurring in the Sierra Vista area geared toward reducing the impact of the community on area water resources and ecosystems. The Partnership includes summaries of past, ongoing, and planned water resources management activities in its annual report (USPP, 2004). Activities reported by the Partnership include effluent reuse and groundwater recharge projects, public education, stormwater detention basins, water conservation incentive programs for private citizens and industries, upgrading water infrastructure, installing water conserving appliances in public buildings, and modification of ordinances to support water conservation. Educational, research, economic, and legislative avenues toward sustainable water resources are being pursued. The Partnership, Fort Huachuca, and BLM have set the goal of eliminating deficit groundwater pumping in the Sierra Vista subbasin by 2011. This goal has been written into legislation governing the Partnership and BLM's management of the SPRNCA.

20.1.2.11 Monitoring and Evaluation

Monitoring is included in the Partnership's ongoing and planned activities. Groundwater levels, San Pedro River flow regime, and vegetation characteristics in the SPRNCA are the indicators that are being monitored to evaluate the effects of water conservation and management efforts in the Sierra Vista subbasin. Fort Huachuca, ADWR, USGS, and BLM monitor groundwater levels and quality in wells they maintain in the Sierra Vista subbasin. The USGS monitors San Pedro River flow at several gaging stations, and The Nature Conservancy organizes an annual effort to map wet and dry portions of the San Pedro River in the SPRNCA. The Corps of Engineers Topographic Engineering Center is mapping vegetation in the SPRNCA every four years, in a project that is funded through 2008. The Partnership plans to use public surveys to evaluate the effectiveness of public outreach and education efforts in the Sierra Vista subbasin (USPP, 2004).

20.1.3 Watershed Scale

The watershed chosen for this case study is the upper San Pedro River watershed (USGS, HUC 15050202) (Figure 20-1). This watershed extends from the headwaters of the San Pedro River in Sonora, Mexico, to the Narrows, a bedrock constriction in the valley located 12 miles north of Benson, Arizona (Lacher, 1994). The watershed is approximately 2900 sq mi, with 2200 sq mi in Arizona, and 700 sq mi in Sonora (Kepner et al., 2003). Approximately 75,000 people lived in the Arizona portion of the watershed during 2000 (www.census.gov). Approximately 35,400 people lived in the two largest cities in the watershed in Sonora, Naco and Cananea, during 2000 (www.world-gazetteer.com/d/d_ms/so.html). The three most prevalent races recorded in the watershed during the 2000 U.S. Census were Whites (69%), Hispanics (19%), and Blacks (6%). Approximately 57% of the people older than 25 had completed high school in 2000; 22% had college degrees. Median household income in the watershed was approximately $33,000, and 12% of people in the watershed had 1999 incomes below the poverty level (www.census.gov). Fort Huachuca in Sierra Vista, Arizona (see Section 20.1.2) is the largest employer in the watershed. In Cochise County, Arizona, tourism associated with the SPRNCA and other historic and natural sites in the watershed also contributes significantly to the local economy (Orr & Colby, 2002; Liverman et al., 1997; Leones et al., 1994). In Sonora, Mexico, the Cananea Copper Mine, located at the headwaters of the San Pedro River, and the associated smelter in Cananea, Mexico, are significant economic forces.

Average rainfall in the upper San Pedro watershed is 8–12 inches, ranging from 6 inches in the valley to 30 inches in the mountains. Historically approximately 75% of rainfall occurs during the summer monsoon, but recent research indicates that this pattern may be changing (BLM, 2000). Annual rainfall varies from 11.8 in to 29.5 inches (SCERP, 1999).

20.1.3.1 Community Beliefs and Values

There are a number of interest groups active in this watershed, both government and nongovernment, on both sides of the U.S.–Mexico border. In the Arizona portion of the watershed, the Partnership is the most active nongovernmental interest group. In the Sonora portion of the watershed, the Asociacon Regional Ambiental de Sonora y Arizona (ARASA), formed in 2002, is the primary watershed interest group. Stakeholder outreach and education activities have occurred in this watershed, including surveys to determine water related beliefs and values. The Nature Conservancy survey conducted in Cochise County, Arizona, in the mid

1990s is discussed in Section 20.1.2.1. The Udall Center for Studies in Public Policy at the University of Arizona has surveyed stakeholders in both Arizona and Sonora. Their findings are summarized in Table 20-3. In Arizona, many believe that Fort Huachuca and Sierra Vista are the primary threat to maintaining the San Pedro River and the SPRNCA. When the SPRNCA was first created, it was believed that irrigation wells along the river were the primary threat, and the pasture lands along the river were placed in land conservation programs. In Sonora, the Cananea mining operations are believed to be the primary threat to the San Pedro River, however, there are suggestions that reducing irrigation use of water would improve the river. Many see reducing irrigation water use as an easier fix than reducing water use in the mining operation, or Sierra Vista, but farmers feel it is unfair for them to bear the brunt of the negative impacts for improving the San Pedro River (www.udallcenter.arizona.edu/sanpedro, Harris et al., 2001).

Table 20-3. Summary of Upper San Pedro Stakeholder Water Issues from Udall Center Surveys.

Arizona*	Sonora*
There is widespread appreciation for the river.	The primary water concerns are water quality, water quantity, economic benefit, and preservation of ejidos (local farms).
Residents of the watershed want to provide substantive input to water management and planning for the river.	Mexicans believe the Cananea Mine is behind many of their problems with water quantity (they are pumping too much groundwater), water quality (discharging mine wastes into San Pedro River).
Opinions are divided over the nature and extent to which the river is in jeopardy.	Mexicans are suspicious that United States is pushing for conservation in Mexico solely to improve conditions in the United States, with little regard for conditions in Mexico.
There is virtually unanimous support for water conservation and aquifer recharge.	Mexicans believe that 30 to 40 years ago the San Pedro River was very different from how it is now – it flowed, and was a place where people went to fish, swim, or walk.
Many people see a need for growth management.	There is concern about potential negative effects on Mexican farmers and ranchers from conservation and restoration efforts. Alternative options for livelihoods need to be provided.
Multiple jurisdictions, both local and national, should share in the river's management.	Mexican stakeholders desire to benefit from protecting the San Pedro River to a similar degree that U.S. stakeholders benefit.
The burdens of water conservation should be equitably shared among water users.	Municipal and industrial water conservation needs to be promoted.
There is a need for more study of the Mexican portion of the watershed.	There is a need for explanations of what conservation areas are, their benefits, and how they will affect ecosystems, and ejidos.
Mexico should be included in cooperative planning and management processes for the watershed.	

*Arizona: http://www.udallcenter.arizona.edu/sanpedro/pubinput.html
*Sonora: http://www.udalcenter.arizona.edu/sanpedro/mexico.html

20.1.3.2 Water Quantity

There is quite a bit of information available regarding the amount of water in the upper San Pedro River watershed:

- Groundwater storage in the watershed is estimated to be 59 million ac-ft (ADWR, 1990).
- Groundwater recharge in the watershed is estimated to be 30,000 ac-ft/yr (ADWR, 1990).

- The groundwater contribution from the Mexican portion of the watershed is estimated to vary from 700 to 3400 ac-ft/yr (ADWR, 1990; Freethey, 1982).
- The annual average flow of the San Pedro River at Reddington, (USGS, HUC 09472000, drainage 2927 sq mi, just downstream of the Narrows), is 29.8 cfs for 1990–1997 (www.usgs.gov).
- The annual average flow of the San Pedro River at Palimonas, (USGS, 09470500, drainage 737 sq mi, near international border), coming into Arizona from Sonora, is 24.4 cfs for 1996–2002 (www.usgs.gov).
- At both of these gaging stations, the month with the lowest average flow was May, and the month with the greatest average flow was August (www.usgs.gov).
- Average annual precipitation in the watershed ranges from 11.4 inches per year in the valley to over 30 inches per year in the Huachuca Mountains (Liverman et al., 1997).
- Precipitation tends to be lowest in April and May, and greatest in July and August (http://www.wrcc.dri.edu).
- The watershed experiences a bimodal precipitation pattern, with periods of higher precipitation in the summer and winter, separated by periods of very low precipitation (Steinetz et al., 2003).
- The San Pedro River is the only perennial stream in the watershed, although it is not perennial along its entire length. The portion of the river in this watershed that is perennial extends from Hereford, Arizona, approximately 18 miles to a point just south of Fairbank, Arizona. (http://www.water.az.gov/adwr/Content/WaterInfo/OutsideAMAs/SoutheasternArizona/Basins/uppersanpedro.html). All tributaries are ephemeral.

20.1.3.3 Water Quality

There are several water quality issues in the upper San Pedro River watershed. The majority of the San Pedro River in Arizona is listed as impaired by ADEQ. Portions of the river are listed as impaired because of copper, *Escherichia coli*, and/or nitrate (ADEQ, 2004). Water quality data for the San Pedro River in Mexico was not found. Anecdotal information suggests there may be, or may have been in the past, problems with Cannanea mine discharges and discharges of inadequately treated sewage to San Pedro River in Sonora, Mexico (www.udallcenter.arizona.edu/sanpedro). Impairments identified in Arizona may be partially the result of conditions in Sonora.

There are also groundwater quality issues in this watershed. Two areas in particular have received a lot of attention in the past. Near St. David, Arizona, groundwater is contaminated with nitrates, lead, and sulfates, possibly from the Apache Powder Company superfund site, or inadequate septic systems and agricultural practices (Hargis & Associates, 1991; http://www.water.az.gov/adwr/Content/WaterInfo/OutsideAMAs/SoutheasternArizona/Basins/uppersanpedro.html). Also, in the Bisbee-Naco area, groundwater is contaminated with sulfate (Hargis and Associates, 1991). Groundwater monitoring by ADEQ has also identified wells in this watershed where concentrations of arsenic, fluoride, hardness, nitrate, and total dissolved solids are greater than the state drinking water standards. Pesticide concentrations in groundwater were below drinking water standards. Groundwater in this watershed was not analyzed for radiochemicals (ADEQ, 2004).

20.1.3.4 Water Demand and Use

The most recent water use data for the watershed that we located was for 1990. The USGS compiled data characterizing human use of water in the Arizona portion of the watershed for 1990 (http://water.usgs.gov/cgi-bin/wuhuc?huc=1505202). In 1990, approximately 44,000 ac-ft of groundwater, and 4900 ac-ft of surface water was withdrawn in the watershed for human uses. Thirty-four percent of the groundwater withdrawn was used for public water supply, 16% was for commercial uses (75% of commercial water was supplied by public water supply), 1% was withdrawn from private wells for residential use, 1% was used by industry, 1% was used for livestock, and 58% was used for irrigation. All of the surface water withdrawn was used for irrigation. Approximately 65% of the water withdrawn was consumed.

ADWR (1991, reported in Liverman et al.,1997), estimated 1990 water demand for the entire San Pedro River basin. In the Mexican portion of the watershed, ADWR estimated that in 1990 approximately 2600 ac-ft of water was utilized by domestic uses, 3100 ac-ft was used by industry (Cananea Mine use estimated at 3000 ac-ft), 5000 ac-ft of surface water was used for irrigation, and 1000 ac-ft of surface water was used for livestock. Domestic and industrial uses were supplied from groundwater (ADWR, 1991). Therefore, water use estimated for the Mexican portion of the watershed totaled 11,700 ac-ft.

Summing the water withdrawals for the Arizona and Sonora portions of the watershed yielded a total of 10,600 ac-ft of water withdrawn for human uses in the watershed in 1990. ADWR (1991) estimated that 44% of the 1990 water demand in the entire San Pedro River basin was consumed by natural processes (evapotranspiration and evaporation). Assuming the same percentage for just the upper San Pedro River watershed, approximately 26,000 ac-ft of water was consumed by natural processes in the watershed in 1990. This would bring the total 1990 water demand for the upper San Pedro River watershed to approximately 87,600 ac-ft. Note that ADWR estimated 1990 water demand for the entire San Pedro River basin to be approximately 139,000 ac-ft.

More recent water use data is likely available from local sources.

20.1.3.5 Water Needs

We need to consider what portion of the water demand for the watershed constitutes water needs, versus desired water uses. Water for support of human health and welfare; to maintain important economic sectors such as mining, agriculture, tourism, and Fort Huachuca; and to maintain the natural systems in the watershed that support human health and economic needs would be considered water needs. Multiplying the population estimate for the watershed (104,400) by Glieck's minimum per capita water requirement (13 gal/day) we calculate that approximately 1500 ac-ft/yr of water would be required for the health and welfare of the permanent population of the watershed. We did not find information on minimum water requirements for mining or agriculture in the watershed. Minimum water requirements for tourism should include water to support the health and welfare needs of tourists as well as water to support the natural systems tourists come to see (i.e., San Pedro River riparian habitat, birds, rare and endangered species of animals and plants). The National Park Service reports that 396,185 people visited the five parks in this area (Coronado National Monument, Ft. Bowie National Historic Site, Tombstone Courthouse State Historic Park, Chiricahua National Monument, and Kartchner Caverns State Park) during 2003 (CCCER, 2004). We can use this total as an estimate of the number of tourists staying in the watershed. Multiplying the number of tourists by Gleik's minimum per capita water requirement (50 L/day) we calculate that approximately 16 ac-ft/yr of water would be required for the health and welfare of tourists in the watershed. Fort Huachuca has implemented a number of projects to reduce its water usage. The Upper San Pedro Partnership reports that Fort Huachuca expects to achieve a minimum annual water usage of approximately 1500 ac-ft/yr (USPP, 2004).

We did not locate any information on recommended San Pedro River flow levels for maintaining the SPRNCA, although this is a question of intense study. There has been quite a bit of research on water requirements for the San Pedro River riparian habitat starting in the mid 1990s (www.tucson.ars.gov/salsa, www.ars.usda.gov/research/projects) that should be helpful in characterizing the water requirements of natural systems in the upper San Pedro River watershed. At least one report is available (Stromberg et al., 1996) that lists hydrologic thresholds for maintaining some of the San Pedro River riparian plant communities. Scott et al. (2002) provides a good summary of research activities in the SPRNCA. There is some information available about aquatic habitat requirements for some endangered species present in the watershed such as the spikedace, loach minnow, Sonoran tiger salamander (www.natureserve.org), and Huachuca water umbel (http://www.gf.state.az.us/w_c/edits/documents/Lilascre.fo.pdf). A request for water rights for protection of entire San Pedro River basin ecosystem that was submitted to ADWR sometime in the past specified 37,000 ac-ft/yr (Liverman et al., 1997). We were not able to determine the result of that request.

San Pedro River flows are under consideration as part of the ongoing adjudication of Gila River water rights (Lacher, 1994), however, we did not locate any information indicating that downstream flow requirements have yet been set for the San Pedro River under this process. Legislated flow requirements would be considered water quantity needs.

Water quality criteria are in place for the San Pedro River, based on its designated uses as assigned by the ADEQ (2004). These are considered water quality needs.

20.1.3.6 Desired Uses

Desired water uses are those not necessary for the survival of the society of the upper San Pedro River watershed and the natural systems on which it depends. Examples of desired water uses that are probably present in the watershed include water for swimming pools, watering landscaping, landscape water features, evaporative coolers, washing pavement, vehicles, and equipment, dust control, and leaks. Desired water use can also refer to the amount of water used over and above the minimum requirement for a necessary use. Current water demand could be considered the desired level of water use (see Section 20.1.3.4).

20.1.3.7 Sustainability Criteria

The next step in our watershed SWRM process is to determine if the four water resources sustainability criteria are being met. The first criterion is that the groundwater elevation must be relatively constant for the last 10 years. Putnam et al. (1988) and the USGS (1999) report that groundwater levels in the watershed are fairly stable, although cones of depression are present around urban areas. Hydrologic modeling by Steinitz et al. (2003) of the period from 1940 through 1997 indicate cones of depression at Sierra Vista, Fort Huachuca, St. David, and Benson in Arizona, at Cananea in Sonora, and at Naco at the U.S.–Mexico border. The cones of depression modeled at Cananea and Sierra Vista and Fort Huachuca are 130 ft and 80 ft deep, respectively. The hydrologic model indicates that by 1997, groundwater withdrawals exceed recharge by approximately 65%, and reduce San Pedro River flows by approximately 20% (Steinitz et al., 2003). These modeled effects suggest that groundwater usage in the watershed is not sustainable.

The second criterion is that there be sufficient surface and groundwater to satisfy existing and projected water demand. The total available water for the watershed could be assumed to be equal to the estimated groundwater recharge rate, 30,000 ac-ft/yr, plus the volume of San Pedro River flow at Reddington, 29.8 cfs, or 51,600 ac-ft/yr. This total is a little more than half of the estimated 1990 water demand for the watershed, 87,600 ac-ft/yr. More recent information for the Sierra Vista subbasin of the upper San Pedro River watershed (USPP, 2004, see Section 20.1.2.7 for discussion) also suggests that existing and projected water demand exceeds available water resources, as does hydrologic modeling performed by Steinitz et al. (2003). This criterion is not met in the upper San Pedro River watershed.

The third criterion is that the timing of water inputs be adequate to satisfy existing and projected water demands. The normal precipitation cycle in the watershed consists of intense, isolated summer monsoon rains from mid July through mid September, and cold late autumn and early winter storms from November through January (BLM, 2000). Fifty to sixty percent of precipitation occurs during the monsoon season, and the driest months, with minimum runoff and flow, are usually April–June (Liverman et al., 1997). The growing season in the watershed is considered to be May–September. Monitoring of 2001 seasonal riparian area water use indicate growing season vegetative water use starts around June and ends around October (Scott et al., 2002). This pattern follows the typical precipitation pattern. We did not find information on seasonal patterns of groundwater pumping, so it is not possible to evaluate if seasonal water supply patterns are adequate to satisfy human uses of water. Therefore, this criterion cannot be evaluated.

The fourth criterion is that water quality be adequate to support existing and project water uses. Based on the most recent state water quality assessments (ADEQ, 2004), there are places in the watershed where surface and groundwater quality are not adequate to support their designated uses.

Three of the four criteria were not met. Therefore, existing water resources management in the watershed is determined to be unsustainable.

20.1.3.8 Alternative Futures

An alternative futures analysis of the upper San Pedro River watershed was completed by Steinitz et al. (2003). Indicators that were evaluated in this analysis included San Pedro River flows, groundwater levels, aesthetics, endangered species habitat, vegetation, and land use. The alternative futures scenarios that were evaluated included scenarios in which existing development plans and restrictions continued, scenarios in which development was only permitted in previously developed areas, and scenarios in which there were fewer restrictions on development than currently exist. Projections of future conditions based on these scenarios yielded the following insights:

- All of the alternative futures analyzed resulted in lowering the groundwater table near Sierra Vista, with a projected drop of 33–49 ft over 20 years (note that the scenarios did include reductions in per capita water use, but not groundwater recharge programs).
- All of the alternative futures analyzed result in reduced areas of perennial flow in the San Pedro River, although increases in river flow in some segments did result from some scenarios.
- Policy decisions affecting irrigated agriculture in Arizona had the greatest effects on hydrology in the upper San Pedro River watershed.

- Policy decisions about development and its control in Arizona had the second most significant impact on hydrology in the upper San Pedro River watershed. These policies also affected habitat and aesthetics indicators.
- Policy decisions about growth/development in Sonora ranked as the third most important scenario element with regard to effects on hydrology in the upper San Pedro River watershed.
- Management of Fort Huachuca significantly affected conditions in Sierra Vista, but had significantly less effect on the rest of the watershed than irrigation and development decisions.
- Critical areas where there were conflicts between demand for development and conservation needs included the lower slopes of the Huachuca Mountains and the arroyos in Sierra Vista. Lands currently owned by Fort Huachuca were also in conflict in scenarios where they became available for development (i.e., the Fort closes).

20.1.3.9 Prioritizing Uses

The first step for prioritizing water uses is to compare available water with required and desired water uses. We have already determined that water resources in the upper San Pedro River watershed appear to be overallocated, so desired water use levels (i.e., current water use levels) cannot be sustained. Some sort of reduction of water demand will be required to achieve sustainability. To get an idea of how much water demand will need to be reduced, we can compare available water to minimum water needs estimated for the watershed.

We did calculate an estimate of minimum water required for the support of the watershed population (1500 ac-ft/yr), tourists (16 ac-ft/yr), and Fort Huachuca (1500 ac-ft/yr), however, we did not locate data on required water uses for other economic sectors such as agriculture and mining. The sum of the available estimates of minimum water requirements with the estimated natural systems water demand for 1990 (26,000 ac-ft/yr) could be considered as a rough estimate of minimum water needs for the watershed. That sum is approximately 29,000 ac-ft/yr. The estimated available water for the watershed is 51,600 ac-ft/yr. This comparison suggests that it could be possible to support human health needs and natural system needs sustainably. Prioritization of water uses and allocation of water at sustainable levels would likely be a lengthy process, requiring conflict resolution skills, and adjustment of mental models and social mindscapes.

20.1.3.10 Management Options

Several vehicles are being used to pursue conservation of water resources and ecosystems in the watershed. There are several land preserves in the watershed in both Arizona and Sonora (http://lastgreatplaces.org/sanpedro/index.html), and several groups (e.g., The Nature Conservancy, Udall Center, ARASA) are working to add land preserves in the watershed. Land use plans are in place for the Arizona portions of the watersheds; for Sierra Vista (Vista 2020, http://www.ci.sierra-vista.az.us/Community%20Development/pdf/VISTA2020Draft3.pdf), and Cochise County (Cochise County Comprehensive Land Use Plan, http://www.cochisecounty.com/ P&Z/Comprehensive.htm; and Southern San Pedro Valley Area Plan, http://www.cochisecounty.com/P&Z/Comprehensive.htm). The Sierra Vista and Fort Huachuca community has set the goal of eliminating deficit groundwater pumping by 2011 (see Section 20.1.2.10). Work is underway in Sonora to improve wastewater treatment facilities in Cananea and Naco, and improve water supply infrastructure in these cities (www.udallcenter.arizona.edu). The Partnership is working with Cochise County to develop and implement several county codes related to reducing water usage, and is pursuing state legislation to limit expansion of irrigated agriculture in the watershed (USPP, 2004). Economic vehicles in place or under consideration include fines for violation of water conservation codes, rebate programs for retrofitting high water use fixtures, and rental payments for land enrolled in conservation programs (USPP, 2004). Stakeholder education and outreach programs are being conducted by Cochise County, The Nature Conservancy, the Partnership, and the Udall Center. Personnel from the University of Arizona and the USDA Agricultural Research Service are collaborating on development of a GIS-based watershed decision support tool (http://www.udallcenter.arizona.edu/sanpedro/dialogues/archives/May04_technical.pdf).

There is also a lot of research occurring in the watershed through universities (e.g., University of Sonora, Arizona State University, University of Arizona), state and national agencies (e.g., ADEQ, USDA, USGS, EPA, BLM, Institue del Medio Ambiente y el Desarrollo del Sonora, Secretaria de Medio Ambiente y Recursos Naturales), local NGOs (e.g., Partnership, ARASA, Udall Center), and international groups (e.g., Semi-Arid Land-Surface-Atmosphere, The Nature Conservancy, North American Commission for Environmental Cooperation). Research addresses a range of topics including surface and groundwater quality, surface and groundwater hydrology, land use and land cover, endangered species, water budgets, vegetation water use, methods for stakeholder education, and water management policy. There is a strong effort to promote collaboration of research and conservation effort across the national border (www.udallcenter.arizona.edu).

To date, water management in the upper San Pedro River watershed has been primarily directed from the local level. International collaboration occurs among interest groups and state and federal agencies, rather than at the level of the executive governments.

20.1.3.11 Monitoring and Evaluation

Groundwater levels, San Pedro River flow regime, and vegetation characteristics in the SPRNCA are the indicators that are being monitored to evaluate the effects of water conservation and management efforts in the upper San Pedro River watershed. Fort Huachuca, ADWR, ADEQ, USGS, and BLM monitor groundwater levels and quality in wells they maintain in the Sierra Vista subbasin. The USGS monitors San Pedro River flow at several gaging stations, and is working to develop collaborative flow monitoring into Mexico (http://www.udallcenter.arizona.edu/sanpedro/dialogues/archives/May04_technical.pdf). The Nature Conservancy organizes an annual effort to map wet and dry portions of the San Pedro River in both Arizona and Sonora (http://www.udallcenter.arizona.edu/sanpedro/dialogues/archives/May04_technical.pdf). The U.S. Army Corps of Engineers Topographic Engineering Center is mapping vegetation in the SPRNCA every 4 years, in a project that is funded through 2008. The Partnership plans to use public surveys to evaluate the effectiveness of public outreach and education efforts in the Sierra Vista subbasin (USPP, 2004). The Nature Conservancy and other interest groups also monitor effectiveness of education and outreach through their interactions with stakeholders.

20.1.4 Larger-scale Influences

There are several programs that operate at larger scales that encompass the upper San Pedro River watershed, thus having the potential to influence management of the watershed as well as the potential to be influenced by management of the watershed. As a tributary of the lower Colorado River, San Pedro River flow contributes to the Colorado River discharge required by the Mexican Water Treaty and the Colorado River Compact, and comes under the consideration of the Colorado River Water User's Organization and the BLM's management of the Colorado River (Carter & Morehouse, 2001). San Pedro River water resources are also being considered in the Gila River water rights adjudication (Gelt, 1994; http://www.supreme.state.az.us/wm/).

National and international economics affect water use in the watershed, most noticeably through the mineral mining interests located in the San Pedro River basin; the Cananea mine in Mexico, and copper mining operations located in the Arizona portion of the San Pedro River basin (www.nationalatlas.gov). As a result of the presence of these mining operations, whatever affects the value of copper and other minerals mined in the watershed, and thus the rate of mineral production in the watershed, has the potential to affect water demand in the watershed.

Since water in this watershed is supplied solely by precipitation, regional weather patterns influence water resources in the watershed. High summer precipitation in the watershed is the result of the "Arizona Monsoon" weather pattern (http://weather.about.com/library/weekly/aa091302a.htm). Precipitation patterns in this region are affected by Gulf of California temperatures, ground temperature, Pacific tropical storms, Pacific Ocean temperatures, ENSO (El Niňo/La Niňa), and the position of the jet-stream (Pagano, 1999; Brown & Comrie, 2002).

20.2 Spring Creek, Pennsylvania

20.2.1 Background

Spring Creek is located in the south central portion of Centre County in central Pennsylvania, in the Ridge and Valley physiographic province with steep slopes, rich and thin soils, and fertile flat valleys (Giddings, 1974). It is part of the Susquehanna River watershed, a tributary of the Chesapeake Bay. The hydrology of this area is precipitation dominated and water quantity has not been a major concern, although drought conditions in Pennsylvania from 1995 to 2002 did raise awareness of water quantity issues. Water quality is currently the primary issue associated with this waterbody. Spring Creek is a spring-fed creek; one study indicates that, on average, more than 85% of the annual flow of Spring Creek is provided by groundwater (Taylor, 1997). The aquifer is bounded by the same ridges that form the watershed boundary, and is recharged by mountain runoff and valley infiltration (Giddings,1974). Spring Creek is a well-known high-quality brown trout fishery. Maintaining the integrity of this fishery, which contributes to the local economy, is a priority with stakeholders.

Because the creek is spring fed, it is threatened by pollution from both surface runoff and underground sources. Concern about the effects of pollution on the fishery has provided the impetus for restoration, management, and planning activities in the watershed. The I-99 Partnership for Sustainable Development was created to proactively plan for the construction of the new I-99 highway through this area. This partnership has created planning guides suitable for local communities, counties, and the region; as well as assisted communities with developing plans that promote thoughtful use and conservation of land and water resources as they respond to the effects of the new highway corridor (www.clearwaterconservancy.org).

20.2.2 Community Scale

The borough of State College is the largest community in the Spring Creek watershed (Figure 20-2). It is a growing community that is home to Pennsylvania State University. The university is a major contributor to the economy of State College and is the primary force driving

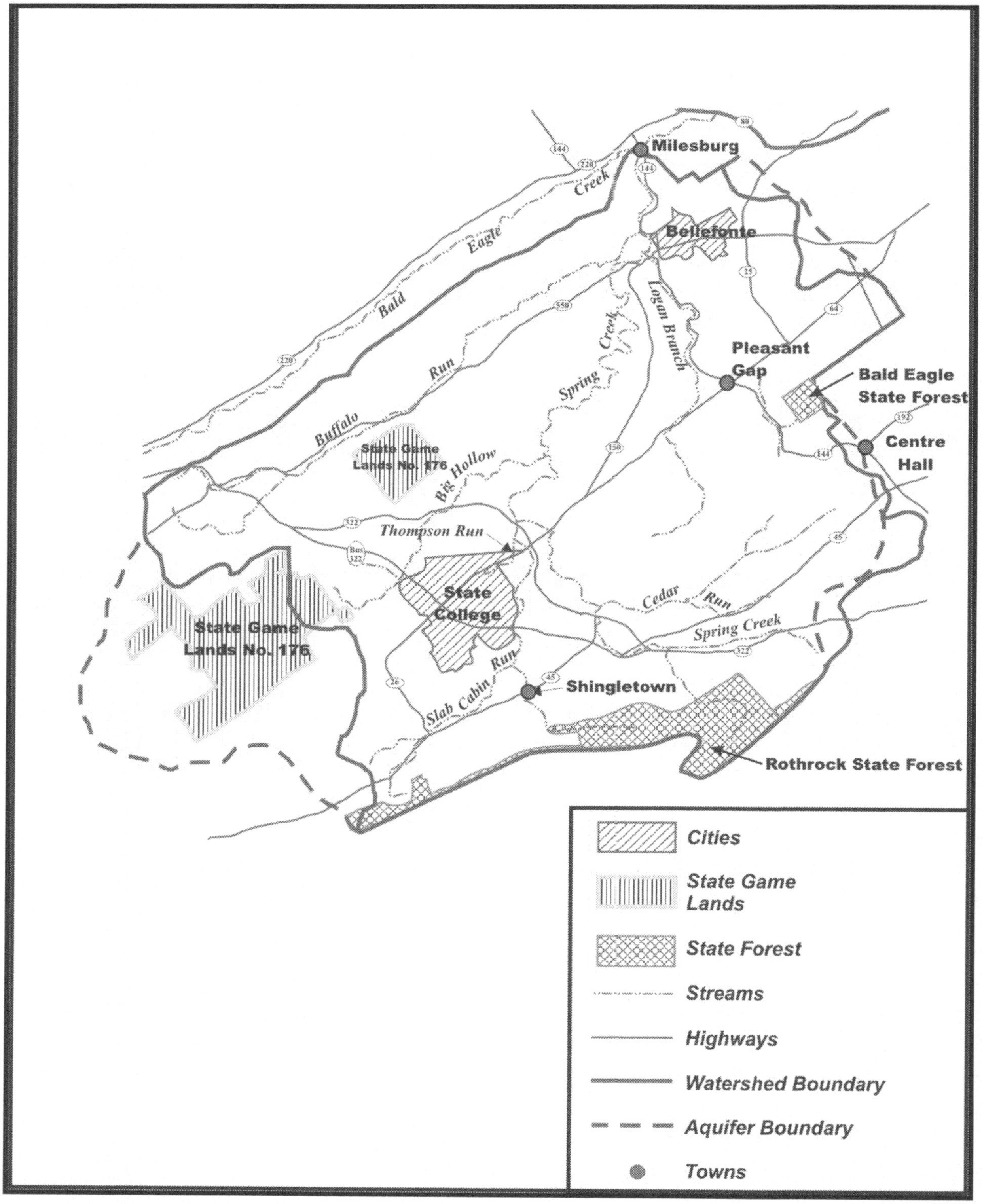

Figure 20-2. Spring Creek, Pennsylvania, Watershed.

its growth. Background information about this community that may be useful for water resources planning includes the following:

- State College covers approximately 4.5 sq mi of land.
- The 2000 census population estimate for this community was 38,420 (www.census.gov).
- The three most prevalent races recorded in State College during the 2000 Census were "White non-Hispanic" at 91.4%, Asian at 4%, and Black at 3%.
- Approximately 88% of the adult population have completed high school; 36% have college degrees.
- Pennsylvania State University is the largest employer in State College and exerts a strong influence on the local economy (Tripp Umbach & Associates, 2004; http://www.getawayguides.com/pennsylvania/statecollege/diversions.htm; http://www.lindalowe.com/l_economy-state-college-bellefonte-centre-county-pennsylvania-real-estate.asp).
- Temperatures in State College usually range between 28 °F and 69 °F.
- The average annual rainfall at State College is 39.1 inches. Rainfall is fairly evenly distributed throughout the year, with average rainfall for most months between 2 and 4 inches. June is the month with the highest average rainfall (Taylor, 1997).
- This area experiences periods of wetter or drier than normal weather as a result of the influences of cycles in Atlantic Ocean surface temperatures (i.e., Atlantic Multidecadal Oscillation).
- This community is located in the Ridge and Valley ecoregion (Omernick, 1987), in the Valley Center Upland hydrogeologic formation (Giddings, 1974).

20.2.2.1 Community Beliefs and Values

There are three locally led interest groups active in the Spring Creek watershed. The missions of these groups include conservation and restoration of natural resources, and protection and enhancement of quality of life, environment, and economy (www.clearwaterconservancy.org, www.springcreekwatershed.org). These groups have solicited local stakeholder input in developing and prioritizing conservation and restoration related activities within the Spring Creek watershed. In Pennsylvania, resource planning authority is concentrated at the county level (e.g., Greenways Program Plans). Therefore, there is a high level of community involvement in planning activities. We did not find specific information about community values, beliefs, interests, and concerns about water resources. However, based on the activities of the interest groups, we would postulate that maintaining existing quality of life with regards to aesthetics and recreational opportunities (e.g., fishing) are priorities for the community. Pennsylvania State University is currently preparing a survey to evaluate attitudes and beliefs about Spring Creek in the watershed (Jim Shortle, personal communication, 2004).

20.2.2.2 Water Quantity

The State College Borough Water Authority (SCBWA) has a permitted water usage of 9 mgd (www.scbwa.org). The SCBWA maintains seven well fields and Shingletown Reservoir to provide water to its customers in State College and surrounding townships. We did not find explicit information on the recharge rate of the aquifer in the State College area, however, Giddings (1974) reported that in the Valley Center Upland, groundwater response to rainfall was slow. We also did not locate information on the capacity of the Shingletown Reservoir.

20.2.2.3 Water Quality

The quality of available water is an important consideration in this area. Portions of Slab Cabin Run and its tributaries adjacent to State College are included on the 2004 303(d) list as impaired by siltation from urban runoff. Portions of Slab Cabin Run adjacent to State College, are also listed as impaired because of thermal modification caused by golf courses [Pennsylvania Department of Environmental Protection (PDEP), 2004c]. We were not able to locate recent groundwater quality data online. Water quality data for the area may be available from the Pennsylvania Department of Conservation of Natural Resources.

The SCBWA 2003 water quality report (http://www.scbwa.org/waterquality3.htm) lists treated water quality. Constituents measured in treated water include nitrate, radioactivity, turbidity, atrazine, and barium. The report indicates that these constituents occur in the raw water, but not at what concentrations. The report also states that the water authority treats for *Cryptosporidum* and *Giardia* in the Shingletown reservoir water. Groundwater contamination reportedly affects springs that feed Slab Cabin Run (PDEP, 2003), but we did not determine the exact location of the contamination.

20.2.2.4 Water Demand and Use

The SCBWA reports that their current average water demand is 6 mgd, with an average per capita use for their system of 92 gal/day (6 mgd/65,000 customers) (www.scbwa.org/history.html). Based on this per capita use, we estimate that water demand from State College is approximately 4000 ac-ft/yr. The water authority does not report water use by use sector, or community, online, but it is possible that there are other local sources for this information. Pennsylvania's Act 220 mandates that all water withdrawals greater than 10,000 gal/day be registered with, and report water usage to, PDEP. Smaller water users are encouraged to register and report their water usage voluntarily (PDEP, 2004a). It would be useful to have an idea of how much water is used in State College by business, industry, private homes, public facilities, and the university.

The water authority does include water conservation tips on its web page. The only edition of its Water Conservation Newsletter online is for the summer of 2002, about the time the drought ended in Centre County (http://www.scbwa.org/conserve/conserve.html). This newsletter lists mandatory water use restrictions that would be imposed under "drought emergency" conditions as defined by the Commonwealth of Pennsylvania. We found no indication of currently active local water conservation programs.

20.2.2.5 Water Needs

We need to consider what portion of the water demand constitutes water needs, versus desired water uses. Water for support of human health and welfare would be considered a need. Multiplying the population estimate for State College by Glieck's minimum per capita water requirement (13 gal/day) we calculate that approximately 600 ac-ft/yr of water would be required for the health and welfare of the population of State College. This water would need to meet Pennsylvania drinking water quality standards (PDEP, 2002).We do not have the information necessary to estimate water quantity and quality needs for maintaining the important economic sectors in State College.

There are some state guidelines related to water requirements for natural resources that apply to the State College area. Slab Cabin Run and its tributaries have been assigned the designated use of cold water fisheries by PDEP, and are subject to water quality criteria associated with this designated use (Title 25 Pennsylvania Code, Chapter 93). The Slab Cabin

Run tributary northeast of State College is designated as a Wild Trout Fishery by the Pennsylvania Fish and Boat Commission. PDEP drafted regulations stating that water allocation from streams with this designation cannot result in more than a 10–15% loss of habitat (PDEP, 2000). We did not determine if this policy is currently in use. Other published guidelines for evaluation of water allocation requests indicate that water allocations can only be such that stream 7Q10 (i.e., the lowest 7-day average flow that occurs, on average, every 10 years) flows are maintained (PDEP, 2001).

20.2.2.6 Desired Uses

Desired water uses are those not necessary for the survival of the society of State College and the natural systems on which it depends. Examples of desired water uses that are probably present in State College include water for pools, watering landscaping, landscape water features, washing pavement, vehicles, and equipment, dust control, and leaks. Desired water use can also refer to the amount of water used over and above the minimum requirement for a necessary use. Current water demand could be considered the desired level of water use.

20.2.2.7 Sustainability Criteria

The next step in our community SWRM process is to determine if the four water resources sustainability criteria are being met. The first criterion is that the groundwater elevation must be relatively constant for the last 10 years. There are indications that there may be groundwater elevation data available that can be used to evaluate this criterion. Water level measurements are available for a USGS observation well located west of State College, which is up gradient. The period of record of this well is 1997 through 2004, less than 10 years. Water levels in this well dropped almost 35 ft between 1999 and 2003, during the time when this area was under drought conditions. Since 2003, the water level has been increasing, and is back to predrought levels (http://nwis.waterdata.usgs.gov). This rapid recovery suggests that under normal conditions water use is sustainable. However, significant adjustments of water use would be necessary under prolonged drought conditions.

The second criterion is that there be sufficient surface and groundwater to satisfy existing and projected water demand. The fact that the SCBWA is currently utilizing less water than its permit allows suggests that there is adequate surface and groundwater to satisfy existing and future water demand. However, the sustainability of the SCBWA permitted water use is unknown. An evaluation of this criterion is not possible with the data that we have.

The third criterion is that the timing of water inputs be adequate to satisfy existing and projected water demands. There is information available on the seasonality of precipitation and flow inputs (Taylor, 1997; www.usgs.gov). However, we did not locate any information on the seasonality of water demand. Therefore, this criterion cannot be evaluated at this time.

The fourth criterion is that water quality be adequate to support existing and project water uses. Based on the most recent state water quality assessments (PDEP, 2004c), surface waters adjacent to State College are not meeting their aquatic life support designated use because of sediment, and flow and temperature modification. There are also indications that there is pesticide groundwater contamination in the State College area, however, the SCBWA does not report any water quality issues with their water supply well fields. Water quality does not support sustainability.

The fourth criterion was not met, therefore, existing water resources management is determined to be unsustainable.

20.2.2.8 Alternative Futures

As mentioned previously, there is a high level of community involvement in watershed, county, and regional resources planning. From what we gather, this involvement often occurs in workshops or meetings and involves identification of issues and/or management options, and prioritizing the same. It does not appear that there has been any formal alternative futures analysis or development of relationships between management activities and ecologic, social, and economic outcomes. Work in this area would enhance and improve the planning, restoration, and conservation efforts already ongoing in the community.

20.2.2.9 Prioritizing Water Uses

The first step for prioritizing water uses is to compare available water with required and desired water uses. A cursory evaluation of the ability to meet existing community uses with available water can be made by comparing the current water use of the SCBWA with its permitted use. Because the current use is less than the permitted use, we could infer that there is adequate water to meet existing uses, and to cover future expansion of water demand. The sustainability of the SCBWA permitted water use needs to be evaluated with regard to potential for effects on the local hydrology, ecosystems, economy, and society.

Community priorities for water use have not been explicitly stated. A borough zoning ordinance does specify that preservation of the community's "natural setting" is a planning goal, and outlines policies for achieving that goal which include suppressing water pollution, and discouraging unplanned changes in contours that would modify natural drainage patterns (State College Zoning Ordinance, 1959).

20.2.2.10 Management Options

Water quality is currently the primary water related issue being addressed by State College Borough. Planning, research, regulatory, and education and outreach methods are being utilized by the community to improve and protect water resources. Although the primary focus of most of these management methods is water quality, there is some work being done that addresses water quantity. Pennsylvania State University has been involved in several water quality restoration and sustainable development planning projects in the community and other parts of the watershed, as well as development of a Spring Creek water budget. The State College Area School District received funding in 2000 to support water quality sampling field trips to stream sites for fifth-grade classes (PDEP, 2003b). Borough zoning goals include encouraging light industry in the community, which appears compatible with goals for maintaining water quality and natural setting (State College Zoning Ordinance, 1959). State College has an NPDES Phase II permit for stormwater management. State College appears to be addressing issues related to improving and maintaining water quality. The next step is for the community to consider issues related to maintaining water quantity.

20.2.2.11 Monitoring and Evaluation

Data collection in the community already includes a number of the sustainable water resources indicators (see Chapter 14.0) including, groundwater level, precipitation, water quality, median household income, migration, and population below poverty level. Pennsylvania State University has done some work on measuring impervious area in the Spring Creek watershed (PDEP, 2003b), so this information may also be being monitored in the State College Borough. PDEP assesses and tracks stream use impairment, and that information is available to the

Table 20-4. Current Land Use for the Spring Creek Watershed (from SCWC, 2000).

Land Use	Acres	Percentage of the Watershed
Agriculture	33,795	36
Commercial	1,117	1
Forest/water	34,767	37
Industrial/mines	1,609	2
Public/institutional	1,350	2
Recreation	1,868	2
Residential	8,755	10
Transportation/utilities	4,996	5
Vacant	4,846	5

community. State College most likely also monitors job creation by economic sectors and waste water treatment discharge. Forest cover measures are not appropriate for a community. Additional indicators that it could be useful to add are water use by sector, and flow monitoring on Slab Cabin Run.

20.2.3 Watershed Scale

The Spring Creek watershed is located in the Ridge and Valley physiographic province and is characterized by steep slopes with thin fertile soils, and fertile flat valleys. The watershed drains approximately 146 sq mi. The extent of the underlying aquifer is only slightly greater than the watershed boundary, approximately 175 sq mi (www.pgwd.org). The watershed is underlain primarily by limestone and dolomite—karst geology. There are six major tributaries to Spring Creek in this watershed; Cedar Run, Slab Cabin Run, Big Hollow, Thompson Run, Logan Branch, and Buffalo Run (Figure 20-2). Communities located in the watershed (partially or completely) include State College, Bellefonte, Centre Hall, Pleasant Gap, and Milesburg.

Predominant land uses in the watershed are forest, agriculture, and urban areas (Table 20-4). Forested land accounts for 37% of the watershed and is mainly located on mountain tops. Major forested areas include Bald Eagle State Forest, Rothrock State Forest, and State Game Lands No. 176. More than one-third (36%) of the watershed is in dairy or crop production. More than 100 farms support animal operations (Alliance for the Chesapeake Bay, 2004). Agricultural land is predominant in the valley regions outside the towns of State College and Bellefonte.

Although Spring Creek watershed is located in a rural county, it has become an urbanized stream with 14 municipalities in the watershed. Approximately 20% of the watershed is urban, including a variety of land uses such as commercial, public and institutional, residential, and utilities. As a result of growing urbanization in the last four decades, not only in the State College area but also in other parts of the watershed, estimated impervious area has increased from 5% in 1960 to 11% in 2000 [Clear Water Conservancy (CWC), 2004].

Approximately two-thirds of the stream corridor is owned privately. Some stretches are completely inaccessible because landowners do not allow trespassing on their property. The other one-third of the Spring Creek floodplain is public property held by Pennsylvania Fish Commission and Rockview State Correctional Institution (Alliance for the Chesapeake Bay, 2004). There are approximately 94,000 people living in the watershed (SCWC & CWC, 2002). The three most prevalent races recorded in the watershed during the 2000 U.S. Census were Whites (87%), Asians (6%), and Blacks (4%). Approximately 43% of the people older than 25 had completed high school in 2000; 26% had college degrees. Median household income in the watershed was approximately $36,000 in 1999, and 26% of people in the watershed had 1999 incomes below the poverty level (www.census.gov). Pennsylvania State University is the major economic force in the watershed (Tripp Umbach & Associates, 2004). The recreational fishing industry associated with the high-quality trout fisheries in the watershed (e.g., Fisherman's Paradise) is also important to the economy of the watershed (PDEP, 2003b).

20.2.3.1 Watershed Community Beliefs and Values

We did not find specific information about community values, beliefs, interests, and concerns about water resources. However, based on the activities of watershed interest and planning groups, we would postulate that maintaining existing quality of life with regards to aesthetics and recreational opportunities (e.g., fishing) are priorities for the watershed community. Pennsylvania State University is currently preparing a survey to evaluate attitudes and beliefs about Spring Creek in the watershed (Jim Shortle, personal communication, 2004).

The watershed community appears to value its natural resources and consider their conservation a priority. There are three locally led interest groups active in the Spring Creek watershed; The Spring Creek Watershed Community, the Clear Water Conservancy, and the Spring Creek Watershed Commission. The missions of these groups include conservation and restoration of natural resources, and protection and enhancement of quality of life, environment and economy (www.clearwaterconservancy.org, www.springcreekwatershed.org). These groups have solicited local stakeholder input in developing and prioritizing conservation and restoration related activities within the Spring Creek watershed. In Pennsylvania, resource planning authority is concentrated at the county level (e.g., Greenways Program Plans). Therefore, there is a high level of local involvement in planning activities.

20.2.3.2 Water Quantity

Average precipitation in the watershed is 39.07 in/yr. Based on this precipitation value, we estimate that approximately 304,000 ac-ft/yr of water enter this watershed. A water budget calculated for the period 1968 through 1997 indicates that approximately 44% of precipitation in the watershed ends up as groundwater (Taylor, 1997). Therefore, we would estimate that the aquifer recharge rate in the watershed is approximately 134,000 ac-ft/yr (44% of 304,000 ac-ft/yr). The same water budget indicates that 6% of precipitation in the watershed runs off, becoming surface water. Therefore we would estimate that approximately 18,000 ac-ft/yr of surface water comes from direct runoff. From 1968 through 1997, the average flow in Spring Creek at Milesburg (the downstream end of the watershed) was 232 cfs (Taylor, 1997). This is equivalent to a volume of approximately 168,000 ac-ft/yr. Taylor's water budget indicates that, on average, 80% of this flow is contributed by groundwater (Taylor 1997). The remainder of the water in the budget, 152,000 ac-ft/yr, leaves the watershed through evapotranspiration (Taylor, 1997).

20.2.3.3 Water Quality

Approximately 17 miles of streams have been identified as impaired by the PDEP (2004c). Causes of impairment noted in the 303(d) listing include siltation, thermal modification, and organic enrichment. Exposure of pyretic rock in 2003 during construction of I-99 resulted in acid drainage to Buffalo Run (www.springcreekwatershed.org). Water quality of stormwater runoff is also a concern. Stormwater runoff tends to have higher concentrations of metals, ammonia, chloride, total organic carbon, orthophosphate, and total suspended solids (SCWC & CWC, 2002).

The researchers did not find any groundwater quality data for the watershed. Public water suppliers monitor the quality of their finished water. State College Borough Water Authority does not report any problems with their groundwater supplies (SCBWA, 2004). There is a known groundwater quality issue in the watershed, however. Groundwater has been contaminated with pesticides at a chemical plant site northwest of State College on Highway 26. The groundwater contamination has affected nearby springs and portions of Slab Cabin Run and Spring Creek, and has resulted in fish consumption advisories for these streams (PDEP, 2003a). There is currently a fish consumption advisory for PCBs in brown trout in Logan Branch (PDEP, 2003a). The chemical plant site has been declared a Superfund Site (PDEP, 2003b; www.epa.gov).

20.2.3.4 Water Demand and Use

There are approximately 25 public water supply systems in the watershed pumping approximately 19,000 ac-ft of groundwater daily (SCWC & CWC, 2002). A water budget calculated for 1968–1997 indicates that approximately 50% of watershed precipitation goes to evapotranspiration (Taylor, 1997). This is equivalent to a volume of 152,000 ac-ft/yr (50% of 304,000 ac-ft/yr). We did not find information on agricultural or industrial water use.

20.2.3.5 Water Needs

We would identify water for the support of human health and welfare, Pennsylvania State University, the watershed recreational fishing industry, and agriculture as water needs for the watershed. Multiplying the population estimate for the watershed (94,000) by Gleik's minimum per capita water requirement (13 gal/day) we calculate that approximately 1400 ac-ft/yr of water would be required for the health and welfare of the watershed population. This water would need to meet state and federal drinking water quality standards. We will assume that this estimated water need includes water needs for the university.

There are some state guidelines related to water requirements for natural resources that apply to the Spring Creek watershed. All of the streams in the watershed have been assigned the designated use of either a Cold Water Fishery or a High Quality Cold Water Fishery by PDEP, and are subject to water quality criteria associated with these designated uses (Title 25 Pennsylvania Code, Chapter 93). A number of streams in this watershed have been designated as Class A Wild Trout Waters by the Pennsylvania Fish and Boat Commission (PFBC). Cedar Run, Thompson Run, Buffalo Run, and portions of Spring Creek and Logan Branch are classified as wild brown trout fisheries (PFBC, 2004). PDEP drafted regulations stating that water allocation from streams designated as High Quality or Cold Water Fisheries by PDEP, or as Class A, B, C, or D Wild Trout Streams by PFBC, cannot result in more than a 10–15% loss of habitat (PDEP, 2000). These regulations were based on a Susquehanna River Basin Commission study for development of a procedure for evaluating impacts of water withdrawals on wild trout fisheries in Pennsylvania (SRBC, 1998). We did not determine if this policy is currently in use. Other published guidelines for evaluation of water allocation requests indicate that water allocations can only be such that stream 7Q10 flows are maintained (PDEP, 2001).

20.2.3.6 Desired Uses

Desired water uses are those not necessary for the survival of the society of Spring Creek watershed and the natural systems on which it depends. Examples of desired water uses that are probably present in the watershed include water for swimming pools, watering landscaping, landscape water features, washing pavement, vehicles, and equipment, dust control, and leaks. Desired water use can also refer to the amount of water used over and above the minimum requirement for a necessary use. Current water demand could be considered the desired level of water use.

20.2.3.7 Sustainability Criteria

The next step in our watershed SWRM process is to determine if the four water resources sustainability criteria are being met. The first criterion is that the groundwater elevation must be relatively constant over the last 10 years. There are indications that there may be groundwater elevation data that could be used to evaluation this criterion, however, we did not find much online. An online publication by Giddings (2004) does show graphs of groundwater elevations in two wells (CE 686 and CE 118) located west of State College for the period from 1996 through 2003, a 7-year period. This graph shows significant drops in groundwater levels during the drought that started in the summer of 1998 and ended early in 2002. The graph also shows the rapid recovery of the groundwater elevations. These graphs do not suggest a significant downward trend in groundwater elevations in this part of the aquifer, so we would say that this criterion is met. However, these graphs do make it apparent that the sustainability of the aquifer in the watershed is fragile and susceptible to changes in climate.

The second criterion is that there be sufficient surface and groundwater to satisfy existing and projected water demand. In Section 20.2.3.2 we estimated that there is approximately 134,000 ac-ft/yr of groundwater recharge and 18,000 ac-ft/yr of surface water available in the watershed, for a total of 152,000 ac-ft/yr of available water. Known water demand for the watershed includes 19,000 ac-ft/yr of water for municipal use. This total is less than the estimated total available water, however, it does not include all water use currently occurring in the watershed (e.g., agricultural, fisheries support). We anticipate that better water demand data will be available in the near future as a result of the implementation of Pennsylvania's Water Resources Planning Act, Act 220 (www.dep.state.pa.us/dep/deputate/watermgt/wc/Subjects/WaterResources/).

The third criterion is that the timing of water inputs be adequate to satisfy existing and projected water demands. There is adequate data available to determine the timing of precipitation inputs to the watershed, as well as flow, and possibly groundwater, response to these inputs. However, we did not find information on seasonal water demand in the watershed. This would be a data gap to consider addressing in future data collection efforts.

The fourth criterion is that water quality be adequate to support existing and projected water uses. As noted in Section 20.2.3.3, there are areas in the watershed where water quality has been determined to be inadequate to support designated water uses. Therefore, this criterion is not met.

The fourth criteria is not met, therefore, existing water resources management is not sustainable.

20.2.3.8 Alternative Futures

The Vision 2020: Living With I-99 project (http://www.clearwaterconservancy.org/vision2020.htm) analyzed alternative futures for the Mid Bald Eagle watershed, which includes Spring Creek watershed. The first step in this project was to determine the stakeholders objectives for the Mid Bald Eagle watershed. Next, an evaluation of the probable effects of the new I-99 on watershed land use and sustainability based on existing management policies was completed (Community Planning Consultants Inc. et al., 2002). Particular impacts considered included land conversion to residential, retail, and office space; roads with potential for traffic flow problems; threats to natural resources and agricultural land; increase of impervious area; and affects on water supply, wastewater treatment, solid waste systems, emergency services, school districts, and recreation lands. It was determined that this probable future did not meet the stakeholder objectives for the watershed. Stakeholders were then asked to envision their ideal future and this input was used to develop the vision of a sustainable future in which controlled development occurs that maintains quality of life and sense of community, including the characteristic natural beauty and agricultural heritage of the watershed. This vision forms the basis of the Land Use and Sustainability Plan (Community Planning Consultants Inc. et al., 2002) for the Mid Bald Eagle watershed.

The USGS is developing a Spring Creek watershed model linking changes in water resources to changes in land use. The purpose of this modeling effort is to give municipal officials in the watershed a tool to use for evaluating the effects of proposed land use changes. The model is expected to be completed sometime around 2008 (Fulton, 2003).

20.2.3.9 Prioritizing Uses

The first step for prioritizing water uses is to compare available water with required and desired water uses. Based on the information we were able to find regarding current water use, groundwater levels, and stream flows, it appears that available water quantity is adequate to meet current water demand, although there are some water quality issues, associated primarily with fisheries uses.

There is strong potential for significant future development in this watershed, because of the influence of Pennsylvania State University, and the new I-99 highway that passes through the watershed. Associated with the potential for future development is a strong potential for conflicts between development and conservation concerns and their associated demands for water. Based on the outcome of the Vision 2020: Living With I-99 project (Community Planning Consultants Inc. et al., 2002), water usage for natural systems and agriculture are priorities in the Spring Creek watershed.

20.2.3.10 Management Options

Management options for Spring Creek watershed water resources have been developed as part of a number of planning efforts occurring in the watershed, including the Act 167 Stormwater Management Plan for the Spring Creek Watershed (Sweetland Engineering and Assoc. and Centre County Planning Office, 2002), Spring Creek Rivers Conservation Plan (PSU and Clear Water Conservancy, 2001), Vision 2020: Living With I-99 Land Use and Sustainability Plan (Community Planning Consultants Inc. et al., 2002), and The Spring Creek Watershed Plan Phase I Final Report (Clear Water Conservancy, 2004). These include zoning ordinances to protect sensitive areas from inappropriate land uses, riparian buffers to protect stream habitat and water quality, conservation easements, creation of parks and reserves to protect sensitive areas and endangered species and provide educational opportunities, water conservation programs, wellhead protection programs, stormwater management practices, education and outreach programs, and research programs. There is recognition of the need to coordinate planning efforts at multiple scales (e.g., amongst communities, within the watershed), within Centre County, and within the Upper and Middle Susquehanna Region.

20.2.3.11 Monitoring and Evaluation

There are water resources monitoring programs in place in the Spring Creek watershed. The Spring Creek Community water resources monitoring project consists of 12 monitoring stations installed in 1998 on five streams in the watershed; Spring Creek, Buffalo Run, Logan

Branch, Slab Cabin Run, Thompson Run, and Cedar Run. Flow, temperature, and water quality are monitored at these stations under baseflow conditions and during storm events (http://www.springcreekwatershed.org/index.php?option=content&task=view&id=45&Itemid=62). The Spring Creek Community water resources monitoring project also includes monitoring of groundwater levels at seven wells in the watershed, which was initiated in 2001 (http://www.springcreekwatershed.org/index.php?option=content&task=view&id=48&Itemid=65). PDEP also maintains one fixed water quality monitoring station on Spring Creek near Benner Township as part of the Surface Water Quality Monitoring Network (WQN). Stations in the WQN are sampled bimonthly for flow and chemical analysis, and annually for biological evaluation (PDEP, 2004b). We did not find information on routine groundwater quality monitoring in the watershed. This would be a useful addition to the existing water resources monitoring programs.

Sustainability indicators for the Mid Bald Eagle watershed are presented in the Vision 2020: Living With I-99 Land Use and Sustainability Plan (Community Planning Consultants Inc. et al., 2002) that are applicable to Spring Creek watershed. Indicators of working together, retaining distinct communities, economic diversity, efficient transportation systems, and a healthy environment are identified in the plan.

20.2.4 Larger-Scale Influences

There are several larger-scale programs that include the Spring Creek watershed, thus having the potential to affect management of the Spring Creek watershed, as well as the potential to be affected by management of the watershed. These programs include the Centre County Comprehensive Plan (Centre County Planning Commission, 2003), Vision 2020 Land Use and Sustainability Plan (Community Planning Consultants, Inc. et al., 2002), PDEP Watershed Restoration Action Strategy for Bald Eagle Creek Watershed (PDEP, 2003b), Upper/Middle Susquehanna Regional Water Plan, Pennsylvania State Water Plan, Pennsylvania Greenways Program, and the Chesapeake Bay Program.

Because water in this watershed is supplied solely by precipitation, regional weather patterns influence water resources in the watershed. Moisture from the Gulf of Mexico tracking north along the Appalachians is the source for much of the precipitation in the Spring Creek area (http://garden.garden.org/regional/report?action=description&region=13). Precipitation patterns in this region are affected by ENSO (El Niño/La Niña), the North Atlantic Oscillation, and the Pacific/North American large-scale weather patterns (http://www.cpc.ncep.noaa.gov/data/teledoc/telecontents.html). It is also possible that precipitation patterns in this area are affected by changes in land cover (i.e., increase of urban areas) (http://www.nasa.gov/centers/goddard/news/topstory/2003/1211urban.html, http://www.nasa.gov/vision/earth/environment/Land_Cover_Change.html).

CHAPTER 21.0

CONCLUSIONS AND RECOMMENDATIONS

21.1 Conclusions

- Sustainable water resources are achievable, and represent an outcome of using a total systems approach to monitor, assess, and adaptively manage water resources through time.
- A conceptual framework and guidelines for SWRM have been formulated. The framework and guidelines incorporate the principles, practices, concepts, and theories established or suggested by national and international agencies, organizations, and coalitions focusing on sustainability.
- There are multiple elements within each of the framework sectors (e.g., hydrology, ecology, economics, social/cultural) that are interactive and reinforce other elements within and among sectors. Because water is essential for life, there is almost no management practice or element in any sector that does not directly or indirectly affect water.
- SWRM must simultaneously pursue multiple management practices such as water conservation, best management practices, increased institutional coordination, water markets and incentives, revised water policies, and public education. SWRM is not "*either-or*", but rather "*both-and*".
- Adaptive management – Plan→Do →Assess→Revise→Plan…– is the only feasible approach for achieving sustainable water resources.
- Existing information is sufficient to begin the process of moving toward sustainable water resources. Critical information for selected elements needed to achieve sustainable water resources, however, is not available for every community or watershed.
- WERF subscribers must become involved not only within their local community, but also in watershed, state, and regional efforts.

21.2 Recommendations

- Testing the framework with real community, watershed, and regional information, identifying key processes, developing knowledge-based rules for these processes, and moving toward decision support systems represent next steps in the process.
- Information inventories should be conducted in each community and watershed to determine what information might be needed that is currently not being monitored or measured. In many communities, water use information is a major deficiency.

- Water requirements to sustain ecological systems, goods, and services is unknown for many aquatic communities. One of the first steps might be an inventory or review of studies that have established minimum water quantity and quality requirements for different types of aquatic systems. Quantifying water requirements to provide and sustain ecological services must receive critical attention.
- Mental models and social mindscapes related to water should be developed for different communities in different geographic regions and hydrologic regimes (e.g., Northeast, Southeast, Upper Midwest, Southwest, Pacific Northwest) and the different water models compared and contrasted. Without an understanding of perceived community beliefs about water, it will be difficult to formulate effective management practices that can be implemented and sustained.
- WERF subscribers should select a water-related issue in their community to initiated the process of moving toward SWRM. Documenting other issues within the community should be a part of this process. Success breeds success; select an issue, resolve it, and move on to related issues, documenting the linkages among the ecological, economic, and social sectors.

APPENDIX A

Table A-1. International Water Management Agencies Organizations and Institutions (after Anderson, 2001).

Name/Web address	Purpose
Blue Planet Project www.canadians.org/blueplanet/index2.html	Address impacts of trade and privatization on water and water services. Protect water as a common resource and basic human right.
Global Applied Research Network (GARNET) www.info.lut.ac.uk/departments/cv/wedc/garnet/grntover.html	Share applied research information among researchers working in the water and sanitation sector worldwide.
Global Environment Monitoring System (GEMS/Water) www.cciw.ca/gems/intro.html	United Nations Environment Programme housed at the Canada Centre for Inland Waters (CCIW) is oriented toward understanding freshwater quality issues throughout the world.
Global Energy and Water Cycle Experiment (GEWEX) www.gewex.org/	World Climate Research Programme observe, understand and model the hydrological cycle and energy fluxes in the atmosphere, at land surface and in the upper oceans.
Global Environment Facility (GEF) www.gefweb.org/	Multilateral financing mechanism, created in 1991. GEF was established to address four critical threats to the global environment; biodiversity loss, climate change, degradation of international waters, and ozone depletion.
Global International Waters Assessment (GIWA) www.giwa.net	Produce a comprehensive and integrated global assessment of international waters, describe the ecological status of and the causes of environmental problems in 66 water areas in the world, and focus on the key issues and problems facing aquatic ecosystems in transboundary waters.
Global Water Partnership www.gwpforum.org	"Support countries in the sustainable management of their water resources," translating this vision into action through regional Technical Action Committees.
Hydrology for the Environment, Life and Policy (HELP) www.nwl.ac.uk/ih/help/index.html	Encourages the water policy, water resources management, and scientific communities to work together within a field-oriented context so that science is closely integrated with policy and management needs.
Inter-American Water Resources Network (IWRN) www.iwrn.net	Build and strengthen water resources partnerships throughout the Americas.
International Association for Hydraulic Engineering and Research (IAHR) www.iahr.nl/	Advance applied and basic research on water-related issues and contribute to the optimization of the world's water resources.
International Association of Hydrological Sciences (IAHS) www.cig.ensmp.fr/~iahs/	Hydrology and water resources

Name/Web address	Purpose
International Commission on Irrigation and Drainage (ICID) www.icid.org/	"Dedicated to enhancing the worldwide supply of food and fiber for all people by improving water and land management and the productivity of irrigated and drained lands through appropriate application of irrigation, drainage, and flood management techniques."
International Commission on Large Dams (ICOLD) http://genepi.louis-jean.com/cigb/anglais.html	Forum for the exchange of knowledge and experience in dam engineering
International Hydrological Programme (IHP) www.unesco.org/water/ihp/	Improve the scientific and technological basis for the development of methods for the rational management of water resources.
International Network of Basin Organizations (INBO) www.oieau.fr/riob/friobang.htm	Develop ongoing dialogue with those responsible for global river basin management.
International Private Water Association (IPWA) www.ipwa.org/	Promote opportunities for private sector investment in water systems, wastewater treatment, desalination, irrigation, flood control, and other water-related infrastructure projects worldwide.
International Rivers Network (IRN) www.irn.org/	"To halt and reverse the degradation of river systems; to support local communities in protecting and restoring the well-being of the people, cultures and ecosystems that depend on rivers; to promote sustainable, environmentally sound alternatives to damming and channeling rivers; to foster greater understanding, awareness and respect for rivers; to support the world wide struggle for environmental integrity, social justice and human rights; and to ensure that our work is exemplary of responsible and effective global action on environmental issues."
International Water and Sanitation Centre (IRC) www.irc.nl/	Community-based technologies; participation and community management; gender awareness; hygiene promotion; operation and maintenance; monitoring and evaluation; information management; community water resources management; and resource center development
International Water Association (IWA) www.iwahq.org.uk	Improvement of urban water management worldwide in an environmentally sustainable way
International Water Management Institute (IWMI) www.cgiar.org/iwmi/	Improve water resources and irrigation management through better technologies, policies, institutions, and management
International Water Resources Association (IWRA) www.iwra.siu.edu	Improve and expand the understanding of water resources issues through education, research, and information exchange between countries and across disciplines.
The Pacific Institute www.pacinst.org/	Research and policy analysis in the areas of environment, sustainable development, and international security
Stockholm International Water Institute (SIWI) www.siwi.org	Venue for the awarding of the Stockholm Water Prize, given annually in honor of outstanding achievements in science, engineering, technology, education, or public policy related to protection of the world's water resources
United Nations Commission on Sustainable Development (UNCSD) www.un.org/esa/sustdev/csd.htm	Follow-up on the implementation of Earth Summit agreements, including Agenda 21 – a plan for achieving sustainable development

Name/Web address	Purpose
United Nations Development Programme (UNDP) www.undp.org	"To help countries in their efforts to achieve sustainable human development by assisting them to build their capacity to design and carry out development programs in poverty eradication, employment creation and sustainable livelihoods, the empowerment of women, and the protection and regeneration of the environment, giving first priority to poverty eradication"
UNDP/World Bank Water and Sanitation Program (WSP) www.wsp.org/english/index.html	Alleviate poverty by helping the poor gain sustained access to improved water and sanitation services
United Nations Environment Programme (UNEP) www.unep.org	Freshwater unit promotes integrated management and use of freshwaters, enhances environmental quality and promotes environmentally sustainable socioeconomic development. Water Branch develops policy-relevant assessments of the state of freshwater and marine resources.
United Nations Food and Agriculture Organization (FAO) www.fao.org	Alleviate poverty and hunger by promoting agricultural development, improved nutrition, and the pursuit of food security.
United Nations Educational, Scientific and\Cultural Organization (UNESCO) www.unesco.org	Promoting collaboration among nations through education, science, culture and communication to further universal respect for justice, for the rule of law and for the human rights and fundamental freedoms which are affirmed for the peoples of the world, without distinction of race, sex, language, or religion, U.N. World Water Development Report 2003 International Year of Freshwater
WaterAid www.wateraid.org.uk/index.html	British water industry work through partner organizations to help poor people in developing countries achieve sustainable improvements in their quality of life by improved domestic water supply, sanitation, and associated hygiene practices
Water Associations Worldwide www.wef.org/conferences/affiliations/waw.jhtml	Seeks to advance international water management through the cooperation of their organizations and members.
Water for People (WFP) www.water4people.org/	Helps people in developing countries obtain safe drinking water.
Water Supply and Sanitation Collaborative Council (WSSCC) www.wsscc.org/	Providing a regular way for water and sanitation sector professionals to exchange views and experiences and develop approaches that will lead to faster achievement of the goal of universal coverage
WaterWeb Consortium www.waterweb.org	Promote the sharing of water-related information using Internet and communication technologies
Wetlands International (WI) www.wetlands.agro.nl/	Sustain and restore wetlands, their resources and biodiversity for future generations through research, information exchange and conservation activities

Name/Web address	Purpose
The World Bank www.worldbank.org	Hydropower; water supply and sanitation; water resources management in rural development (including agricultural irrigation and drainage); and environmental dimensions of water resources management (such as freshwater, costal, and marine resource management)
The World Bank Institute: Water Policy Reform Program www.worldbank.org/wbi/topics.html	Helping countries prepare and implement policy reforms leading to SWRM, through capacity building and learning activities
World Commission on Dams www.dams.org/	Review the development effectiveness of large dams and assess alternatives for water resources and energy development; and develop internationally acceptable criteria, guidelines and standards, where appropriate, for the planning, design, appraisal, construction, operation, monitoring, and decommissioning of dams
The World Conservation Union (IUCN) www.uicn.org/	Influence, encourage, and assist societies throughout the world to conserve the integrity and diversity of nature and to ensure that any use of natural resources is equitable and ecologically sustainable
World Health Organization (WHO) www.who.org	Improve global understanding of health problems and lead the efforts to solve them
World Meteorological Organization (WMO) www.wmo.ch	Increase prompt delivery of accurate weather information. Seeks to apply hydrology to meet the needs for sustainable development and use of water and related resources; to mitigate water-related disasters; and to effectively manage the environment at national and international levels.
World Resources Institute (WRI) www.wri.org/	Move human society to live in ways that protect the Earth's environment and its capacity to provide for the needs and aspirations of current and future generations
World Water Council (WWC) www.worldwatercouncil.org	"Devoted to the critical issues of long-term global water policy and to advocating solutions to problems of water resource management"
World Water Forum (WWF) www.worldwaterforum.org	Triennial meeting sponsored by the World Water Council

APPENDIX B

Table B-1. Some of the Links Available at Hydrology Web for Government Sites, Data Publications, and Bibliographies Related to Hydrology.

Resource	Description
U.S. Army Corps of Engineers Institute for Water Resources www.iwr.usace.army.mil/iwr/atlas/Atlasintro.htm	The Institute was formed to provide the Corps with the capability to anticipate changes in water resources conditions and to provide the agency with forward-looking insights and analysis of emerging national water resources issues.
Department of Energy Climate Change Research Division www.science.doe.gov/ober/CCRD_top.html	Supports basic research in chemical, physical, and biological processes of the Earth's atmosphere, land, and oceans and how these processes may be affected by energy production and use, primarily the emission of carbon dioxide from fossil fuel combustion.
U.S. Climate Change Science Program www.climatescience.gov/	A collaborative interagency program, under a cabinet-level organization designed to improve the government wide management of climate science and climate-related technology development. The CCSP incorporates and integrates the U.S. Global Change Research Program (USGCRP) with the Administration's U.S. Climate Change Research Initiative (CCRI).
U.S. Environmental Protection Agency Office of Water www.epa.gov/OW/	Charged with implementing the water-based goals of the EPA Strategic Plan: clean and safe water, and healthy communities and ecosystems.
U.S. Environmental Protection Agency Groundwater and Ecosystem Restoration Division www.epa.gov/ada/kerrlab.html	Conducts EPA-investigator led laboratory and field research to provide the scientific basis to support the development of strategies and technologies to protect and restore ground and surface water quality, and ecosystems impacted by man-made and natural processes.
U.S. Environmental Protection Agency Office of Research and Development Urban Watershed Management Branch www.epa.gov/ednnrmrl/	Focuses on the drainage and treatment systems of the urban environment, the structural integrity of such systems and the effects of discharges to receiving waters.
National Oceanic and Atmospheric Administration www.noaa.gov	Information on precipitation and climate
National Oceanic and Atmospheric Administration National Weather Service www.nws.noaa.gov/	Provides weather, hydrologic, and climate forecasts and warnings for the United States, its territories, adjacent waters and ocean areas, for the protection of life and property and the enhancement of the national economy.
National Oceanic and Atmospheric Administration National Weather Service Hydrology Laboratory www.weather.gov/oh/hrl/	Conducts studies, investigations, and analyses leading to the application of new scientific and computer technologies for hydrologic forecasting and related water resources problems.
U.S. Department of the Interior Bureau of Reclamation Flood Hydrology Group www.usbr.gov/pmts/flood/	Performs technical investigations, designs, reviews, and related work in flood hydrology and hydrometeorology, involved in the hydrological aspects of Reclamation's Dam Safety Program and provides services to other Department of the Interior agencies as well. Standards and criteria are developed and coordinated with other federal agencies concerned with flood hydrology, hydrometeorology, and dam safety.

Resource	Description
U.S. Department of the Interior Bureau of Reclamation River Systems and Meteorology Group www.usbr.gov/pmts/rivers/	Provides expertise in the areas of watershed and river system management, water demand and conservation, water management and accounting, hydrologic modeling, water supply analysis and forecasting, and meteorology.
U.S. Department of the Interior Bureau of Reclamation Water Resources Research Laboratory www.usbr.gov/pmts/hydraulics_lab/	Performs applications of hydraulic modeling, analysis, and field testing expertise to the solution of water resources, hydraulics, and fluid mechanics problems. We solve site-specific problems and pursue applied research that develops new water resource management tools for Reclamation projects.
U.S. Department of the Interior Bureau of Reclamation Water Supply, Use, and Conservation Group www.usbr.gov/pmts/water_use/	Provides technical assistance and conducts comprehensive water supply investigations associated with water resource planning and watershed management and project operation activities.
U.S. Department of Agriculture Hydrology and Remote Sensing Laboratory www.ars.usda.gov/main/site_main.htm?modecode=12-65-06-00	Conducts nationally orientated basic and applied research on water resources and remote sensing concerns related to the production of food and fiber and the conservation of natural resources.
U.S. Geological Survey Water Resources of the United States water.usgs.gov/	Collects data to evaluate the quantity, quality, and use of the nation's water resources; conducts water-resources appraisals describing the occurrence, availability, and physical, chemical, and biological characteristics of surface and ground water; and conducts research that aids in alleviating water resources problems and provides an understanding of hydrologic systems sufficient to predict their response to natural or human-caused stress.
U.S. Geological Survey Water Use in the United States water.usgs.gov/watuse/	Estimates of national water use
U.S. Geological Survey Water Resources Local Offices water.usgs.gov/local_offices.html	Local offices collect data and conduct studies in their area. Best sources for local information.
Hydrological Processes www3.interscience.wiley.com/cgi-bin/jhome/4125?CRETRY=1&SRETRY=0	International journal devoted to the publication of scientific and technical papers in hydrology. The editors welcome submission of research papers on the physical, chemical and mathematical aspects of hydrological processes.
Journal of Environmental Hydrology www.hydroweb.com/journal-hydrology.html	A journal covering all areas of the environment and fresh water resources, including the fields of hydrology, groundwater, environmental hydrology, urban hydrology, groundwater pollution, groundwater contamination, and groundwater remediation.
Stormwater: The Journal for Surface Water Quality Professionals www.stormh2o.com/sw.html	Publishes information on NPDES Phase II compliance, TMDLs, and other issues related to surface water quality.
Hydrological Processes www.interscience.wiley.com/jpages/0885-6087/	International journal devoted to the publication of scientific and technical papers in hydrology. The editors welcome submission of research papers on the physical, chemical, and mathematical aspects of hydrological processes.
Journal of Environmental Hydrology www.hydroweb.com/journal-hydrology.html	A journal covering all areas of the environment and fresh water resources, including the fields of hydrology, groundwater, environmental hydrology, urban hydrology, groundwater pollution, groundwater contamination, and groundwater remediation.
Texas Water Resources Institute http://twri.tamu.edu/twripubs/	Publishes four online newsletters on water-resources-related topics.

Resource	Description
Journal of Water Science www.rse.uquebec.ca/index_english.html	A multidisciplinary scientific journal publishing recent research in hydrology, hydrogeology, water resources management, groundwater and surface water quality, taxonomy, distribution and ecology of aquatic organisms, aquatic toxicology, ecotoxicology, potable water quality and treatment, wastewater treatment, and socioeconomic and legal aspects of water resources management.
Water Resources Publications LLC www.wrpllc.com/	Publisher and distributor of water resources books, related publications, and computer software
National Weather Service Glossary of Hydrologic Related Terms www.crh.noaa.gov/hsd/hydefa-c.html	Online glossary of hydrology related terms
United Nations Educational Scientific and Cultural Organization www.unesco.org/water/	International water issues, primarily safe water supply
International Office for Water www.oieau.fr/anglais/index.htm	Purpose is to gather public and private partners involved in water resources management and protection in France, Europe and in the world to set up a partners network.
Water Resources Abstracts www.uwin.siu.edu/databases/wrsic/index.html	Collection of international water research compiled by the Water Resources Scientific Information Center of the USGS
Groundwater Atlas of the United States http://capp.water.usgs.gov/gwa/	USGS publication describing the location, extent, and geologic and hydrologic characteristics of important U.S. aquifers.

APPENDIX C

Table C-1. Summary of the Major Hydrologic Effects of Land Use Change. (Reprinted with permission from Calder, 1992. Copyright 1992, McGraw-Hill, New York.)

Land-use change	Component affected	Principal hydrologic processes involved	Geographic scale and likely magnitude of effect
Afforestation (deforestation has converse effect except where disturbance caused by forest clearance may be of overriding importance)	Annual flow	Increased interception in wet periods Increased transpiration in dry periods through increased water availability to deep root systems	Basin scale; magnitude proportional to forest cover, world average is 34 mm/yr^{-1} reduction for 10% increase in forest cover
	Seasonal flow	Increased interception and increased dry period transpiration will increase soil moisture deficits and reduce dry season flow	Basin scale; can be of sufficient magnitude to stop dry season flows
		Drainage activities associated with planting may increase dry season flows through initial dewatering and also through long-term effects of the drainage system	Basin scale; drainage activities will increase dry season flows
		Cloud water (mist or fog) deposition will augment dry season flows	High-altitude basins only; increased cloud water deposition may have a significant effect on dry season flows
	Floods	Interception reduces floods by removing a proportion of the storm rainfall and by allowing buildup of soil moisture storage	Basin scale; effect is generally small but greatest for small storm events
		Management activities; cultivation; drainage; road construction, all increase floods	Basin scale; increased floods for all sizes of storm events
	Water quality	Leaching of nutrients is less from forests through reduced surface runoff and reduced fertilizer application	Basin scale; variable but leaching can be an order of magnitude less than from agricultural land
		Deposition of most atmospheric pollutants is higher to forests because of reduced aerodynamic resistance	Basin scale; leads to acidification of catchments and runoff
	Erosion	High infiltration rates in natural, mixed forests reduce surface runoff and	Basin scale; reduces erosion

Land-use change	Component affected	Principal hydrologic processes involved	Geographic scale and likely magnitude of effect
		erosion	
		Slope stability is enhanced by reduced soil pore water pressure and binding of forest roots	Basin scale; reduces erosion
		Windthrow of trees and weight of tree crop reduce slope stability	Basin scale; increases erosion
		Soil erosion, through splash detachment, is increased from forests without an understory of shrubs or grass	Basin scale; increases erosion
		Management activities; cultivation, drainage, road construction, felling, all increase erosion	Basin scale; management activities are often more important than the direct effect of the forest
	Climate	Increased evaporation and reduced sensible heat fluxes from forests affect climate	Micro, meso, and global scale, forests generally cool and humidify the atmosphere; a 2°C increase in regional temperature is predicted for Amazonia if deforestation continues
Agricultural intensification	Water quantity	Alteration of transpiration rates affects runoff	Basin scale; effect is marginal
		Timing of storm runoff altered of storm runoff altered through land drainage	Basin scale; significant effect
	Water quality; fertilizers	Application of inorganic fertilizers	Basin scale; increased nutrient concentration in surface and groundwaters
	Pesticides	Application of nonselective and persistent pesticides poses health risks to humans and animal life	Basin, regional, and global scale; effects can be long-lasting
	Farm wastes	Inadequate disposal of farm organic water pollutes surface and groundwater bodies	Basin scale; effects are very site dependent
	Erosion	Cultivation without proper soil conservation measures and uncontrolled grazing on steep slopes increase erosion	Basin scale; effects are very site dependent
Draining wetlands	Seasonal flow	Upland peat bogs, groundwater fens, and African dambos have little effect in maintaining dry season flows	Basin scale; drainage or removal of wetland will not reduce, and may increase dry season flows
		Lowering of the water table may induce soil moisture stress, reduce transpiration,	Basin scale; a reduction of water-table depth to minimum of 30 cm below surface is

Land-use change	Component affected	Principal hydrologic processes involved	Geographic scale and likely magnitude of effect
		and increase dry season flows	required
		Initial dewatering following drainage will increase dry season flows	Basin scale; effect may be from 1 to 2 years to decade
		The deeper flow outlet of the drainage system will lead to increased dry season flows	Basin scale; effects will be long term
	Annual flow	Initial dewatering following drainage will increase annual flow	Basin scale; effect may last from 1 to 2 years to decade
		Afforestation following drainage will reduce annual flow	Basin scale; effects in the afforestation
	Floods	Drainage method; soil type, channel improvement, all affect flood response	Basin scale; open drain increase, closed drain reduce, flood peak
	Water quality	Redox potentials are altered leading to decomposition of peat, acidification, and increased organic loads in runoff	Basin scale; increased acidity
		Drainage systems intercepting mineral horizons will reduce acidity	Basin scale; reduce acidity
	Carbon balance	Accumulating peat bogs are a sink for atmospheric CO_2	Global scale effect; disturbance and drainage of bogs in such that the carbon balance of the world's peat bogs in neutral at present

APPENDIX D

Table D-1. Statutory and Programmatic Drivers Affecting Several Critical Ecosystems Types and Processes (from D. Norton, personal communication).

Ecosystem Type/Process	Statutory/Programmatic Drivers
Critical Ecosystems in General	♦ National Environmental Policy Act (NEPA) ♦ Clean Water Act Sections 101, 303d, 319, 401, 404 and others ♦ Endangered Species Act ♦ Safe Drinking Water Act ♦ Clean Air Act ♦ State heritage programs ♦ Smart Growth programs, open space programs ♦ Numerous federal/state/local laws protecting sensitive areas
Wetlands	♦ Clean Water Act (Section 404, regulates discharge of dredged and fill material into waters of the United States, including wetlands; Section 402, National Pollutant Discharge Elimination System; Section 401, state certification of Water Quality) ♦ Rivers & Harbors Appropriations Act of 1899 (Section 10-regulates activities affecting navigation in U.S. waters, including wetlands) ♦ National Environmental Policy Act of 1969 (NEPA) ♦ Federal Agriculture Improvement and Reform Act of 1996 (Farm Bill) ♦ Endangered Species Act ♦ Transportation Equity Act for the 21st Century (TEA-21) ♦ Coastal Wetland Planning, Protection & Restoration Act (CWPPRA) ♦ North America Wetlands Conservation Act (NAWCA) ♦ Government Performance Results Act of 1993 Goal: +100,000 acres of wetlands annually by 2005 (U.S. EPA) ♦ Wetlands and Floodplains Executive Orders ♦ Recent court cases (Solid Waste Agency of North Cook County, Illinois, Tulloch, others) ♦ State wetlands legislation
Stream Corridors/ Riparian Zones/ Floodplains	♦ Clean Water Act ♦ Safe Drinking Water Act ♦ National Flood Insurance Act of 1968 ♦ Flood Disaster Protection Act of 1973 ♦ Executive Order 11988, "Floodplain Management" ♦ Executive Order 11990, "Protection of Wetlands" ♦ Threatened & Endangered species listings ♦ Federal Energy Regulatory Commission relicensing ♦ State best management practices ♦ Forest practice acts ♦ Numerous buffer/restoration laws, initiatives, and incentives ♦ Programs and policies affecting hyporheic zones

Ecosystem Type/Process	Statutory/Programmatic Drivers
Headwaters and Springs	♦ Clean Water Act ♦ "Mountaintop mining" policy development ♦ See also all stream/riparian floodplain drivers
Estuaries	♦ National Estuary Program Comprehensive Conservation & Management Plans ♦ Estuary Restoration Act of 2000 ♦ Estuaries and Clean Waters Act of 2000 ♦ Clean Water Act Sections 301h, 303, 312, 403, 404 ♦ Invasive Species Executive Order, Management Plan, Aquatic Nuisance Species Task Force ♦ National Coastal Condition Report, Coastal Research and Monitoring Strategy
Coral Reefs	♦ Florida Keys National Marine Sanctuary Water Quality Protection Program ♦ Executive Order 13089 (Coral Reef Protection) ♦ Executive Order 13158 (Marine Protected Areas) ♦ International Coral Reef Initiative ♦ Clean Water Act Sections 403, 303, 301h, and 312 ♦ Proposed revisions to CWA Section 403 (especially Special Ocean Site establishment provisions) ♦ U.S. Coral Reef Task Force ♦ Coastal Zone Management Act of 1972 ♦ Coral Reef Conservation Act of 2000
Submerged Aquatic Vegetation Beds	♦ National Estuary Program ♦ Clean Water Act ♦ Invasive Species Executive Order ♦ Invasive Species Management Plan ♦ See also estuaries drivers
Shellfish Beds	♦ See also coral reefs drivers ♦ U.S. EPA geographic initiatives (Chesapeake Bay) ♦ Harvest regulations ♦ National Estuary Program ♦ Clean Water Act Sections 301h, 303, 312, 403 ♦ Invasive Species Executive Order ♦ Invasive Species Management Plan ♦ Clean Vessel Act
Unique Pelagic Habitats and Special Ocean Sites	♦ Clean Water Act Sections 301h and 403 ♦ Proposed revisions to CWA Section 403 (especially Special Ocean Site establishment provisions) ♦ Clean Water Act Section 312(n) – the Uniform National Discharge Standards (UNDS) for vessels of the Armed Forces ♦ Oceans Act of 2000 ♦ Marine Protection, Research and Sanctuaries Act (MPRSA) ♦ Fisheries Conservation and Management Act ♦ Convention on the Prevention of Marine Pollution by Dumping of Wastes and Other Matter (London Dumping Convention)
Aquifer Recharge Areas	♦ Safe Drinking Water Act Sections below: - Section 1453 - source water assessments that assess the threats to Public Water Systems from all sources of contamination (whether groundwaters or surface waters) - Section 1428 - wellhead protection programs that assess and protect Public Water Systems from all sources of contamination

Ecosystem Type/Process	Statutory/Programmatic Drivers
	(only groundwaters, but not entire recharge areas) - Sections 1422 and 142 - protecting portions of recharge areas that are contiguous to underground injection wells and protecting the groundwaters not only from contaminants from those wells, but across a wide spectrum of contamination sources and contaminants - Section 1424 (e) - protecting officially designated "sole source aquifers" [i.e., certain entire aquifers (about 71 nationwide) are protected from contamination from "Federally-assisted projects."] - Section 1452 - Drinking Water State Revolving Fund has set-asides of funding for various source water protection efforts from buying easements to implementing best management practices for sources of contamination for protecting groundwaters and surface waters Note: These statutes also assess and protect wetlands, stream corridors, floodplains, estuaries, headwaters and springs, and hyporheic zones, in addition to recharge areas. ♦ National Source Water Contamination Prevention Strategy is in draft–for groundwaters and surface waters that are sources of drinking water
Threatened & Endangered Species Habitats/Other Unusual Habitats	♦ Endangered Species Act ♦ Clean Water Act Section 403 ♦ Proposed revisions to CWA Section 403 (especially Special Ocean Site establishment provisions) ♦ OW's memorandum of understanding on consulting, with U.S Fish & Wildlife Service, National Marine Fisheries Service, U.S. EPA/Office of Water ♦ State Heritage programs ♦ Regional ecosystem programs ♦ Migratory bird international treaties

REFERENCES

Abler, D.G.; Shortle, J.S. Environmental and Farm Commodity Policy Linkages in the U.S. and E.C. *Eur. Rev. Agric. Econ.* **1992**, *19*, 197–217.

Adams, R.M.; Crocker, T.D. Materials Damages. In *Measuring the Demand for Environmental Quality*, Braden, J.B.; Kolstad, C.D., Eds.; North-Holland: Amsterdam, 1991.

Arizona Department of Environmental Quality. *Second Draft – Status of Water Quality in Arizona, 2004.* Arizona Department of Environmental Quality, 2004. (Available at http://www.adeq.state.az.us/environ/water/assessment/2004.html.)

Arizona Department of Water Resources. *Preliminary Hydrographic Survey Report for the San Pedro Watershed, Vol.1: General Assessment.* Arizona Department of Water Resources: Phoenix, Ariz., 1990.

Arizona Department of Water Resources. *Hydrographic Survey Report for the San Pedro River Watershed, Volume 1, General Assessment.* Arizona Department of Water Resources: Phoenix, Ariz., 1991.

Agarwal, A.; delos Angeles, M.S.; Bhatis, R.; Cheret, I.; Davila-Poblete, S.; Falkenmacle, M.; Vallarreal, F.G.; Jonch-Clausen, T.; Kodi, M.A.; Kindler, J.; Rees, J.; Roberts, P.; Rogers, P; Solanes, M.; Wright, A. *Technical Advisory Committee Background Paper No. 4: Integrated Water Resources Management*; Global Water Partnership: Stockholm, Sweden, 2000.

Air University. *SpaceCast 2020*, Vol. 1. Air University: Maxwell Air Force Base, Ala., 1994.

Alderfer, C. An Empirical Test of a New Theory of Human Needs. *Org. Behav. Hum. Perform.* **1969**, *4*, 142–175.

Alderfer, C. *Existence, Relatedness, & Growth.* Free Press: New York, 1972.

Alliance for the Chesapeake Bay. *Spring Creek Fact Sheet 2004.* Alliance for the Chesapeake Bay: Camp Hill, Pa., 2004. (Available at www.acb-online.org/pubs.cfm.)

American Society of Civil Engineers Task Committee on Sustainability Criteria. *Sustainability Criteria for Water Resource Systems.* American Society of Civil Engineers: Reston, Va., 1998.

Anderson, F. Understanding the International Water Management Arena: A Newcomer's Guide to the Major Players. *Water Resour. Impact.* **2001**, *3*(4), 32–38.

Anderson, J. Current Water Recycling Initiatives in Australia: Scenarios for the 21st Century. *Water Sci. Technol.* **1996**, *33*(10-11), 37–43.

Anderson, R.C.; Lohof, A.Q.; Carlin, A. *The United States' Experience with Economic Incentives in Environmental Pollution Control Policy.* U.S. Environmental Protection Agency: Washington, D.C., 1997.

Anderson, R.C.; Rusin, M.; Hoffman, L. *The Use of Economic Incentive Mechanisms in Environmental Management,* Research Paper No. 051; American Petroleum Institute: Denver, Colo., 1990.

Anderson, T.R. *Water Crisis: Ending the Policy Drought*. Johns Hopkins University Press: Baltimore, 1982.

Andrade, A.J. Reclaimed Water Public Education and Awareness. Paper presented at American Water Works Association (AWWA) Conserv 96 Conference, Orlando, Fla., January 4–8, 1996.

Andreasen, A.R. *Marketing Social Change: Changing Behavior to Promote Health, Social Development, and the Environment*. Jossey-Bass: San Francisco, Calif., 1995.

Angelakis, A.N.; Marecos Do Monte, M.H.F.; Bontoux, L.; Asano, T. The Status of Wastewater Reuse Practice in the Mediterranean Basin: Need for Guidelines. *Water Res.* **1999**, *33*(10), 2201–2217.

Antich, A.; Gobas, H.R.; Salganokar, J. The Santa Monica Urban Runoff Recycling Facility and the Sustainable Environment. *Stormwater 5*(2). (Available at www.forester.net/sw_0207_santa.html.)

Antich, T.; Borboa, G.; Shapiro, N. *Urban Runoff Reuse: A Paradigm Shift.* Government West: Nov./Dec. 2001. (Available at www.govwest.com/urbanrunoff_wat.asp.)

Asano, T. Proposed California Regulations for Groundwater Recharge with Reclaimed Municipal Wastewater. *Water Sci. Technol.* **1993**, *27*(7–8), 157–164.

Asano, T.; Levine, A. Wastewater Reclamation, Recycling, and Reuse: An Introduction. In *Wastewater Reclamation and Reuse*, Asano, T., Ed.; Technomic Publishing Co., Inc.: Lancaster, Pa., 1998.

Bailey, R.G. Delineation of Ecosystem Regions. *Environ. Manag.* **1976**, *7*, 365–373.

Bailey, R.G. *Description of Ecoregions of the United States*, 2nd ed.; Misc. Publ. No. 1391 (rev); USDA Forest Service: Washington, D.C., 1995; p 108.

Bailey, R.G. *Ecoregion-Based Design for Sustainability*. Springer-Verlag: New York, 2002.

Baird, B.L.; Waller, P.L. The Tampa Bay Water Resource Recovery Project – Addressing Implementation, The Next Step. *Proceedings of the AWWA Annual Conference and Exposition, Atlanta, Ga., June 15–19, 1997*; American Water Works Association: Alexandria, Va., 1997; p 43.

Barbour, M.T.; Norton, S.B.; Preston, H.R.; Thornton, K.W., Eds. *Ecological Assessment of Aquatic Resources, Linking Science to Decision-Making.* Society of Environmental Toxicology and Chemistry: Pensacola, Fla., 2004.

Barlow, P.M.; Hirsch, R.M.; Alley, W.M.; Dettinger, M.D.; Gain, W.S.; Hamilton, P.A.; Hutson, S.S.; Kirby, W.H.; Konikow, L.F.; Leake, S.A.; Lins, H.F.; Maupin, M.A.; Schefter, J.E.; Schwarz, G.E.; Slack, J.R.; Wiche, G.J. *Concepts for National Assessment of Water Availability and Use;* U.S. Geological Survey Circular 1223; U.S. Geological Survey: Reston, Va., 2002.

Bastian, M.B.; Cash, K.; Dorn, P.; Smith, K.; van Derveer, W.D.; Wrona, F.J. Engaging the Local Community in the Stakeholder Network. In *Ecological Assessment of Aquatic Resources, Linking Science to Decision-Making;* Barbour, M.T.; Norton, S.B.; Preston, H.R.; Thornton, K.W., Eds.; Society of Environmental Toxicology and Chemistry: Pensacola, Fla., 2004; pp 39–54.

Bazerman, M.H.; Curhan, J.R.; Moore, D.A.; Valley, K.L. Negotiation. *Annu. Rev. Psychol.* **2000**, *51*, 279–314.

Beattie, B.R. Irrigated Agriculture on the Great Plains: Problems and Policy Alternatives. *Western J. Agr. Econ.* **1981**, *6,* 289–300.

Beavis, M.; Walker, M. Achieving Environmental Standards with Stochastic Discharges. *J. Environ. Econ. Manag.* **1983**, 10, 103–111.

Benton, G.S.; Blackburn, R.T.; Snead, V.O. The Role of the Atmosphere in the Hydrologic Cycle. *EOS Trans. Am. Geophys. Union* **1950**, *31*, 61–73.

Berkes, F.; Folke, C., Eds. *Linking Social and Ecological Systems. Management Practices and Social Mechanisms for Building Resilience*. Cambridge University Press: Cambridge, U.K., 1998.

Bernados, B. The Importance of Reliability in Potable Reuse. Paper presented at Joint AWWA/WEF Water Reuse Conference, San Diego, Calif., February 25–28, 1996.

Blancas, C.; Castillo, A.; Picazo, J.; Lopez, J. Guidelines for the Safe Reuse of Municipal Wastewater in Andalusia (Spain). In *Proceedings of the Beneficial Reuse of Water and Biosolids Speciality Conference,* Marabella, Malaga, Spain, April 6–9, 1997; Water Environment Federation: Alexandria, Va., 1997.

Bloomquist, G.C.; Berger, M.C.; Hoehn, J.P. New Estimates of Quality of Life in Urban Areas. *Am. Econ. Rev.* **1988**, *78*(1), 89–107.

Bobertz, B. *Public Participation in Environmental Regulation*. Environmental Law Institute: Washington, D.C., 1981.

Booth, D. Urbanization and the Natural Drainage System—Impacts, Solutions and Prognoses. *Northeast Environ. J.* **1991**, *7*(1), 93–118.

Bord, R.J., A. Fisher, A.; O'Connor, R.E.; Wheeler, W.J. *Freshwater Quality, Quantity, and Availability: An American Policy Perception.* Environmental Resources Research Institute, The Pennsylvania State University: University Park, Pa., 1993.

Borisova, T.; Shortle, J.; Horan, R.; Abler, D. The Value of Information for Water Quality Management. *Water Resour. Res.* **2005**, *41*(6).

Botkin, D.B. *Discordant Harmonies: A New Ecology for the Twenty-First Century*. Oxford University Press: New York, 1990.

Bovee, K.D.; Lamb, B.L.; Bartholow, J.M.; Stalnaker, C.B.; Taylor, J.G.; Henriksen, J. *Stream Habitat Analysis Using the Instream Flow Incremental Methodology*; Biological Resources Division Information and Technology Report; USGS/BRD-1998-004; U.S. Geological Survey: Fort Collins, Colo., 1998.

Bowles, S. Endogenous Preferences: The Cultural Consequences of Markets and Other Economic Institutions. *J. Econ. Lit.* **1998**, *36*, 75–111.

Bowlin, W.F. Measuring Performance: An Introduction to Data Envelopment Analysis (DEA). *J. Cost Anal*. **1998**, *Fall*, 3–27.

Braden, J.B., R. Larson, E. Herricks. Impact Targets vs. Discharge Standards in Agricultural Pollution Management. *Am. J. Agr. Econ.* **1991**, *73*, 388–397.

Braden, J.B.; Segerson, K. Information Problems in the Design of Nonpoint Pollution. In *Theory, Modeling and Experience in the Management of Nonpoint-Source Pollution,* Russell, C.S.; Shogren, J.F., Eds.; Kluwer Academic Publishers: Dordrecht, The Netherlands, 1993.

Broeker, W.S. The Great Ocean Conveyor. *Oceanography* **1991**, *4,* 79–84.

Brookshire, D.; Burness, S.; Chermak, J.; Krause, K. *Urban Residential Water Demand.* Working Paper. Department of Economics, University of New Mexico: Albuquerque, 2001.

Brown, D.P.; Comrie, A.C. Sub-regional Precipitation Linkages to SOI and PDO in the Southwest United States. *Atmos. Sci. Lett.* 2002, 3, 94–102.

Brown, T.C. Projecting U.S. Freshwater Withdrawals. *Water Resour. Res.* **2000**, *36*, 769–780.

Brubaker, E. *The Promise of Privatization.* Energy Probe Research Foundation: Toronto, Ontario, 2001. (Available at http://www.environmentprobe.org/enviroprobe/pubs/ev548.htm.)

Brubaker, K.L.; Entekhabi, D.; Eagleson, P.S. Estimation of Continental Precipitation Recycling. *J. Climatol.* **1993**, *6*, 1077–1089.

Brundtland, G.H. *Population, Environment and Development*, Rafael M. Salas Memorial Lecture, United Nations Population Fund, New York, September 28, 1993.

Bureau of Land Management. *Miracle of a Desert River*. Bureau of Land Management: Boise, Idaho, 2000. [Multimedia CD-ROM].

Byström, O.; H. Andersson, H.; Gren, I.-M. Economic Criteria for Restoration of Wetlands under Uncertainty. *Ecol. Econ.* **2000**, *35*, 35–45.

Calder, L.R. Hydrological Effects of Land-use Change. In *Handbook of Hydrology;* Maidment, D.R., Ed.; McGraw-Hill: New York, 1992; p 50.

California Biodiversity News. Wildlife "Corridors" Save Species from "Islands" of Habitat. *California Biodiversity News* **1997**, *5*(1) (Available at http://ceres.ca.gov/biodiv/newsletter/v5n1/wildlife_corridors.html.)

Carpenter, S.R.; Walker, B.; Anderies, J. M.; Abel, N. From Metaphor to Measurement: Resilience of What to What? *Ecosystems* **2001**, *4*, 765–781.

Carpenter, S.R. Alternate States of Ecosystems: Evidence and Some Implications. In *Ecology: Achievement and Challenge*, M.C. Press, M.C.; Huntly, N.J.; Levin , S., Eds.; Blackwell Science: Oxford, U.K., 2001; pp 357–383.

Carpenter, S.R.; Ludwig, D.; Brock, W. Management of Lakes Subject to Potentially Irreversible Change. *Ecol. Applic.* **1999**, *9*(3), 751–771.

Carter, R.H.; Morehouse, B.J. *An Examination of Arizona Water Law and Policy from the Perspective of Climate Impacts;* CLIMAS Report Series CL2-01; Institute for the Study of Planet Earth, University of Arizona: Tucson, 2001. (Available at http://www.ispe.arizona.edu/climas/pubs/CL2-01.html.)

Case, P.; Alward, G. *Patterns of Demographic, Economic and Value Change in the Western U.S., Implications for Water Use and Management*; Report to the Western Water Policy Review Advisory Commission. U.S. Department of Agriculture Forest Service: Washington, D.C., 1997.

Cavallina, J.M.; Young, L.E. Integrated Systems for the Treatment and Recycling of Wastewater in Latin America: Reality and Potential. *Urban Agric.* **2002**, *8*, 18–19.

Cayan, D.R.; Redmond, K.T.; Riddle, L.G. ENSO and Hydrologic Extremes in the Western U.S. *J. Climatol.* **1999**, *12*, 2881–2893.

Centre County Planning Commission. *Centre County Comprehensive Plan, Phase I.* Centre County Planning Commission: Bellefonte, Pa., 2003.

Chahine, M.T. The Hydrologic Cycle and its Influence on Climate. *Nature (London)* 1992, 359, 373–380.

Charnes, A.; Cooper, W. ; Lewin, A.Y.; Seiford, L. M. *Data Envelopment Analysis: Theory, Methodology and Applications.* Kluwer Academic Press: Boston, Mass., 1994.

Chase, T.N.; Pielke, R.A.; Kittel, T.G.F.; Nemani, R.R.; Running, S.W. Simulated Impacts of Historical Land Cover Changes on Global Climate in Northern Winter. *Clim. Dynam.* **2000**, *16*, 93–105.

Chorley, R.J. The Hillslope Hydrological Cycle. In *Hillslope Hydrology*, Kirkby, M.J., Ed.; Wiley & Sons: Chichester, U.K., 1978; pp. 1–42.

Christensen, N.L.; Bartuska, A.M.; Brown, J.H.; Carpenter, S.; D'Antonio, C.; Francis, R.; Franklin, J.F.; MacMahon, J.A.; Noss, R.E.; Parsons, D.J.; Peterson, C.H.; Turner, M.G.; Woodmansee, R.G. The Report of the Ecological Society of America Committee on the Scientific Basis for Ecosystem Management. *Ecol. Appl.* **1996**, *6*, 665–691.

Cialdini, R.B. Influence: *The Psychology of Persuasion.* Quill: New York, 1998.

Cifuentes, E.; Siebe, C. Environmental Impact of Wastewater Irrigation in Central Mexico. In *Proceedings of the Beneficial Reuse of Water and Biosolids Speciality Conference,* Marabella, Malaga, Spain, April 6–9, 1997; Water Environment Federation: Alexandria, Va., 1997.

Ciracy-Wantryp, S.V. Water Economic: Relations to Law and Policy. In *Water and Water Rights,* Clark, R.E., Ed.; Allen Smith Co.: Indianapolis, Ind., 1967.

Clear Water Conservancy. *The Spring Creek Watershed Plan Phase I Final Report, Our Challenges and Direction for the Future*. Clear Water Conservancy: State College, Pa., 2004. (Available at http://www.clearwaterconservancy.org/wsplan%20files/Phase%201%20Final%20Repor.pdf.)

Clinton, W. Executive Order 12866 – Regulatory Planning and Review. September 30, 1993. *Fed. Regist.* 1993, 58, 51735.

Coburn, J. Cleaning Up Urban Stormwater: The Storm Drain Stenciling Approach (or Getting to the Nonpoint Source). *J. Soil Water Conserv.* **1994**, *49*, 312–316.

Cochise College Center for Economic Research. *Sierra Vista Economic Focus 2003–2004.* Cochise College Center for Economic Research: Sierra Vista, Ariz., 2004.

Colby, B.G. Regulation, Imperfect Markets and Transaction Costs: The Elusive Quest for Efficiency in Water Allocation. In *Handbook of Environmental Economics*, Bromley, D., Ed.; Blackwell Publishers: Oxford, U.K., 1995; pp 475–502.

Coleman, W.T. Legal Barriers to the Restoration of Aquatic Systems and the Utilization of Adaptive Management. *Vermont Law Rev.* **1998**, *23*, 177.

Community Planning Consultants Inc., KELLERCO Inc., Land Studies Inc., LD Harper and Associates, Thomas Point Associates Inc., CET Engineering Services Inc. *Vision 2020: Living with I-99 Land Use and Sustainability Plan.* Clear Water Conservancy: State College, Pa., 2002. (Available at http://www.clearwaterconservancy.org/ Vision%202020%20Web%20Final%20Plan/ Table_of_Contents_for_web.htm)

Costa, A.L.; Kallick, B., Eds. *Assessment in the Learning Organization: Shifting the Paradigm.* Association for Supervision and Curriculum Development: Alexandria, Va., 1995.

Craik, K. *The Nature of Explanation.* Cambridge University Press: Cambridge, U.K., 1943.

Crook, J.; Surampali, R.Y. 1996. Water Reclamation and Reuse Criteria in the U.S. *Water Sci. Technol.* **1996**, *33*(10-11), 451–462.

Crook, J., Mosher, J.J.; Casteline, J.M. Status and Role of Water Reuse: An International View. Global Water Research Coalition: London, U.K., August 2005,

Cropper, M.L.; Evans, W.N.; Beradi, J.J.; Ducclas Soares, M.M.; Portney, P.R. The Determinants of Pesticide Regulations: Statistical Analysis of EPA Decision-making. *J. Polit. Econ.* **1992**, *100*, 175–197.

Curran, T.M.; Kiss, S.K. Operational Experiences with the World's Largest Residential Dual Water System. *Proceedings Water Environment Federation 68th Annual Conference*, Miami Beach, Fla., 1995; pp 6,81.

D'Angelo S., Chair. *Using Reclaimed Water to Augment Potable Water Resources*, Public Information Outreach Programs (Special Publications). Water Environment Federation and American Water Works Association: Alexandria, Va., 1998.

Daily, G.C., Ed. *Nature's Services, Societal Dependence on Natural Ecosystems.* Island Press: Washington, D.C., 1997.

Daly, H.E; Farley, J. *Ecological Economics: Principles and Applications.* Island Press: Washington, D.C., 2004.

Dator, J. 1993. From Future Workshops to Envisioning Alternative Futures. *Future Res. Q.* 1993, Fall.

Davies, J.; Mazurek, J. *Pollution Control in the U.S.: Evaluating the System.* Resources for the Future Inc.: Washington, D.C., 1998.

Del Piero, M. Water Reuse in South America: A Chilean Study. *Proceedings of the National Conference Environmental Engineering,* American Society of Civil Engineers: Chicago, 1998.

Deller, S.C.; Hai, S.T.; Marcouiller, D.; English, D.B.K. The Role of Amenities and Quality of Life in Rural Economic Growth. *Am. J. Agr Econ.* **2001**, *83*(2):352–365.

Dendauw, I. The Great Lakes Region and Bulk Water Exports: Issues in International Trade in Water. *Water Intl.* **2000**, *25*(4), 565–571.

Dickinson, R.E.; Kennedy, P. Impacts on Regional Climate of Amazon Deforestation. *Geophys. Res. Lett.* **1992**, *19*(19), 1947–1950.

Dinar, A.; Knapp, K.C.; Letey, J. Irrigation Water Pricing to Reduce and Finance Subsurface Drainage Disposal. *Agr. Water Manage.* **1989**, *16*, 155–172.

Dinham, B. *The Pesticide Hazard: A Global Health and Environmental Audit.* Zed Books: London, 1993.

Doba, R.; Pellatz, R. The Flagstaff, Arizona, Experience: Year Number One. Paper presented at Joint AWWA/WEF Water Reuse Conference, San Diego, Calif., February 25–28, 1996.

Drucker, P.F. *1999 Year in Review.* The Peter F. Drucker Foundation for Nonprofit Management: New York, 1999.

Drucker, P.F.; Keeney, R.L.; Hammond, J.S. *Harvard Business Review on Decision Making.* Harvard Business School Press: Cambridge, Mass., 2001.

Dynesius, M.; Nilsson, C. Fragmentation and Flow Regulation of River Systems in the Northern Third of the World. *Science* **1994**, *266,* 753–762.

Eagleson, P.S. The Emergence of Global-scale Hydrology. *Water Resour. Res.* **1986**, *22*, 6s–14s.

Easum, B. 1998. *We Are Our Mental Models.* (Available at www.easumbrandy.com/resources/index.php?acton=details&record=344.)

Eltahir, E.A.B.; Bras, R.L. Precipitation Recycling in the Amazon Basin. *Q. J. Roy. Meteor. Soc.* **1994**, *120*, 861–880.

Eltahir, E.A.B.; Bras, R.L. Precipitation Recycling. *Rev. Geophys.* **1996**, *34*, 367–378.

Enfield, D.B.; Mestas-Nuñez, A.M.; Trimble, P.J. The Atlantic Multidecadal Oscillation and Its Relation to Rainfall and River Flows in the Continental U.S. *Geophys. Res. Lett.* **2001**, *28*(10), 2077–2080.

Eshleman, K.N. Predicting Regional Episodic Acidification of Surface Waters Using Empirical Models. *Water Resour. Res.* **1988**, *24,* 1118–1126.

Estes, J.A.; Duggins, C.O. Sea otters and kelp forests in Alaska: Generality and Variation in a Community Ecological Paradigm. *Ecol. Monogr.* **1995**, *65*, 75–100.

Esty, D.C.; Srebotnjak, T.; Goodall, M.; Andonov, B.; Campbell, K.; Gregg, K.; Kim, C.; Li, Q.; Martinez, M.; Townsend, J.; Zhang, B.; Levy, M.; de Sherbinin, A.; Anderson, B.; Saltelli, A.; Saisana, M.; Nardo, M.; Dahl, A. *2005 Environmental Sustainability Index.* Yale Center for Environmental Law and Policy: New Haven, Conn., 2005.

European Commission. *Towards Sustainable and Strategic Management of Water Resources*, European Commission: Brussels, 2000.

Faeth, P. *Fertile Ground: Nutrient Trading's Potential to Cost-Effectively Improve Water Quality.* World Resources Institute: Washington, D.C., 2000.

Falkenmark, M. *Water Management and Ecosystems: Living with Change*, TEC Background Paper No. 9; Global Water Partnership: Stockholm, Sweden, 2003.

Farid M.S.; Atta, S.; Rashid, M.; Munnick, J.O.; Platenburg, R. Impact of the Reuse of Domestic Wastewater for Irrigation on Groundwater Quality. *Water Sci. Technol.* **1993**, *27*(9):147–157.

Feachem, R.G.; Bradley, H.; Garelick, H.; Mara, D.D. *Sanitation and Disease – Health Aspects of Excreta and Wastewater Management.* World Bank, John Wiley & Sons: Chichester, England, 1983.

Fekete, B.M.; Vörösmarty, C.J.; Grabs, W. *Global, Composite Runoff Fields Based on Observed River Discharge and Simulated Water Balances*; WMO – Global Runoff Data Center Report #22; Global Runoff Data Center: Koblenz, Germany, 1999.

Firth, P.; Fisher, S.G., Eds. *Global Climate Change and Freshwater Ecosystems*. Springer-Verlag: New York, 1992.

Fisher, A.; Raucher, R. Intrinsic Benefit of Improved Water Quality: Conceptual and Empirical Perspectives. In *Advances in Applied Micro-Economics,* Smith, V.K., Ed.; JAI Press, Inc.: Greenwich, Conn., 1984.

Fisher, A.N.; L. Appel, L.; Johnson, F.R.; Rogers, C.; Swietlik, W.; Wallace, J.B. Valuing Ecological Resources. In *Ecological Assessment of Aquatic Resources, Linking Science to Decision-Making;* Barbour, M.T.; Norton, S.B.; Preston, H.R.; Thornton, K.W., Eds.; Society of Environmental Toxicology and Chemistry: Pensacola, Fla., 2004; pp 113–139.

Federal Interagency Stream Restoration Working Group. *Stream Corridor Restoration: Principles, Processes, and Practices;* GPO Item No. 0120-A; National Technical Information Service, U.S. Department of Commerce: Springfield, Va., 1998.

Folke, C.; Carpenter, S.; Elmquist, T.; Gunderson, L.; Holling, C.S.; Walker, B.; Bengtsson, J.; Berkes, F; Colding, J.; Danell, K.; Falkenmark, M.; Gordon, L.; Dasperson, R.; Dautsky, N.; Kinzig, A.; Levin, S.; Maler, K.; Moberg, F.; Ohlsson, L.; Olsson, P.; Ostrom, E.; Reid, W.; Rockstrom, J.; Savenije, H.; Svedin, U. *Resilience and Sustainable Development: Building Adaptive Capacity in a World of Transformations*; Environmental Advisory Council of the Swedish Ministry of the Environment: Stockholm, 2002.

Frederick, K.D. *Scarce Water and Institutional Change*. Resources for the Future, Inc.: Washington, D.C., 1985

Freeman, A.M., III. *Air and Water Pollution Control: A Benefit–Cost Assessment*. John Wiley & Sons: New York, 1983.

Freeman, A.M., III. Clean Water Act Reauthorization: How Far Have We Come? *Water Resour. Bull.* **1994**, *30*(5), 793–798.

Freeman, A.M., III. Environmental Policy since Earth Day I: What Do We Know about Costs and Benefits? *Agr. Resour. Econ. Rev.* **2002**, *31*(1), 1–4.

Freethey, G.W. Hydrologic Analysis of the Upper San Pedro Basin from the Mexico-Unites States International Boundary to Fairbanks, Arizona; Open-File Report 82-752; U.S. Geologic Survey: Herndon, Va., 1982.

Frost, T.M.; Montz, P.K.; Kratz, T.K.; Badillo, T.; Brezonik, P.L.; Gonzalez, M.J.; Rada, R.G.; Watras, C.J.; Webster, K.E.; Wiener, J.G.; Williamson, C.E.; Morris, D.P. Multiple Stresses from a Single Agent: Diverse Responses to the Experimental Acidification of Little Rock Lake, Wisconsin. *Limnol. Oceangr*. **1999**, *44*, 78–794.

Fulton, J. Spring Creek Watershed Model. *Springs & Sinks*, September 2003.

Galloway, D.; Jones, D.R.; Ingebritson, S.E. *Land Subsidence in the U.S.;* U.S. Geologic Survey Circular 1182; U.S. Geological Survey: Reston, Va., 1999; p 175.

Galloway, Jr., G.E.; Clamen, M. The International Joint Commission: A Model of Cooperation in Dealing with Boundary Water and Transboundary Environmental Issues. *Water Resour. Impact*. **2001**, *3*(2), 12–15.

Garcia, L.; Rosenblum, E.; Newby, J.; Grant, F.; Lindow, D. Getting the Job Done: A Resource Partnership on the South Bay Water Recycling Program. In *Proceedings of the Beneficial Reuse of Water and Biosolids Speciality Conference,* Marabella, Malaga, Spain, April 6–9, 1997; Water Environment Federation: Alexandria, Va., 1997.

Garnham, A.; Oakhill, J.V. The Mental Models Theory of Language Comprehension. In *Models of Understanding Text;* B.K. Britton, B.K.; Graesser, A.C., Eds.; Lawrence Erlbaum Associates: Hillsdale, N.J., 1995; pp 313–339.

Gelt, J. Managing the Interconnecting Waters: The Groundwater—Surface Water Dilemma. *Arroyo* **1994**, *7*(4).

Georgia Basin Futures Project. *Georgia Basin Futures Project Newsletter*, January 2004. (Available at http://www.basinfutures.net/pdfs/GBFP_Jan_2004_Newlsetter.pdf.)

Geselbracht, J.; Requa, D.; Lopez, D.; Henderson, R. Advanced Water Recycling for Disposal of Peak Wet-Weather flows. *Proceedings Joint AWWA/WEF Water Reuse Conference.* Lake Buena Vista, Fla., February 1–4, 1988; p 349.

Gibbons, D. *The Economic Value of Water*. Resources for the Future, Inc.: Washington, D.C., 1986.

Giddings, M.T., Jr. *Hydrologic Budget of Spring Creek Drainage Basin, Pennsylvania.* Ph.D. Dissertation. The Pennsylvania State University: State College, Pa., 1974.

Giddings, T. *2002 Marked the Beginning of the End of the Invisible Drought.* Online publication. Pennsylvania Groundwater Association: Hummelstown, Pa., 2004. (Available at http://www.pgwa.org/invisible%20drought.htm.)

Gillian, D.M.; Brown, T.C. *Instream Flow Protection: Seeking a Balance in Western Water Use.* Island Press: Washington, D.C., 1997.

Gilliom, R.J.; Alley, W.M.; M.E. Gurtz. *Design of the National Water Quality Assessment Program—Occurrence and Distribution of Water-Quality Conditions*; U.S. Geological Survey Circular 1112; U.S. Geological Survey, Sacramento, Calif., 1995.

Gleick, P.H. Basic Water Requirements for Human Activities: Meeting Basic Needs. *Water Intl.* **1996**, *21*, 83–92

Gleick, P.H. Water in Crisis: Paths to Sustainable Water Use. *Ecol. Appl.* **1998a**, *8*, 571–579.

Gleick, P.H. *The World's Water, 1998–1999*. Island Press: Washington, D.C., 1998b.

Gleick, P.H. *The World's Water, 2000–2001*. Island Press: Washington, D.C., 2000.

Gleick, P.H.; Haasz, D.; Henges-Jeck, C.; Srinivasan, V.; Wolff, G.; Cushing, K.K.; Mann, A. 2003. *Waste Not, Want Not: The Potential for Urban Water Conservation in California.* http://www.pacinst.org/reports/urban_usage/.Pacific Institute for Studies in Development, Environment and Security: Oakland, Calif., 2003.

Gleick, P.H.; Wolff, G.; Chalecki, E.L.; Reyes, R. *The New Economy of Water, The Risks and Benefits of Globalization and Privatization of Fresh Water.* Pacific Institute for Studies in Development, Environment, and Security: Oakland, Calif., 2002a.

Gleick, P.H.; Wolff, G.; Chalecki, E.L.; Reyes, R. The Privatization of Water and Water Systems. In *The World's Water, The Biennial Report of Freshwater Resources, 2002–2003*; Gleick, P.H., et al., Eds.; Island Press: Washington, D.C., 2002b.

Global Environmental Management Initiative. *Connecting the Drops Toward Creative Water Strategies, A Water Sustainability Tool.* Global Environmental Management Initiative: Washington, D.C., 2002.

Global Water Partnership. *Catalyzing Change: A Handbook for Developing Integrated Water Resources Management (IWRM) and Water Efficiency Strategies*. Global Water Partnership: Stockholm, Sweden, 2004.

Goldvarg, Y.; Johnson-Laird, P.N. Illusions in Modal Reasoning. *Memory Cogn.* **2000**, *28*, 282–294.

Goodland, R. The Concept of Environmental Sustainability. *Annu. Rev. Ecol. Syst.* **1995**, *26*, 1–24.

Gordon, A.L. Interocean Exchange of Thermohaline Water. *J. Geophys. Res.* **1986**. *91*, 5037–5046.

Gore, J.A.; Nestler, J.M. Instream Flow Studies in Perspective. *Regul. Riv.* **1988**, *2*, 93–101.

Gosselink, J.G.; Touchet, B.A.; Beek, J.V.; Hamilton, D. Bottomland Hardwood Forest Ecosystem Hydrology and the Influence of Human Activities: The Report of the Hydrology Workgroup. In *Ecological Processes and Cumulative Impacts,* Gosselink, J.G.; Lee, L.C.; Muir, T.A., Eds.; Lewis Publishers: Chelsea, Mich., 1990; pp 347–387.

Graczyk, D.J.; Krug, W.R.; Gebert, W.A. *A History of Annual Streamflows from the 21 Water-resources Regions in the U.S. and Puerto Rico*, U.S. Geological Survey Open File Report 86-128; U.S. Geological Survey, Reston, Va., 1986.

Graf, W.L. Landscapes, Commodities, and Ecosystems: The Relationship between Policy and Science for American Rvers. In *Sustaining Our Water Resources,* Water and Science Technology Board, National Research Council, Eds.; National Academy Press: Washington, D.C., 1992; pp 11–42.

Graf, W.L. Dam Nation: A Geographic Census of American Dams and Their Large-scale Hydrologic Impacts. *Water Resour. Res.* **1999**, *35*, 1305–1311.

Gray, W.M. Strong Association between West African Rainfall and U.S. Landfall of Intense Hurricanes. *Science*. **1990**, *249*, 1251–1256.

Gray, W.M.; Landsea, C.W.; Knaff, J.A.; Mielke Jr., P.W.; Berry, K.J. *Summary of 1999 Atlantic Tropical Cyclone Activity and Verification of Author's Seasonal Activity Prediction*. Department of Atmospheric Science, Colorado State University: Ft. Collins, Colo., 1999; p 32.

Gray, W.M.; Landsea, C.W.; Mielke Jr., P.W.; Berry, K.J. *Early April Forecast of Atlantic Seasonal Hurricane Activity and US Landfall Strike Probabilities for 2000*. Colorado State University: Fort Collins, Colo., 2000. (Available at http://hurricane.atmos.colostate.edu/Forecasts/2000/april2000/.)

Griffin, R.C.; Bromley, D.W. Agricultural Runoff as a Nonpoint Externality: A Theoretical Development. *Am. J. Agr. Econ.* **1982**, *64*, 547–552.

Grobicki, A.; Cohen, B. *Water Reclamation for Direct Re-use in Urban and Industrial Applications in South Africa, and Its Projected Impact on Water Demand*, Water Research Technical Commission No. KV118/99; Water Research Commission: Pretoria, South Africa, 1998.

Gunderson, L.H. Managing Surprising Ecosystems in Southern Florida. *Ecol. Econ.* **2001**, *37*, 371–378.

Gunderson, L.H.; Holling, C.S. *Panarchy: Understanding Transformations in Systems of Humans and Nature*. Island Press: Washington, D.C., 2002.

Hahn, R. Economic Prescriptions for Environmental Problems: How the Patient Followed the Doctor's Orders. *J. Econ. Perspect.* **1989**, *3*, 95–114.

Hahn, R. The Impacts of Economics on Environmental Policy. *J. Environ. Econ. Manag.* **2000**, *39*, 375–400.

Halme, M.; T. Joro, T.; Korhonen, P.; Salo, S.; Wallenius, J. A Value Efficiency Approach to Incorporating Preference Information in Data Envelopment Analysis. *Manag. Sci.* **1999**, *45*(1), 103–115.

Hamlin, M. From Here to Where? Managing Water in a Sustainable Way. In *Proceedings AWRA–University of Dundee, Globalization and Water Management: The Changing Value of Water*, Dundee, Scotland, August 6–8, 2001; American Water Resources Association: Middleburg, Va., 2001.

Hammond, J.S.; Keeney, R.L.; Raiffa, H. The Hidden Traps in Decision Making. *Harvard Bus. Rev.* **1998**, *75*, 25–34.

Hamon, W.R. Estimating Potential Evapotranspiration. *J. Hydraul. Div., Proc. Am. Soc. Civil Engin.* **1961**, *87*, 107–120.

Hanley, N. Policy on Agricultural Pollution in the European Union. In *Environmental Policies for Agricultural Pollution Control,* Shortle, J; Abler, D., Eds.; CAB International: Wallingford, U.K., 2001

Hanley, N.; Shogren, J.F.; White, B. *Environmental Economics in Theory and in Practice*. Oxford University Press: New York, 1997.

Hargis and Associates, Inc. *Results of Quarterly Monitoring Program, October 1991*. Apache Nitrogen Products: Benson, Ariz., 1991.

Harrington, P. *Pricing of Water Services*. Organization for Economic Cooperation and Development: Paris, 1987.

Harris, E.; Huntley, C.; Mangle, W.; Ran, N. Transboundary Collaboration in Ecosystem Management: Integrating Lessons from Experience. Master's Project. University of Michigan School of Natural Resources and Environment, Ann Arbor, Mich., 2001.

Harrison, D.E.; Larkin, N.K. El Niño–Southern Oscillation Sea Surface Temperature and Wind Anomalies, 1946–1993. *Rev. Geophys.* **1998**, *36*(3), 353–399.

Hartley, T.W. *Public Perception & Participation in Water Reuse: White Papers.* RESOLVE, Inc.: Washington, D.C., 2001.

Hartley, T.W. *Water Reuse: Understanding Public Perception and Participation;* Project Report 00-PUM-1; Water Environment Research Foundation: Alexandria, Va., 2003.

Harvey, C. The Reasonableness of Non-Constant Discounting. *J. Public Econ.* **1994**, *53*, 31–51.

Hayami, Y.; Ruttan, V.W. *Agricultural Development: An International Perspective*. Johns Hopkins University Press: Baltimore, Md., 1985.

Heberling, M.T. Valuing Public Goods Using the Stated Choice Method. Ph.D. Dissertation, The Pennsylvania State University, State College, Pa., 2000.

Herzberg, F. *Work and the Nature of Man.* Staple Press. London, U.K., 1968.

Hicks, A.L.; Larson, J.S. *The Impact of Urban Stormwater Runoff on Freshwater Wetlands and the Role of Aquatic Invertebrate Bioassessments*. The Environment Institute, University of Massachusetts: Amherst, Mass., 1997.

Holdren, C.; Jones, W.; Taggart, J. *Managing Lakes and Reservoirs*. North American Lake Management Society and Terrene Institute, in cooperation with Office of Water Assessment, Watershed Protection Division, U.S. Environmental Protection Agency: Madison, Wisc., 2001.

Holling, C.S. The Resilience of Terrestrial Ecosystems: Local Surprise and Global Change. In *Sustainable Development of the Biosphere*, Clark, W.C.; Munn, R.E., Eds.; Cambridge University Press. Cambridge, U.K., 1986.

Holling, C.S. Theories for Sustainable Futures. *Conserv. Ecol.* **2000**, *4*(2), Article 7.

Holling, C.S. Understanding the Complexity of Economic, Ecological, and Social Systems *Ecosytems* **2001**, *4*, 390–405.

Holling, C.S. *From Complex Regions to Complex Worlds. Ecol. Soc*. **2004**, *9*(1), Article 11.

Holling, C.S., Ed. *Adaptive Environmental Assessment and Management*. John Wiley: Chichester, U.K., 1978.

Holling, C.S.; Gunderson, L.H. Resilience and Adaptive Cycles. In *Panarchy: Understanding Transformations in System of Humans and Nature*, Holling, C.S.; Gunderson, L.H., Eds.; Island Press: Washington, D.C., 2002.

Holtermann, S. Alternative Tax-Systems to Correct Externalities and the Efficiency of Paying Compensation. *Economica* **1976**, *46*, 1–16.

Horan, R.; Shortle, J. Environmental Instruments for Agriculture. In *Environmental Policies for Agricultural Pollution Control*, Shortle J.; Abler, D. Eds.; CAB International: Wallingford, U.K., 2001.

Horan, R.; Abler, D.; Shortle, J.; Carmichael, J. Cost-Effective Point–Nonpoint Trading: An Application to the Susquehanna River Basin. *J. Am. Water Resour. Assn.* **2002a**, *38*(2), 467–477.

Horan, R.D. Cost-effective and Stochastic Dominance Approaches to Stochastic Pollution Control. *Environ. Res. Econ.* **2001**, *18*, 373–389.

Horan, R.D.; Shortle, J.S.; Abler, D.G. Point-Nonpoint Nutrient Trading in the Susquehanna River Basin. *Water Resour. Res.* **2002b**, *38*(5), 1050.

Houghton, J.T.; Filho, L.G.; Callandar, B.A; Harris, N.; Kattenberg, A.; Maskell, K., Eds. *Climate Change 1995: Contribution of Working Group 1 to the Second Assessment Report of the Intergovernmental Panel on Climate Change*. Cambridge University Press: New York, 1995.

Houghton, R.A. The Worldwide Extent of Land-Use Change. *BioScience.* **1994**, *44*, 305–313.

Howe, C.W.; Salles, J.M. Seeking an Optimum Mix of Property Rights, Regulation, and Liability: Theory and Applications to Water Resources. Annual meeting of the European Association of Environmental and Resource Economists, Lisbon, Portugal, 1996.

Howe, C.W.; Schurmeir, D.R.; Shaw, W.D. Innovative Approaches to Water Allocation: The Potential for Water Markets. *Water Resour. Res.* **1986**, *22*, 439–445.

Hultquist, R.H. Augmentation of Ground and Surface Drinking Water Sources with Reclaimed Water in California. Water Quality 91. Proceedings American Water Works Association's 1995 Annual Conference, Anaheim, Calif., 1995.

Hurby, T. Assessments of Wetland Functions: What They Are and What They Are Not. *Environ. Manag.* **1999**, *23*, 75–85.

Hutson, S.S.; Barber, N.L.; Kenny, J.F; .Linsey, K.S.; Lumia, D.S.; Maupin, M.A. *Estimated Use of Water in the Water in the United States in 2000*; USGS Circular 1268; U.S. Geological Survey: Herndon, Va., 2004. (Available at www.USGS.gov/pubs/circ/2004/circl268.)

Hynes, H.B.N. The Stream and Its Valley. *Int. Ver. Theor. Agnew. Limnol. Verh.* **1975**, *19*, 1–15.

Idaho Department of Environmental Quality. *Summary of Participant Recommendations for a Trading Framework in the Lower Boise River.* Idaho Department of Environmental Quality: Boise, Idaho, 2000.

Jackson, R.B., Carpenter, S.R.; Dahm, C.N.; McKnight, D.M.; Naiman, R.J.; Postel, S.L.; Running, S.W. Water in a Changing World. *Ecol. Appl.* **2001**, 11, 1027–1045.

Jimenez, B.; Ramos, J. Reuse Priorities for the Establishment of a National Program on Efficient Use of Water for Agriculture. In *Proceedings of the Beneficial Reuse of Water and Biosolids Speciality Conference,* Marabella, Malaga, Spain, April 6–9, 1997; Water Environment Federation: Alexandria, Va., 1997.

Johnson, E. A Study in the Economics of Water Quality Management. *Water Resour. Res.*, **1967**, *3*.

Johnson-Laird, P.N. *Mental Models: Towards a Cognitive Science of Language, Inference, and Consciousness*. Cambridge University Press. Cambridge, U.K.; and Harvard University Press. Harvard, Mass., 1983.

Jonch-Clausen, T. *Integrated Water Resources Management (IWRM) and Water Efficiency Plans by 2025, Why, What and How?;* TEC Background Paper No. 10; Global Water Partnership: Stockholm, Sweden, 2004.

Jones, R.; Clark, C. Impact of Watershed Urbanization on Stream Insect Communities. *Water Resour. Bull.* **1987**, *15*(4), 1047–1055.

Juanico, M.; Friedler, E. Wastewater Reuse for River Recovery in Semi-arid Israel. *Water Sci. Technol.* **1999**, *40*(4–5), 43–50.

Just, R.E.; Antle, J.M. Effects of Commodity Program Structure on Resource Use and the Environment. In *Commodity and Resource Policies in Agricultural Systems*; Just, R.E.; Bockstael, N., Eds.; Springer-Verlag: Berlin, Germany, 1991.

Karr, J.R.; Schlosser, I.J. Water Resources and the Land-Water Interface. *Science* **1978**, *201*, 229–234.

Kasperson, J.X.; Kasperson, R.E.; Turner, B.L. *Regions at Risk: Comparisons of Threatened Environments.* United Nations University Press: New York, 1995.

Kates, R.W.; Clark, W.C.; Corell, R.; Hall, J.M.; Jaeger, C.C.; Lowe, I.; McCarthy, J.J.; Schellnhuber, H.J.; Bolin, B.; Dickson, N.M.; Faucheux, S; Gallopin, G.C.; Grubler, A.; Huntley, B.; Jager, J.; Jodha, H.S.; Kasperson, F.E.; Mabogunje, A.; Matson, P.; Mooney, H.; Moore III, B.; O'Riordan, T.; Svedin, U. Sustainability Science. *Science* 2001, 292, 641–642.

Katz, S.M.; Tennyson, P. Public Education Is the Key to Water Repurification's Success. In *Proceedings of the Beneficial Reuse of Water and Biosolids Speciality Conference,* Marabella, Malaga, Spain, April 6–9, 1997; Water Environment Federation: Alexandria, Va., 1997.

Kenney, D.S. Historical and Sociopolitical Context of the Western Watershed Movement. *Water Resour. Bull.* **1999**, *35*, 493–503.

Kepner, W.G.; Semmens, D.J.; Goodrich, D.C.; Mouat, D.A.; Bassett, S.D. Scenario Analysis for the San Pedro River, Analyzing Hydrological Consequences for a Future Environment. Presentation at the ReVA-MAIA Conference. May 13–15, 2003, King of Prussia, Pa., 2003.

KezziahWatkins. Upper San Pedro Partnership Community Connector Outreach Project Summary Report. KezziahWatkins: Colorado Springs, Colo., 2004.

Kingsolving, A.D.;Bain, M.E. Fish Assemblage Recovery along a Riverine Disturbance Gradient. *Ecol. Appl.* **1993**, *3*, 531–544.

Klein, R. Urbanization and Stream Quality Impairment. *Water Resour. Bull.* **1979**, *15*(4) 948–963.

Klessig, L.L.; Hagengruber, J. Eleven Necessary Conditions for Societal Sustainability. *J. Hum. Values* **1999**, *5*(1), 33–52.

Korhonen, P.; Tainio, R.; Wallenius, J. Value Efficiency Analysis of Academic Research. *Eur. J. Oper. Res.* 2001, 130, 121–132.

Kotler, P.; Armstrong, G.; Saunders, J.; Wong, V. *Principles of Marketing,* European Edition. Prentice Hall: London, U.K., 1996.

Kotler, P.; Roberto, N.; Lee, N. *Social Marketing: Improving the Quality of Life.* Sage Publications: Thousand Oaks, Calif., 2002.

Lacher, L.J. Hydrologic and Legal Issues of the Upper San Pedro River Basin. Semi-Arid Land-Surface Atmosphere Program online publication, 1994. (Available at http://www.tucson.ars.ag.gov/salsa/archive/publications/lacher/lacher0.html.)

Lauer, W.C.; Rogers, S.E. The Demonstration of Direct Potable Water Reuse: Denver's Pioneer Project. Paper presented at Joint AWWA/WEF Water Reuse Conference, San Diego, Calif., February 25–28, 1996.

Laurance, W.F.; Laurance, S.G.; Delamonica, P. Tropical Forest Fragmentation and Greenhouse Gas Emissions. *Forest Ecol. Manag.* **1998**, *110*, 173–180.

Lawrence, D.B. *The Economic Value of Information*. Springer: New York, 1999.

Lean, J.; Rowntree, P.R. A GCM Simulation of the Impact of Amazonian Deforestation on Climate Using Improved Canopy Representation. *Q. J. Roy. Meteor. Soc.* **1993**, *119*, 509–530.

Lee, K.N.; Lawrence, J. Adaptive Management: Learning from the Columbia River Basin Fish and Wildlife Program. *Environ. Law* **1986**, *16*, 431.

Leighton, J.P.; Sternberg, R.J. Reasoning and Problem Solving. In *Experimental Psychology.* Healy A.F.; Proctor, R.W., Eds.; Wiley & Sons: New York, 2003; pp 623–648.

Leones, J.; Dunn, D.; Worden, M.; Call, R.E. *Agricultural Tourism in Chochise County, Arizona, Characteristics and Economic Impacts*. Michigan State University Extension: East Lansing, Mich., 1994. (Available at http://www.msue.msu.edu/msue/imp/modtd/33839801.html.)

Leserman, J.R.; Ferry, W.K. Economics of Water Reclamation for Groundwater Recharge and Injection. In *Proceedings of the AWWA Annual Conference and Exposition, Atlanta, Ga., June 15–17, 1997*; American Water Works Association: Denver, Colo., 1977; p 33.

Letson, D. Point/Nonpoint Source Trading: An Interpretive Survey. *Nat. Resour. J.* **1993**, *32*, 219–232.

Levin, S. *Fragile Dominion: Complexity and the Commons*. Perseus Books, New York, 1999.

Levin, S.A. Physical and Biological Scales and the Modeling of Predator–Prey Interactions. In *Large Marine Ecosystems: Patterns, Processing, and Yields,* Sherman, K.; Alexander, L.M.; Gold, B.D., Eds.; American Association for the Advancement of Science: Washington, D.C., 1990.

Levine, B.B.; Madireddi, K.; Kahn, E.; Stenstrom, M.K.; Ye, Q.F.; Suffet, I.H. Treatment of Trace Organic Compounds by Ozone/BAC for Wastewater Reuse, The Lake Arrowhead Pilot Plant. In *Proceedings of the Beneficial Reuse of Water and Biosolids Speciality Conference,* Marabella, Malaga, Spain, April 6–9, 1997; Water Environment Federation: Alexandria, Va., 1997.

Levinson, B.; Thornton, K.W. Managing Interdisciplinary Research: Lessons Learned from the EPA/NSF/USDA Water and Watersheds Research Program. In *First Interagency Conference on Research in the Watersheds, October 27-30, 2003, Benson, Arizona*; Renard, K.G.; McElroy, S.A.; Gburek, W.J.; Canfield, H.E.; Scott, R.L., Eds.; U.S. Department of Agriculture: Washington, D.C., 2003; pp 187–193.

Liapis, P.S. Environmental and Economic Implications of Alternative EC Policies. *J. Agric. Appl. Econ*. **1994**, *26*, 241–251.

Lichtenberg, E. Alternative Approaches to Pesticide Regulation. *Northeast. J. Agr. Resour. Econ.* **1992,** *21*, 83–92.

Lichtenberg, E.; Zilberman, D. The Welfare Economics of Price Supports in U.S. Agriculture. *Am. Econ. Rev.* **1986,** *76*, 1135–1141.

Lichtenberg, E.; Zilberman, D. Efficient Regulation of Environmental Health Risk. *Q. J. Econ.* **1988**, *49*, 167–168.

Lichtenberg, E.; Zilberman, D.; Bogen, K.T. Regulating Environmental Health Risks under Uncertainty: Groundwater Contamination in Florida. *J. Environ. Econ. Manag.* **1989**, *17*, 22–34.

Lindblom, C.E. The Science of "Muddling Through." *Public Admin. Rev.* **1959**, *19*(2), 79–88.

Lindblom, C.E. Inquiry and Change: The Troubled Attempt to Understand and Shape Society. Yale University Press: New Haven, Conn., 1992.

Liverman, D.; Meredith, R.; Holdsworth, A. *Climate Variability and Social Vulnerability in the U.S.–-Mexico Border Region: An Integrated Assessment of the Water Resources of the San Pedro River and Santa Cruz River Basins.* U.S. Geological Survey: Tucson, Ariz., 1997. (Available at http://geochange.er.usgs.gov/sw/impacts/society/border/.)

Livezey, R.E.; Masutani, M.; Leetmaa, A.; Rui, H.; Ji, M.; Kumar, A. Teleconnective Response of the Pacific-North American Region Atmosphere to Large Central Equatorial Pacific SST Anomalies. *J. Climatol.* 1997, 10, 1787–1820.

Loreau, M.; Naeem, S.; Inchausti, P.; Bengtsson, J.; Grime, J.P.; Hector, A.; Hooper, D.U.; Huston, M.A.; Raffaelli, D.; Schmid, B.; Tilman, D.; Wardle, D.A. Biodiversity and Ecosystem Functioning: Current Knowledge and Future Challenges. *Science* **2001**, *294*, 804–808.

Loucks, D.P.; Stakhiv, E.Z.; Martin L.R. Sustainable Water Resources Management. *J. Water Resour. Plan. Manag.* **2000**, *126*, 43–47.

Lubchenco, J.; Olson, A.M.; Brubaker, L.B.; Carpenter, S.R.; Holland, M.M.; Hubbell, S.P.; Levin, S.A.; MacMahon, J.A.; Matson, P.A.; Melillo, J.M.; Mooney, H.A.; Peterson, C.H.; Pulliam, H.R.; Real, L.A.; Regal, P.J.; Risser, P.G. The Sustainable Biosphere Initiative: An Ecological Research Agenda. *Ecology* **1991**, *72*(2), 371–412.

Lund, E. Health Problems Associated with Re-Use of Sewage: I. Bacteria, II. Viruses, III. Protozoas and Helminths. Working papers prepared for the World Health Organization (WHO) Seminar on Health Aspects of Treated Sewage Re-Use, Algiers, Algeria, June 1–5, 1980.

Lynch, S.; Smith, K. *Lean, Mean, and Green…Designing Farm Support Programs in a New Era.* Henry Wallace Institute for Alternative Agriculture: Greenbelt, Md., 1994.

MacFadyen, L.; Stead, M.; Hastings, G. *Social Marketing: A Synopsis.* Centre for Social Marketing: Stirling, U.K., 1999. (Available at www.marketing.strath.ac.uk/csm/print/about/synopsis.html.)

Malanczuk, P. Sustainable Development: Some Critical Thoughts in the Light of the Rio Conference. In *Sustainable Development and Good Governance,* Ginther, K.; Denters, E.; de Waart, P.J.I.M., Eds.; Martinus Nijhoff Publishers: Dordrecht. 1995; pp 23–52.

Malik, A.S.; Letson, D.; Crutchfield, S.R. Point/Nonpoint Source Trading of Pollution Abatement: Choosing the Right Trading Ratio. *Am. J. Agric. Econ.* **1993**, *75*, 959–967.

Manabe, S.; Wetherald, R.T. Thermal Equilibrium of the Atmosphere with a Given Distribution of Relative Humidity. *J. Atmos. Sci.* **1967**, *24*(3), 241–259.

Marshall, E.; Shortle, J. Using DEA and VEA to Evaluate Quality of Life in the Mid-Atlantic States. *Agr. Resour. Econ. Rev.* **2005**, *34*(2), 185–203.

Martin, W.E.; Ingram, H.M.; Laney, N.K.; Griffin, A.H. *Saving Water in a Desert City.* Resources for the Future, Inc.: Washington, D.C., 1984.

Maslow, A. A Theory of Human Motivation. *Psychol. Rev.* 1943. 50:381

Maslow, A. *Motivation and Personality.* Harper & Row: New York, 1954.

Master, L.L.; Flack, S.R.; Stein, B.A., Eds. *Rivers of Life: Critical Watersheds for Protecting Freshwater Biodiversity*. The Nature Conservancy: Arlington, Va., 1998.

Matson, P.A.; Parton, W.J.; Power, A.G.; Swift, M.J. Agricultural Intensification and Ecosystem Properties. *Science* **1997**, *277*, 504–508.

May, C.W.; Horner, R.R.; Karr, J.R.; Welch, E.B. Effects of Urbanization on Small Streams in the Puget Sound Lowland Ecoregion. *Watershed Protect. Techniques* **1997**, *2*(4), 483–494.

McClelland, D.C. *Human Motivation*. Cambridge University Press: New York, 1987.

McCreary, A.M. Storytelling that Moves People: A Conversation with Screenwriter Coach, Robert McKee. *Harvard Bus. Rev.* **2003**, *80*, 16–21.

McEwen, B.; Richardson, T. Indirect Potable Reuse: Committee Report. Paper presented at Joint AWWA/WEF Water Reuse Conference, San Diego, Calif., February 25–28, 1996.

McIntosh, R.J., Tainter, J.A.; McIntosh, S.K.; Eds. *The Way the Wind Blows: Climate, History and Human Action.* Columbia University Press: New York, 2000.

McSweeny, W.T.; Shortle, J.S. Probabilistic Cost Effectiveness in Agricultural Nonpoint Pollution Control. *Southern J. Agric. Econ.* **1990**, *22*, 95–104.

Menell, P. The Limitations of Legal Institutions for Addressing Environmental Risk. *J. Econ. Perspect.* **1990**, *5*, 93–114.

Meyer J.L.; Wallace, J.B. Lost Linkages and Lotic Ecology: Rediscovering Small Streams. In *Ecology: Achievement and Challenge,* Press, M.; Hanley N.; Levin, S., Eds.; Blackwell Sciences: Oxford, U.K., 2001.

Meyer, J.L. Changing Concepts of System Management. In *Sustaining our Water Resources*, Tarlock, A.D., Ed.; National Academy Press: Washington, D.C., 1993.

Miceli, T.; Segerson, K. Joint Liability in Torts and Infra-marginal Efficiency. *Intl. Rev. Law Econ.* **1991**, *11*, 235–249.

Michigan Department of Environmental Quality. *Water Quality Trading Program: Kalamazoo Project.* Michigan Department of Environmental Quality: Lansing, Mich., 2004. (Available at www.Michigan.gov/deq/0,1607,7-135-3313_3682_3719-14305--,00.html.

Miller, K. *Property and Water: A Historical Analysis*. National Center for Atmospheric Research: Boulder, Colo., 1995.

Miller, K.; Yates, D. *Climate Change and Water Resources: A Primer for Utilities*. The National Center for Atmospheric Research: Boulder, Colo., 2004.

Mills, R.A.; Asano, T. A Retrospective Assessment of Water Reclamation Projects. *Water Sci. Technol.* **1996**, *33*(10–11), 59–70.

Milon, J.W. Optimizing Nonpoint Source Controls in Water Quality. *Water Resour. Bull.* **1987**, *23*, 387–396.

Montgomery, W.D. Markets in Licenses and Efficient Pollution Control. *J. Econ. Theory*. **1972**, *5*, 395–418.

Moore, M.R.; A. Mulville, A.; Weinberg, M. Water Allocation in the American West: Endangered Fish versus Irrigated Agriculture. *Nat. Resour. J.* **1996**, *36*, 319–357.

Moore, R. *Nitrogen Credit Trading in the Long Island Sound Watershed*, Project No. 97-IRM-5B; Water Environment Research Foundation: Alexandria, Va., 2000.

Morgan, G.; Henrion, M. *Uncertainty, a Guide to Dealing with Uncertainty in Quantitative Risk and Policy Analysis.* Cambridge University Press: New York, 1990.

Mukhamer, S.; Ibrahim, J. Egyptian Experience in Utilizing Recycled Water for Irrigation Purposes. *Desalination* 1993, 93, 633.

Murtagh, F. *Multidimensional Clustering Algorithms*. Physica-Verlag: Wurzberg, Germany, 1985.

Muszkat, L.; Magaritz, M.; Ronen, D. Unsaturated Zone and Groundwater Contamination by Organic Pollutants in Sewage-Effluent-Irrigated Site. *Groundwater* **1993**, *31*(4), 556–565.

Myers, N.; Mittermeier, R.; Mittermeier, C.; da Fonseca, G.; Kents, J. Biodiversity Hotspots for Conservation Priorities. *Nature (London)* **2000**, *403*, 853–858.

Naiman, R.J., Decamps, H. The Ecology of Interfaces: Riparian Zones. *Annu. Rev. Ecol. Syst.* **1997**, *28*, 621–658.

Naiman, R.J.; Turner, M.G. A Future Perspective on North America's Freshwater Ecosystems. *Ecol. Appl.* **2000**, *10*, 958–970.

Naiman, R.J.; Bilby, R.E.; Bisson, P.A. Riparian Ecology and Management in the Pacific Coastal Rain Forest. *BioScience* **2000**, *50*, 996–1011.

National Research Council. *Issues in Potable Reuse: The Viability of Augmenting Drinking Water Supplies with Reclaimed Water*. National Academy Press: Washington, D.C., 1988.

National Research Council. *Managing Coastal Erosion*. National Academy Press: Washington, D.C., 1990.

National Research Council. *Global Environmental Change: Understanding the Human Dimensions*. National Academy Press: Washington, D.C., 1991.

National Research Council. *Sustaining Our Water Resources*. National Academy Press: Washington, D.C., 1992.

National Research Council. *Soil and Water Quality: An Agenda for Agriculture*. National Academy Press: Washington, D.C., 1993.

National Research Council. *Issues in Potable Reuse: The Viability of Augmenting Drinking Water Supplies with Reclaimed Water*. National Academy Press: Washington, D.C., 1998.

National Research Council. *Our Common Journey, A Transition Toward Sustainability*. National Academy Press: Washington, D.C., 1999.

National Research Council. *Watershed Management for Potable Water Supply: Assessing the New York City Strategy*. National Academy Press: Washington, D.C., 2000a.

National Research Council. *Envisioning the Agenda for Water Resources Research in the Twenty-first Century*. National Academy Press: Washington, D.C., 2000b; p 61.

National Research Council. *Assessing the TMDL Approach to Water Quality Management.* National Academy of Sciences: Washington, D.C., 2001.

National Weather Service. *Southeast Arizona Flood History.* National Weather Service Forecast Office, Tucson, Ariz., 2003. (Available at http://www.wrh.noaa.gov/Tucson/hydro/floodhis.htm, accessed 6/21/04.)

The Nature Conservancy. *Miracle in the Desert, San Pedro River.* The Nature Conservancy: Arlington, Va., 2004. (Available at http://www.lastgreatplaces.org/sanpedro/index.html.)

Neal, F.C.; Kubick, K.S.; Jurasek, M. Planning for Water Reuse at the San Francisco Zoo. Paper presented at Joint AWWA/WEF Water Reuse Conference, San Diego, Calif., February 25–28, 1996.

Neuman, J.C. Adaptive Management: How Water Law Needs to Change. In *Proceedings AWRA–University of Dundee, Globalization and Water Management: The Changing Value of Water*, Dundee, Scotland, August 6–8, 2001; American Water Resources Association: Middleburg, Va., 2001; p 5.

Newell, R.; Pizer, W. Discounting the Benefits of Climate Change Policies Using Uncertain Rates. *Resources* **2002**, *146*, 15–20.

Nieswiadomy, M.L. The Demand for Irrigation Water in the High Plains of Texas. *Am. J. Agric. Econ.* **1985**, *67*, 619–626.

Nigam, S.; Barlow, M.; Berbery, E.H. Analysis Links Pacific Decadal Variability to Drought and Stream Flow in U.S. *EOS Trans. Am. Geophys. Union* **1999**, *80*(51), 621–622, 625.

Nobre, C.A.; Sellers, P.J.; Shukla, J. Amazonian Deforestation and Regional Climate Change. *J. Climatol.* **1991**, *4*, 957–988.

Norgaard, R.B.; Howarth, R.B. 1991. Sustainability and Discounting the Future. In *Ecological Economics: The Science and Management of Sustainability*, Costanza, R., Ed.; Columbia University Press: New York, 1991; pp 88–101.

Norton, S.B.; Barbour, M.T.; Thornton, K.W. Injecting Ecological Knowledge into the Decision-making Process. In *Ecological Assessment of Aquatic Resources, Linking Science to Decision-making,* Barbour, M.T.; Norton, S.B.; Preston, R.H.; Thornton, K.W.; Eds.; Society of Environmental Toxicology and Chemistry: Pensicola, Fla., 2004; pp 169–178.

Novotny, V.; Olem, H. *Water Quality: Prevention, Identification, and Management of Diffuse Pollution.* Van Nostrand Reinhold: New York, 1994.

Nystrom, M.; Folke, C. Moberg, F. Coral Reef Disturbance and Resilience in a Human-Dominated Environment. *Trends Ecol. Evol.* **2000**, *15*(10), 413–417.

Obradovic, M.M.; Sklash M.J. An Isotopic Land Geochemical Study of Snowmelt Runoff in a Small Arctic Watershed. *Hydrol. Process.* **1986**, *1*(1), 15–30.

Odendaal, P.E.; van der Westhuizen, J.L.J.; Grobler, G.J. Wastewater Reuse in South Africa. In *Wastewater Reclamation and Reuse,* Asano, T., Ed.; Technomic Publishing Company, Inc.: Lancaster, Pa., 1998; pp 1163–1192.

Odum, E.P. *Fundamentals of Ecology*, 3rd Ed. W.B. Saunders Co.: Philadelphia, Pa., 1971.

Omernik, J.M. Ecoregions of the Conterminous U.S. (Map Suppl.). *Ann. Assoc. Am. Geogr.* **1987**, *101*(1), 118–125.

Omernik, J.M. Ecoregions: A Spatial Framework for Environmental Management. In *Biological Assessment and Criteria: Tools for Water Resource Planning and Decision-making*, Davis, W.S.; Simon, T.P., Eds.; Lewis Publishers: Boca Raton, Fla., 1995; pp 49–63.

O'Neill, R.V.; DeAgelis, D.L.; Waide, J.B.; Allen, T.F.H. A Hierarchical Concept of Ecosystems. *Monographs in Population Biology, 23*. Princeton University Press: Princeton, N.J., 1986.

O'Neill, W.B. Pollution Permits and Markets for Water Quality. Ph.D. Dissertation, University of Wisconsin–Madison, 1980.

Opaluch, J.J.; Kashmanian, R. Assessing the Viability of Marketable Permit Systems: An Application in Hazardous Waste Management. *Land Econ.* **1985**, *61*, 441–447.

Opshoor, J.B.; de Savornin Lohman, A.F.; Vos, H.B. *Managing the Environment: The Role of Economic Instruments.* Organization for Economic Cooperation and Development: Paris, 1994.

Organization for Economic Cooperation and Development (OECD).*Water Pollution by Fertilizers and Pesticides*. OECD: Paris, 1986.

OECD. *Agricultural and Environmental Policies: Opportunity for Integration*. OCED: Paris, 1989.

OECD. *Agricultural and Environmental Policies Integration: Recent Progress and New Directions*. OCED: Paris, 1993.

OECD. *Agriculture and the Environment Issues and Policies*. OCED: Paris, 1998.

Original Jurisdiction. Judiciary and Judicial Procedure, Part IV– Jurisdiction and Venue, Chapter 81–The Supreme Court, *U.S. Code,* Section 1251, Title 28, 2005.

Orr, P.; Colby, B. *Nature Oriented Visitors and Their Expenditures: Upper San Pedro River Basin.* University of Arizona College of Agriculture and Life Sciences Department of Agriculture and Resource Economics: Tucson, Ariz., 2002.

Pagano, T.C. *Southwest Climate in a Nutshell.* Climate Assessment for the Southwest: Tucson, Ariz., 1999. (Available at http://www.ispe.arizona.edu/climas/learn/swnutshell/.)

Palaniappan, M.; Gleick, P.H.; Hunt, C.; Srinivasan, V. Water Privatization Principles and Practices. In *The World's Water 2004–2005*, P. Gleick, P., Ed.; Island Press: Washington, D.C., 2004; pp 45–77.

Paret, M.; Eisner, M. Reclaimed Water Perspectives. *Water Environ. Technol.* **1993**, *5*(2), 46–49.

Pennsylvania Department of Environmental Protection (PDEP). *Draft Policy for Protecting Aquatic Resources and Related Stream Uses in Processing Approvals for Water Rights Acquisitions in Selected Waters of the Commonwealth*. PDEP: Harrisburg, Pa., 2000. (Available at http://www.dep.state.pa.us/dep/subject/advcoun/wrac/2000/ifimforwrac913.pdf.)

PDEP. *Water Allocation Application Instructions*; 3900-PM-WM0001; Commonwealth of Pennsylvania Department of Environmental Protection Bureau of Watershed Management: Harrisburg, Pa., 2001.

PDEP. *Pennsylvania Department of Environmental Protection Division of Drinking Water Management Maximum Contaminant Levels (MCLs), Maximum Residual Disinfectant Levels (MRDLs).* PDEP: Harrisburg, Pa., 2002. (Available at http://www.dep.state.pa.us/dep/deputate/watermgt/WSM/WSM_DWM/PA-MCLs.pdf.

PDEP. *Commonwealth of Pennsylvania Fish Consumption Advisories – 2004.* Commonwealth of Pennsylvania Department of Environmental Protection: Harrisburg, Pa., 2003a. (Available at http://www.dep.state.pa.us/dep/deputate/watermgt/wqp/wqstandards/FishAdvis/FishAdvisory04-Tbl.pdf.)

PDEP. *Watershed Restoration Action Strategy (WRAS), State Water Plan Subbasin 09C, Bald Eagle Creek Watershed, Centre and Clinton Counties*. PDEP: Harrisburg, Pa., 2003b. (Available at http://www.dep.state.pa.us/dep/deputate/watermgt/wc/subjects/nonpointsourcepollution/initiatives/WRASLISTINFO/WrasPlans/WRAS-09C.pdf.)

PDEP. *Fact Sheet: Questions & Answers, Act 220 – The Water Resources Planning Act*; 3920-FS-DEP3104; Commonwealth of Pennsylvania Department of Environmental Protection: Harrisburg, Pa., 2004a.

PDEP. *Pennsylvania's Surface Water Quality Monitoring Network (WQN)*. Commonwealth of Pennsylvania Department of Environmental Protection: Harrisburg, Pa., 2004b.

PDEP. *2004 Pennsylvania Integrated Water Quality Monitoring and Assessment Report*. Commonwealth of Pennsylvania Department of Environmental Protection: Harrisburg, Pa, 2004c.

Peterson, G.D. Estimating Resilience across Landscapes. *Conserv. Ecol*. **2002**, *6*(1), 17. (Available at http://www.consecol.org/vol6/iss1/art17/.)

Pennsylvania Fish and Boat Commission. *2004 Class A Wild Trout Waters*. Pennsylvania Fish and Boat Commission, Bureau of Fisheries, Fisheries Management Division: Harrisburg, Pa., 2004.

Pimm, S.; Raven, P. Extinction by Numbers. *Nature (London)* **2000,** *403*, 843–845.

Pizarro, G.; Lall, U. El Niño-induced Flooding in the U.S. West: What Can We Expect Next? *EOS Trans. Am. Geophys. Union* **2002**, *83*(32), 341–352.

Plantinga, A.J. The Effect of Agricultural Policies on Land Use and Environmental Quality. *Am. J. Agr Econ*. **1996**, *78*, 1082–1091.

Plott, C.R. 1966. Externalities and Corrective Taxes. *Economica* **1966**, *33*, 84–87.

Pool, D.R.; Coes, A.L. *Hydrogeologic Investigations of the Sierra Vista Subwatershed of the Upper San Pedro Basin, Cochise County, Southeast Arizona*; Water Resources Investigation Report 99-4197; U.S. Geological Survey: Tucson, Ariz., 1999. (Available at http://www.earlham.edu/~parkero/swfs/Pool%20and%20Coes%201999%20Hydrogeologic%20Sierra%20Vista.pdf.)

Posner, M.I. *Cognition: An Introduction*. Scott Foresman Publication: Glenview, Ill., 1973.

Postel, S.L. *The Last Oasis, Facing Water Scarcity*; Worldwatch Environmental Alert Series; W.W. Norton & Company: New York, 1992.

Postel, S.L. *Pillar of Sand, Can the Irrigation Miracle Last.* W.W. Norton & Company: New York, 1999.

Postel, S.L. Water and World Population Growth. *J. Am. Wastewater Assn.* **2000a**, *92*(4), 131–138.

Postel, S.L. Entering an Era of Water Scarcity: The Challenges Ahead. *Ecol. Appl.* **2000b**, *10*, 941–948.

Postel, S.L.; Daily, G.C.; Ehrlich, P.R. Human Appropriation of Renewable Freshwater. *Science* **1996**, *271*, 785–788.

Preston, H.R.; Culp, J.; DeMoss, T.; Frydenborg, R.; Kenaga, E.E.; Pittinger, C.A.; Schofield, N. Translating Ecological Science. In *Ecological Assessment of Aquatic Resources, Linking Science to Decision-Making;* Barbour, M.T.; Norton, S.B.; Preston, H.R.; Thornton, K.W., Eds.; Society of Environmental Toxicology and Chemistry: Pensacola, Fla., 2004; pp 141–168.

Pringle, C.M. Exploring How Disturbance is Transmitted Upstream: Going Against the Flow. *J. N. Am. Benthol. Soc.* **1997**, *16,* 425–438.

Profeta, T.H. Beyond the Balance of Nature: Environmental Law Faces the New Ecology: Managing Without a Balance: Environmental Regulation in Light of Ecological Advances. *Duke Environ. Law Policy F* **1996**, *7*, 71–103.

Putnam, F.; Mitchell, K.; Bushner, G. *Water Resources of the Upper San Pedro Basin, Arizona.* Report for the Special Studies Division, Hydrology Division. Arizona Dept. of Water Resources: Phoenix, 1988.

Queensland Department of Natural Resources. *Draft Queensland Water Recycling Strategy, Consultation Document*. Queensland Department of Natural Resources: Queensland, Australia, November 2000.

Radcliffe, J.C. *Water Recycling in Australia.* Australian Academy of Technological Sciences and Engineering: Parkville, Victoria, Australia, 2004

Randall, A. Property Rights and Social Microeconomics. *Nat. Resour. J.* **1975**, *15*, 727–745.

Randall, A. Property Entitlements and Pricing Policies for a Maturing Water Economy. *Aust. J. Agr. Econ.* **1981**, *25*, 195–212.

Rasmusson, E.M.; Dickinson, R.E.; Kutzbach, J.E.; Cleaveland, M.K. Climatology. In *Handbook of Hydrology*; Maidment, D.R., Ed.; McGraw-Hill: New York, 1993.

Raval, A.; Ramanathan, V. Observational Determination of the Greenhouse Effect. *Nature (London)* **1989**, *342*, 758–762.

Reagan, R. Executive Order 12291 of February 17, 1981. *Fed. Regist.* **1987**, *46*, 13193.

Recycled Water Task Force. White Paper of the Public Information, Education, and Outreach Workgroup on Better Public Involvement in the Recycled Water Decision Process. Recycled Water Task Force: Sacramento, Calif., 2003. (Available at http://www.owue.water.ca.gov/recycle/docs/PubInfoDraftPaper.pdf.)

Reuss, M., Ed. *Water Resources Administration in the U.S.: Policy, Practice, and Emerging Issues*. Michigan State University Press: East Lansing, Mich., 1993; p 314.

Revenga, C.; Brunner, J.; Henninger, N.; Kassem, K.; Payne, R. *Pilot Analysis of Global Ecosystems: Freshwater Systems*. World Resources Institute: Washington, D.C., 2000; p 65.

Rhode, A. Spring Flood: Meltwater or Groundwater. *Nordic Hydrol.* **1981**, *12*, 2130.

Ribaudo, M.O. Nonpoint Source Pollution Control Policy in the U.S. In *Environmental Policies for Agricultural Pollution Control;* Shortle, J.; Abler, D., Eds.; CAB International: Wallingford U.K., 2001.

Richter, B.D.; Baumgartner, J.V.; Powell, J.; Braun, D.P. A Method for Assessing Hydrologic Alteration within Ecosystems. *Conserv. Biol.* **1996**, *10*, 1163–1174.

Richter, B.D.; Baumgartner, J.V.; Wigington, R.; Braun, D.P. How Much Water Does a River Need? *Freshwater Biol.* **1997**, *37*, 231–249.

Richter, B.D.; Baumgartner, J.V.; Braun, D.P.; J. Powell, J. A Spatial Assessment of Hydrologic Alteration within a River Network. *Regul. River.* **1998**, *14*, 329–340.

Riebsame, W.E., Westcoat, J.; Morrisette, P. Western Land Use Trends and Policy: Implications for Water Resources. Report to the Western Water Policy Review Advisory Commission, 1997.

Riitters, K.; Wickham, J.; O'Neill,R.; Jones, B.; Smith, E. Global-Scale Patterns of Forest Fragmentation. *Conserv. Ecol.* **2000**,*4*(2), 3.

Roeske, R.H.; Werrell, W.R. Hydrologic Conditions in the San Pedro River Valley, Arizona. *Ariz. Water Comm. Bull.* **1973**, *4*.

Rogers, M. A Model Drought Response Plan. *Government West*, July/August 2003.

Rogers, P.; Hall, A.W. *Effective Water Governance,* TEC Background Paper No. 7; Global Water Partnership: Stockholm, Sweden, 2003.

Rogers, P.; R. Bhatia, R.; Huber, A. *Water as a Social and Economic Good: How to Put the Principle into Practice;* Technical Advisory Committee Background Paper No.2; Global Water Partnership: Stockholm, Sweden, 1998.

Rogers, Y.; Rutherford, Bibby, P.A., Eds. *Models in the Mind: Theory, Perspective and Application*. Academic Press: London, 1992.

Rosegrant, M.W.; Gazmuri, R. *Reforming Water Allocation Policy Through Markets In Tradable Water Rights: Lessons From Chile, Mexico, and California*; EPTD Discussion Paper No. 6; International Food Policy Institute: Washington, D.C., 1994.

Rosenbaum, J. New York State's Project 2015: Preparing for the Impact of an Aging, Diverse Population. In *Proceedings of the Aging Americans: Impacts on Ecology and Environmental Quality Workshop.* U.S. Environmental Protection Agency, Office of Research and Development: Research Triangle Park, N.C., 2005.

Rosenberg, D.M.; Berkes, F.; Bodaly, R.A.; Hecky, R.E.; Kelly, C.A.; Rudd, J.W.M. Large-scale Impacts of Hydroelectric Development. *Environ. Rev*. **1997**, *5*(1), 27–54.

Rosenberg, D.M.; McCully, P.; Pringle, C.M. Global-scale Environmental Effects of Hydrological Alternations: Introduction. *BioScience* **2000**, *50*, 746–752

Rosgen, D. *Applied River Morphology*. Wildland Hydrology Publishers: Ft. Collins, Colo., 1996.

Ross, H., Abel, N.; Herbert, A.; Manning, M.; Walker, P.; Wheeler, H. *Mental Models and Communication in Agriculture*; RIRDC Short Report No 43; Rural Industries Research and Development Corporation: Canberra, Australia, 1998.

Russell, C.S.; Powell, P.T. Practical Considerations and Comparison of Instruments of Environmental Policy. In *Handbook of Environmental and Resource Economics*; van den Berg, N.J., Ed.; Edward Elgar Publishing, Inc.: Cheltenham, U.K., 1999.

Saliba, B. Do Water Market Prices Appropriately Measure Water Values? *Nat. Resour. J.* **1987**, *27*(3), 617–651.

Saliba, B.C.; Bush, D.B. *Water Markets in Theory and Practice.* Westview Press: Boulder, Colo., 1987.

Scheffer, M.; Carpenter, S.R.; Foley, J.; Folke, C.; Walker, B. Catastrophic Shifts in Ecosystems. *Nature (London)* **2001**, *413*, 591–696.

Schkade, D.A.; Payne, J.W. How People Respond to Contingent Valuation Questions: A Verbal Protocol Analysis of Willingness to Pay for an Environmental Regulation. *J. Environ. Econ. Manag.* **1994**, *26*, 88–109.

Schumpeter, J.A. *Capitalism, Socialism and Democracy*; 3rd enlarged ed.; Harper. New York, 1950.

Scott, R.; Goodrich, D.; Williams, D.; Stromberg, J.; Leenhouts, J. *San Pedro Riparian National Conservation Area (SPRNCA) Water Needs Study, Year 2001 Progress Report,* 2002. (Available at http://www.usppartnership.com/documents/USPP-WN%20Project%20Report%202001%20with%20Figures.doc.)

Seaber, P.R.; Kapinos, F.P.; Knapp, G.L. *Hydrologic Unit Maps*: U.S. Geological Survey Water Supply Paper 2294; U.S. Geological Survey: Washington, D.C., 1987.

Segerson, K. Uncertainty and Incentives for Non-Point Source Pollution. *J. Environ. Econ. Manag.* 1988, 15, 87–98.

Sepp, E. *The Use of Sewage for Irrigation: A Literature Review.* California Department of Public Health, Bureau of Sanitary Engineering, Berkley, Calif., 1971.

Serageldin, I. Water Resources Management: A New Policy for a Sustainable Future. *Water Intl.* **1995**, *20*, 15–21.

Shabman, L. Water Use Forecasting—Benefits and Capabilities. In *National Water Summary 1987—Hydrologic Events and Water Supply and Use*; U.S. Geological Survey Water Supply Paper 2350; U.S. Geological Survey, Herndon, Va., 1990; pp 117–122.

Shabman, L.; Stephenson, K. Nitrogen Allowance Trading for Point Sources in the Connecticut Region of the Long Island Sound Watershed, Department of Agricultural and Applied Economics, Virginia Polytechnic Institute and State University: Blacksburg, Va., 1998.

Shavell, S. *Economic Analysis of Accident Law.* Harvard University Press: Cambridge, Mass., 1987.

Sherk, G.W. Interstate Water Conflicts and Individual Rights: Perspectives from the U.S. Supreme Court. *Water Intl.* **2000**, *25*, 519–525.

Shiklomanov, I.A. World Water Resources and Water Use: Present Assessment and Outlook for 2025. In *World Water Scenarios;* Rijsberman, F.R., Ed.; Earthscan: London, 2000; pp 160–203.

Shortle, J. Allocative Implications of Comparisons between the Marginal Costs of Point and Nonpoint Source Pollution Abatement. *Northeast. J. Agr. Resour. Econ.* **1987**, *16*, 17–23.

Shortle, J. The Allocative Efficiency Implications of Water Pollution Abatement Cost Comparisons. *Water Resour. Res*. **1990**, *26*(5), 793–797.

Shortle, J.; Horan, R. The Economics of Nonpoint Pollution Control. *J. Econ. Surv.* **2001**, *15*(3), 255–290.

Shortle, J.S.; Abler, D.G. Incentives for Agricultural Nonpoint Pollution Control. In *The Economics of Nonpoint Pollution Control: Theory and Issues,* Graham-Tomasi, T.; Dosi, C., Eds.; Kluwer Academic Press: Dortrecht, Netherlands, 1994.

Shortle, J.S.; Faichney, R.; Hanley, N.; Munro, A. Least-Cost Pollution Abatement for Probabilistic Water Quality Targets to Protect Salmon on the Forth Estuary. In *Pollution For Sale: Emissions Trading and Joint Implementation,* Sorrel, S.; Skea, J., Eds.; Edward Elgar: Northampton, Mass., 1999.

Shortle, J.S.; Horan, R.D.; Abler, D.G. Research Issues in Nonpoint Pollution Control. *Environ. Res. Econ*. **1998**, *11*(3/4), 571–585.

Shumway, R. Recent Duality Contributions in Production Economics. *J. Agr. Resour. Econ.* **1995**, *20*, 178–194.

Shuval, H.I.; Adin, A.; Fattal, B.; Rawitz, E.; Yekutiel, P. *Wastewater Irrigation in Developing Countries – Health Effects and Technical Solutions*;World Bank Technical Paper Number 51; The World Bank: Washington, D.C., 1986.

Simon, H.A. The Organization of Complex Systems. In *Hierarchy Theory: The Challenge of Complex Systems,* Pattee, H.H., Ed.; Braziller: New York, 1974: pp 3–27.

Simonovic, S.P. Decision Support Systems for Sustainable Management of Water Resources: 1. General Principles. *Water Intl.* **1996**, *21*, 223–232.

Simpson, J.M. Changing Community Attitudes to Potable Re-use in Southeast Queensland. *Water Sci. Technol.* **1999**, *40*(4–5), 59–66.

Smith, K.R. Time to "Green" U.S. Farm Policy. *Issues Sci. Technol.* **1995**, 71–78.

Smith, V.K. Pricing What Is Priceless: A Status Report on Nonmarket Valuation. In *The International Yearbook of Environmental and Resource Economics 1997/1998: A Survey of Current Issues,* Folmer , H.; Tietenberg, T., Eds.; Edward Elgar: Cheltenham, U.K., 1997.

Solanes, M.; Gonzalez-Vallarreal, F. *The Dublin Principles for Water as Reflected in a Comparative Assessment of Institutional and Legal Arrangements for Integrated Water Resources Management*; Technical Advisory Committee Background Paper No. 3; Global Water Partnership: Stockholm, Sweden, 1999.

Solley, W.B.; Pierce, R.R.; Perlman, H.A. *Estimated Use of Water in the U.S. in 1995*; U.S. Geological Survey Circular 1200; U.S. Geological Survey: Denver, Colo., 1998.

Sorenson, D. Planning and Marketing a Valley-wide Reuse Program for the Salt Lake Valley. *Proceedings of the 1994 AWWA/WEF Water Reuse Symposium*, Dallas, Texas, Dallas, Texas, February 27–March 2, 1994; p 645.

Southwest Center for Biological Diversity. *San Pedro River Factsheet*. Southwest Center for Biological Diversity: Phoenix, Ariz., 2004.(Available at http://www.sw-center.org/swcbd/factsheets/sanpedro.html.)

Southwest Center for Environmental Research and Policy. *Transborder Watersheds Research Program: The Upper San Pedro River Watershed Project.* Southwest Center for Environmental Research and Policy: San Diego, Calif., 1999. (Available at http://typhoon.sdsu.edu/research/projects/twrp/sanpedro.html.)

Sparks. R.E. Need for Ecosystem Management of Large Rivers and Their Floodplains. *Bioscience.* **1995**; *45,* 168–182.

Spring Creek Watershed Commission. *Stormwater Management Plan for the Spring Creek Watershed.* Spring Creek Watershed Commission: State College, Pa., 2000

Spring Creek Watershed Commission and Clear Water Conservancy. *Water Resources Monitoring Program Annual Report.* Spring Creek Watershed Commission: State College, Pa., 2002.

Stanford, J.A. *Instream Flows to Assist the Recovery of Endangered Fishes of the Upper Colorado River System*; Biological Report 24; U.S. Fish and Wildlife Service: Denver, Colo., 1994.

Stanford, J.A.; Ward, J.V.; Liss, W.J.; Frissell, C.A.; Williams, R.N.; Lichatowich, J.A.; Coutant, C.C. A General Protocol for Restoration of Regulated Rivers. *Regul River* **1996**, *12*, 391–413.

State College Borough Water Authority. *State College Water Authority 2003 Water Quality Report.* State College Borough Water Authority: State College, Pa., 2004.

State College Zoning Ordinance. Ordinance 559, June 20, 1959, Section 104. (Available at http://www.gov.state-college.pa.us/.)

Steedman, R.J. Modification and Assessment of an Index of Biotic Integrity to Quantity Stream Quality in Southern Ontario. *Can. J. Fish. Aquat. Sci.* **1988**, *45*, 492–501.

Steinitz, C. Scenario-Based Studies of Alternative Futures: Types and Consideration. In *Proceedings of Landscape Futures: An International Symposium on Advances in Research for Natural Resource Planning and Management Across Regional Landscapes*; Armidale, New South Wales, Australia, Sept. 22–25, 1999.

Steinitz, C.; Rojo, H.M.A.; Bassett, S.; Flaxman, M.; Goode, T.; Maddock, III, T.; Mouat, D.; Peiser, R.; Shearer, A. *Alternative Futures for Changing Landscapes, The Upper San Pedro River Basin in Arizona and Sonora.* Island Press: Washington, D.C., 2003.

Stevenson, R.J.; Wiley, M.J.; Hyndman, D.; Pijanowski, B.; Seelbach, P. Developing Relations among Human Activities, Stressors, and Stream Ecosystem Responses for Integrated Regional, Multi-stressor Models. Presented at U.S. Environmental Protection Agency's Global Change and Ecosystem Protection Research STAR Progress Review Workshop, Alexandria, Va., June 16–18, 2004. (Available at http://www.scgcorp.com/global2004/files/Mike%20Wiley.pdf).)

Stillwell, S.A.; Clark, J.H.; Wilson, M.D. Los Angeles Zoo Reclaimed Water Feasibility Study. Paper presented at 69th Annual Water Environment Federation Technical Exposition Conference, Dallas, 1996.

Stommel, H. The Abyssal Circulation. *Deep Sea Res.* **1958**, *5*, 80–82.

Stromberg, J.; Tiller, R.; Richter, B. Effects of Groundwater Decline on Riparian Vegetation of Semiarid Regions: The San Pedro, Arizona. *Ecol. Appl.* **1996**, *6*(1).

Subcommittee on Global Change Research. *Our Changing Planet: The FY2000 U.S. Global Change Research Program. Implementation Plan and Budget Overview, Supplement to the President's FY2000 Budget*; National Science and Technology Council: Washington, D.C., 1999.

Sun, R.J.; Weeks, J.B.; Grubb, H.F. *Bibliography of Regional Aquifer-System Analysis Program of the U.S. Geological Survey;* USGS Water Resources Investigation Report 97-4074; U.S. Geological Survey: Washington, D.C., 1997.

Surtevant, P. Reclaimed Water Could Save Ski Resort. *Water Environ. Technol.* **1993**, *5*(2), 23.

Susquehanna River Basin Commission. *Instream Flow Studies, Pennsylvania and Maryland.* Susquehanna River Basin Commission: Harrisburg, Pa., 1998. (Available at http://www.dep.state.pa.us/dep/DEPUTATE/Watermgt/Wc/subjects/InstreamFlow.htm.)

Sweetland Engineering and Associates, Inc., and Centre County Planning Office. *Act 167 Stormwater Management Plan for the Spring Creek Watershed.* Centre County Commissioners, Bellefonte, Pa., 2002. (Available at http://www.dep.state.pa.us/dep/deputate/watermgt/wc/Subjects/StormwaterManagement/SpringCreek_Centre/SpringCreek.htm.)

Swetnam, T.W.; Betancourt, J.L. Fire – Southern Oscillation Relations in the Southwestern U.S. *Science* **1990**, *249*, 1017–1020.

Swinton, S.M.; Clark, D.S. Farm-level Evaluation of Alternative Policy Approaches to Reduce Nitrate Leaching from Midwest Agriculture. *Agr. Resour. Econ. Rev.* **1994**, *23*, 66–74.

Tarlock, A.D. The Nonequilibrium Paradigm in Ecology and the Partial Unraveling of Environmental Law. *Loy. L. A. Law Rev.* **1994**, *27*, 1121.

Taylor, L.E. *Water Budget for the Spring Creek Basin.* Susquehanna River Basin Commission: Harrisburg, Pa., 1997.

Thabaraj, G.J. Reclaimed Water Use in Florida. Proceedings of the Water Environment Federation's 71st Annual Conference & Exposition, Orlando, Fla. Water Environment Federation: Alexandria, Va., 1998.

Theon, J.; Fugono, N., Eds. *Tropical Rainfall Measurements;* Deepak: Hampton, Va., 1988.

Thomas, K.; Baker, P.A.; Allender, E.B. Design Methods for the Development of Wastewater Land Disposal Systems. *Water Sci. Technol.* (G.B.) **1993**, *27*(1), 77.

Thompson, S.A. *Water Use, Management, and Planning in the U.S.* Academic Press: New York, 1999.

Thornton, K.; Laurin, C. 50 ft Sciences and the Nard Reality of Lake Management. *Lake Reserv. Manag.* **2005**, *21*(2), 203–208.

Thornton, K.W., Laurin, C.; Datanzaro, D. Shortle, Fisher, A.; Sobrinho, J.; Stevens, K. *Moving Toward Sustainable Water Resources Management (SWRM): A Framework and Guidelines for Implementation. Synthesis Report;* 00-WSM-6A; Water Environment Research Foundation. Alexandria, Va., 2006.

Tietenberg, T.H. Economic Instruments for Environmental Protection. *Oxford Rev. Econ. Pol.* **1990**, *6*, 17–33.

Tietenberg, T.H. Tradable Permits for Pollution Control when Emission Location Matters: What Have We Learned? *Environ. Res. Econ.* **1995**, *5*, 95–113.

Trelease, F.J. Policies for Water Law: Property Rights, Economic Forces, and Public Regulation. *Nat. Resour. J.* 1968, 5, 1–48.

Tripp Umbach & Associates. *The Pennsylvania State University Economic Impact Statement 2003*. Tripp Umbach & Associates, Inc.: Pittsburgh, Pa., 2004.

Trush, W.J.; McBain, S.M.; Leopold, L.B. Attributes of an Alluvial River and Their Relation to Water Policy and Management. *Proc. Natl. Acad. Sci.* **2000**, *97*(22), 11858–11863.

Tucker, G.E.; Bras, R.L. Hillslope Processes, Drainage Density, and Landscape Morphology. *Water Resour. Res.* **1998**, *34*(10), 2751–2764.

United Nations Educational, Scientific and Cultural Organization (UNESCO) Working Group M.I.V. *Sustainability Criteria for Water Resource Systems*. Cambridge University Press: Cambridge, U.K., 1999.

UNESCO. Water Use in the World: Present Situation/Future Needs. An online publication of UNESCO prepared for World Water Day 2000. UNESCO: Paris, France, 2000. (Available at http://www.unesco.org/science/waterday2000/water_use_in_the_world,htm.)

United Nations, Economic and Social Council Committee on Economic, Social, and Cultural Rights. *Substantive Issues Arising in the Implementation of the International Covenant on Economic, Social, and Cultural Rights, General Comment No. 15, The Right to Water*; Articles 11 and 12 of the International Covenant on Economic, Social, and Cultural Rights; E/C.12/2002/11; United Nations: Geneva, Switzerland, 2002.

United Nations. *Road Map Towards the Implementation of the United Nation Millennium Declaration;* A/56/326; United Nations: Geneva, Switzerland, 2001.

United Nations. *Earth Summit Agenda 21: The United Nations Program of Action from Rio.* United Nations: New York, 1992.

U.S. Army Corps of Engineers. *Water Control Infrastructure: National Inventory of Dams* [CD-ROM]. Federal Emergency Management Agency: Washington, D.C., 1996.

U.S. Council on Environmental Quality. *Environmental Trends*. Executive Office of the President: Washington, D.C., 1981.

U.S. Department of Agriculture. *State Soil Geographic (STATSGO) Data Base. Data Use Information;* Miscellaneous Publication Number 1492; U.S. Department of Agriculture: Washington, D.C., 1994.

U.S. Department of Agriculture;U.S. Environmental Protection Agency. *Clean Water Action Plan: Restoring and Protecting America's Water*; U.S. Department of Agriculture: Washington, D.C., 1998.

U.S. Environmental Protection Agency (U.S. EPA). *National Water Quality Inventory: 1990 Report to Congress*; EPA-503-9-92-006; U.S. Environmental Protection: Washington, D.C., 1992.

U.S. EPA. *Watershed Protection: A Statewide Approach*; EPA 841-R-95-004; U.S. Environmental Protection Agency Office of Water: Washington, DC, 1995.

U.S. EPA. *Waquoit Bay Watershed. Ecological Risk Assessment Planning and Problem Formulation* (draft); EPA-630-R-96-004a; Risk Assessment Forum, U.S. Environmental Protection: Washington, D.C., 1996.

U.S. EPA. *Principles for Effective Communication with Communities about Ecological Issues;* EPA-236-F-96-001; U.S. Environmental Protection Agency: Washington, D.C., 1997a.

U.S. EPA. *Top 10 Watershed Lessons Learned*; EPA-840-F-97-001; U.S. Environmental Protection Agency: Washington, D.C., 1997b.

U.S. EPA. Risk Assessment Forum. *Guidelines for Ecological Risk Assessment;* EPA-630-R-095-002F. U.S. Environmental Protection Agency: Washington, D.C., 1998.

U.S. EPA. *EPA's Framework for Community Based Environmental Protection;* EPA-237-K-99-001; U.S. Environmental Protection Agency: Washington, D.C., 1999.

U.S. EPA. *The U.S. Experience With Economic Incentives for Protecting the Environment;* EPA-240-R-01-001; U.S. Environmental Protection Agency: Washington, D.C., 2001.

U.S. EPA. *National Water Quality Inventory: 2000 Report;* EPA-8941-R-02-001; U.S. Environmental Protection Agency Office of Water: Washington, D.C., 2002a.

U.S. EPA. *Willamette Basin Alternative Futures Analysis*; EPA 600/R-02/045(b); U.S. Environmental Protection Agency: Washington, D.C., 2002b.

U.S. EPA. *Guidelines for Water Reuse;* EPA-625-R-04-108; U.S. Environmental Protection Agency, 2004a.

U.S. EPA. Indian Tribes; Eligibility for Program Authorization, December 14, 2004. *Fed. Regist.* **2004b**, *59*, 64344.

U.S. Geological Survey. *Ecoregions and Subregions of the United States, Puerto Rico, and the U.S. Virgin Islands;* U.S. Geological Survey: Reston, Va., 2004. (Available at www.nationalatlas.gov.)

U.S. Geological Survey. *Strategic Directions for the U.S. Geological Survey Ground-Water Resources Program. A Report to Congress*. U.S. Geological Survey: Herndon, Va., November 30, 1998.

U.S. Geological Survey. *Hydrogeologic Investigations of the Sierra Vista Subbasin of the Upper San Pedro River Basin, Arizona.* U.S. Geological Survey: Herndon, Va., 1999. (Available at http://az.water.usgs.gov/pubs/pdfs/WRIR%2099-4197%20WEB.pdf.)

Upper San Pedro Partnership. *Upper San Pedro Conservation Plan 2002 Progress Report*. Upper San Pedro Partnership: Sierra Vista, Ariz., 2002, (Available at http://www.usppartnership.com/documents/2002ConservPlan.pdf,)

Upper San Pedro Partnership. *Water Conservation and Management Plan.* Upper San Pedro Partnership: Sierra Vista, Ariz., 2004. (Available at http://www.usppartnership.com/documents.html#consplan.)

U.S. Water Resources Council. *The Nation's Water Resources*. U.S. Government Printing Office: Washington, D.C., 1968.

U.S. Water Resources Council. *Economic and Environmental Principles and Guidelines for Water and Related Land Resources Implementation Studies*. U.S. Government Printing Office: Washington, D.C., 1983.

Vaux, H.J.; Howitt, R.E. Managing Water Scarcity: An Evaluation of Interregional Transfers. *Water Resources Res.* **1984**, *20*, 785–792.

Verdin, K.L.; Greenlee, S.K. Development of Continental Scale Digital Elevation Models and Extraction of Hydrographic Features. *Proceedings of the Third International Conference/Workshop on Integrating GIS and Environmental Modeling*; Santa Fe, NM, 1996.

Vitousek, P.; Mooney, H.; Lubchenco, J.; Melillo, J. Human Domination of Earth's Ecosystems. *Science*. **1997**, *277*, 494–499.

Vörösmarty, C.J.; Sahagian, D. Anthropogenic Disturbance of the Terrestrial Water Cycle. *Bioscience.* **2000**, 50, 753–765.

Walker, B.H. Rangeland Ecology: Understanding and Managing Change. *Ambio*, **1993**, *22*(2–3): 80–87.

Walker, K.F.; Sheldon, F.; Puckridge, J.T. A Perspective on Dryland River Ecosystems. *Regul. River*. **1995**, *11*, 85–104.

Walters, C. *Adaptive Management of Renewable Resources*. Macmillan: New York, 1986.

Walters, C. Large Scale Management Experiments and Learning by Doing. *Ecology* **1990**, *71*, 2060–2068.

Water Environment Research Foundation (WERF). New WERF Initiative Adds Value to Research. *Progress Newsletter* Winter 2004. (Available at http://www.werf.org/press/winter04/winter04_initiative.cfm.)

Weaver, L.A. Low-Intensity Watershed Alteration Effects on Fish Assemblage Structure and Function in a Virginia Piedmont Stream. Unpublished Masters Thesis, Virginia Commonwealth University, 1991.

Webster, J. The Role of Hydrological Processes in Ocean–Atmosphere Interactions. *Rev. Geophys.* **1994**, *32*(4), 427–476.

Weersink, A., Livernois, J.; Shogren, J.F.; Shortle, J.S. Economic Instruments and Environmental Policy in Agriculture. *Can. Public Policy*, **1998**, *XXIV*, 309–327.

Weinberg, M.; Kling, C.L. Uncoordinated Agricultural Policy Making: An Application to Irrigated Agriculture in the West. *Am. J. Agr. Econ.* **1996**, *78*, 65–78.

Weinberg, M.; Kling, C.L.; Wilen, J.E. Water Markets and Water Quality. *Am. J. Agr. Econ.* **1993a.**, *75*, 278–291.

Weinberg, M.; Kling, C.L.; Wilen, J.E. Analysis of Policy Options for the Control of Agricultural Pollution in California's San Joaquin Basin. In *Theory, Modeling and Experience in the Management of Nonpoint-Source Pollution;* Russell, C.S.; Shogren, J.F., Eds.; Kluwer Academic Publishers: Dordrecht, Netherlands, 1993b.

Weinreich, N.K. *Hands-On Social Marketing: A Step-by-Step Guide*. Sage Publications: Thousand Oaks, Calif., 1999.

Wetzstein, M.E.; Centner, T.J. Regulating Agricultural Contamination of Groundwater through Strict Liability and Negligence Legislation. *J. Environ. Econ. Manag.* **1992**, *22*(1), 1–11.

Western Governors Association. *Summary of 2002 Meeting*. Western Governors Association: Salt Lake City, Utah, 2002.

Whittington, J. Working Rivers. *Watershed* **2002**, *24*, 3–4.

Wilcove, D.; Rothstein, D; Dubow, J.; Phillips, A.; Losos, E. Quantifying Threats to Imperiled Species in the United States. *BioScience* **1998**, *48*, 607–615.

Williams, J.G. 1996. Lost in Space: Minimum Confidence Intervals for Idealized PHABSIM Studies. *Trans. Am. Fish. Soc.* **1996**, *125*, 485–465.

Wilson, M.A.; Carpenter, S.R. Economic Valuation of Freshwater Ecosystem Services in the U.S.: 1971–1997. *Ecol. Appl.* **1999**, *9,* 772–793.

Winter, T.C. The Concept of Hydrologic Landscapes. *J. Am. Water Resour. Assn.* **2001**, *37*(2): 335–349.

Woodard & Curran. Water Reclamation Helps Quench Thirsty Region. *Water Environ Technol.* **1999**, *11*(10), 56–57.

World Bank. *Water Resources Management*. World Bank: Washington, D.C., 1993.

World Commission on Dams. *Dams and Development: A New Framework for Decision-Making. The Report of the World Commission on Dams.* World Commission on Dams: Capetown, South Africa, 2000. (Available at http://www.dams.org.)

World Commission on Environment and Development. *Our Common Future* (aka *The Brundtland Report*). Oxford University Press: Oxford, U.K., 1987; p 383.

World Economic Forum (World Economic Forum Global Leaders for Tomorrow Environment Task Force), Yale-Columbia University Center for Environmental Law and Policy, Center for International Earth Science Information Network. *2002 Environmental Sustainability Index*. Yale Center for Environmental Law and Policy: New Haven, Conn., 2002a.

World Economic Forum, Yale Center for Environmental Law and Policy, Center for International Earth Science Information Network. *Pilot Environmental Performance Index.* Yale Center for Environmental Law and Policy: New Haven, Conn., 2002b.

Wouters, P. National and International Water Law: Achieving Equitable and Sustainable Use of Water Resources. *Water Intl.* **2000**, *25*(4), 499–512.

Wouters, P. The Role of Water Law in the Development of an Integrated Water Resources Management Strategy. In *Proceedings AWRA–University of Dundee, Globalization and Water*

Management: The Changing Value of Water, Dundee, Scotland, August 6–8, 2001; American Water Resources Association: Middleburg, Va., 2001; p 5.

Western Water Policy Review Advisory Commission (WWPRAC). *Water in the West: The Challenge for the Next Century*. Report of the Western Water Policy Review Advisory Commission: Denver, Colo., June 1998.

Young, R.A. Why Are There So Few Transactions between Water Users? *Am. J. Agr. Econ.* **1986**, *68*, 1151–1153.

Young, R.A.; Havemann, R.H. Economics of Water Resources: A Survey. In *Handbook of Natural Resource and Energy Economics Vol. II*; Kneese, A.V.; J.L. Sweeny, J.L., Eds.; North Holland: Amsterdam, 1985.

Young, R.E., Thompson, K.A.; McVicker, R.R.; Diamond, R.A.; Gringas, M.B.; Ferguson, D.; Johannesse, J.; Herr, G.K.; Parsons, J.J. Irvine Ranch Water District's Reuse Today Meets Tomorrow's Conservation Needs. In *Wastewater Reclamation and Reuse*, T. Asano, Ed.; Technomic Publishing: Basel, Switzerland, 1998; pp 941–1036.

Young, T.F.; Congdon, C.H. *Plowing New Ground.* Environmental Defense: New York, 1994.

Young, T.F.; Sanzone, S., Eds. *A Framework for Assessing and Reporting on Ecological Condition;* Prepared by the Ecological Reporting Panel Ecological Processes and Effects Committee, U.S. EPA Science Advisory Board: Washington, D.C., 2002

Zacerkowny, O. *Not a Drop Leaves the Plant: Daimler Chrysler's Hi-tech Solution for Industrial Environmental Conservation at Toluca, Mexico, Complex*. Case Study Report 2001. US Filter: Warrendale, Pa., 2001.

Zerubavel, E. *Social Mindscapes, An Invitation to Cognitive Sociology*. Harvard University Press: Cambridge, Mass., 1998.

Zimov, S.A.; Chuprynin, V.I.; Oreshko, A.P.; Chapin, F.S.; Reynolds, J.F.; M.C. Chapin, M.C. Steppe-Tundra Transition: A Herbivore-driven Biome Shift at the End of the Pleistocene (Review). *Am. Natur.* **1995**, *146*, 765–794.

WASTEWATER UTILITY

Alabama
Montgomery Water Works & Sanitary Sewer Board

Alaska
Anchorage Water & Wastewater Utility

Arizona
Gila Resources
Glendale, City of, Utilities Department
Mesa, City of
Peoria, City of
Phoenix Water Services Department
Pima County Wastewater Management

Arkansas
Little Rock Wastewater Utility

California
Calaveras County Water District
Central Contra Costa Sanitary District
Corona, City of
Crestline Sanitation District
Delta Diablo Sanitation District
Dublin San Ramon Services District
East Bay Dischargers Authority
East Bay Municipal Utility District
Eastern Municipal Water District
El Dorado Irrigation District
Fairfield-Suisun Sewer District
Fresno Department of Public Utilities
Inland Empire Utilities Agency
Irvine Ranch Water District
Las Virgenes Municipal Water District
Livermore, City of
Lodi, City of
Los Angeles, City of
Napa Sanitation District
Orange County Sanitation District
Palo Alto, City of
Riverside, City of
Sacramento Regional County Sanitation District
San Diego Metropolitan Wastewater Department, City of
San Francisco, City & County of
Sanitation Districts of Los Angeles County
San Jose, City of
Santa Barbara, City of
Santa Cruz, City of
Santa Rosa, City of
South Bayside System Authority
South Coast Water District
South Orange County Wastewater Authority
Stege Sanitary District
Sunnyvale, City of
Union Sanitary District
West Valley Sanitation District

Colorado
Aurora, City of
Boulder, City of
Colorado Springs Utilities
Greeley, City of
Littleton/Englewood Water Pollution Control Plant
Metro Wastewater Reclamation District, Denver

Connecticut
The Mattabassett District
Greater New Haven WPCA

District of Columbia
District of Columbia Water & Sewer Authority

Florida
Broward, County of
Fort Lauderdale, City of
JEA
Miami-Dade Water & Sewer Authority
Orange County Utilities Department
Reedy Creek Improvement District
Seminole County Environmental Services
St. Petersburg, City of
Stuart Public Utilities
Tallahassee, City of
Tampa, City of
Toho Water Authority
West Palm Beach, City of

Georgia
Atlanta Department of Watershed Management
Augusta, City of
Clayton County Water Authority
Cobb County Water System
Columbus Water Works
Fulton County
Gwinnett County Department of Public Utilities
Savannah, City of

Hawaii
Honolulu, City & County of

Idaho
Boise, City of

Illinois
American Bottoms Wastewater Treatment Plant
Greater Peoria Sanitary District
Kankakee River Metropolitan Agency
Metropolitan Water Reclamation District of Greater Chicago
Wheaton Sanitary District

Iowa
Ames, City of
Cedar Rapids Wastewater Facility
Des Moines, City of
Iowa City

Kansas
Johnson County Unified Wastewater Districts
Unified Government of Wyandotte County/ Kansas City, City of

Kentucky
Louisville & Jefferson County Metropolitan Sewer District

Louisiana
Sewerage & Water Board of New Orleans

Maine
Bangor, City of
Portland Water District

Maryland
Anne Arundel County Bureau of Utility Operations
Hagerstown, City of
Howard County Department of Public Works
Washington Suburban Sanitary Commission

Massachusetts
Boston Water & Sewer Commission
Upper Blackstone Water Pollution Abatement District

Michigan
Ann Arbor, City of
Detroit, City of
Holland Board of Public Works
Lansing, City of
Saginaw, City of
Wayne County Department of Environment
Wyoming, City of

Minnesota
Rochester, City of
Western Lake Superior Sanitary District

Missouri
Independence, City of
Kansas City Missouri Water Services Department
Little Blue Valley Sewer District
Metropolitan St. Louis Sewer District

Nebraska
Lincoln Wastewater System

Nevada
Henderson, City of
Reno, City of

New Jersey
Bergen County Utilities Authority
Ocean, County of
Passaic Valley Sewerage Commissioners

New York
New York City Department of Environmental Protection

North Carolina
Charlotte/Mecklenburg Utilities
Durham, City of
Metropolitan Sewerage District of Buncombe County
Orange Water & Sewer Authority

Ohio
Akron, City of
Butler County Department of Environmental Services
Columbus, City of
Metropolitan Sewer District of Greater Cincinnati
Northeast Ohio Regional Sewer District
Summit, County of

Oklahoma
Tulsa, City of

Oregon
Clean Water Services
Eugene/Springfield Water Pollution Control
Water Environment Services

Pennsylvania
Philadelphia, City of
University Area Joint Authority

South Carolina
Charleston Commissioners of Public Works
Mount Pleasant Waterworks & Sewer Commission
Spartanburg Sanitary Sewer District

Tennessee
Cleveland, City of
Knoxville Utilities Board
Murfreesboro Water & Sewer Department
Nashville Metro Water Services

Texas
Austin, City of
Dallas Water Utilities
Denton, City of
El Paso Water Utilities
Fort Worth, City of
Gulf Coast Waste Disposal Authority
Houston, City of
San Antonio Water System
Trinity River Authority

Utah
Salt Lake City Corporation

Virginia
Alexandria Sanitation Authority

Arlington, County of
Fairfax County
Hampton Roads Sanitation District
Henrico, County of
Hopewell Regional Wastewater Treatment Facility
Loudoun County Sanitation Authority
Lynchburg Regional WWTP
Prince William County Service Authority
Richmond, City of
Rivanna Water & Sewer Authority

Washington
Everett, City of
King County Department of Natural Resources
Seattle Public Utilities
Sunnyside, Port of
Yakima, City of

Wisconsin
Green Bay Metro Sewerage District
Kenosha Water Utility
Madison Metropolitan Sewerage District
Milwaukee Metropolitan Sewerage District
Racine, City of
Sheboygan Regional Wastewater Treatment
Wausau Water Works

Australia
South Australian Water Corporation
Sydney Water Corporation
Water Corporation of Western Australia

Canada
Greater Vancouver Regional District
Regina, City of, Saskatchewan
Toronto, City of, Ontario
Winnipeg, City of, Manitoba

New Zealand
Watercare Services Limited

United Kingdom
Yorkshire Water Services Limited

STORMWATER UTILITY

California
Los Angeles, City of, Department of Public Works
Monterey, City of
San Francisco, City & County of
Santa Rosa, City of

Colorado
Aurora, City of
Boulder, City of

Georgia
Griffin, City of

Iowa
Cedar Rapids Wastewater Facility
Des Moines, City of

Kansas
Overland Park, City of

Maine
Portland Water District

Minnesota
Western Lake Superior Sanitary District

North Carolina
Charlotte, City of, Stormwater Services

Pennsylvania
Philadelphia, City of

Tennessee
Chattanooga Stormwater Management

Washington
Bellevue Utilities Department
Seattle Public Utilities

STATE

Arkansas Department of Environmental Quality
Fresno Metropolitan Flood Control District
Kansas Department of Health & Environment
Kentucky Department of Environmental Protection
Ohio River Valley Sanitation Commission
Urban Drainage & Flood Control District, CO

CORPORATE

Abt. Associates
ADS Environmental Services
Alan Plummer & Associates
Alden Research Laboratory Inc.
American Water
Aqua-Aerobic Systems Inc.
Aquateam–Norwegian Water Technology Centre A/S
ARCADIS
Associated Engineering
Black & Veatch
Blasland, Bouck & Lee Inc.
Boyle Engineering Corporation
Brown & Caldwell
Burns & McDonnell
CABE Associates Inc.
The Cadmus Group
Camp Dresser & McKee Inc.
Carollo Engineers Inc.
Carpenter Environmental Associates Inc.
CDS Technologies Inc.
Chemtrac Systems Inc.
CH2M HILL
Construction & Environmental Consulting, Inc.
D&B/Guarino Engineers, LLC
Damon S. Williams Associates, LLC
Dewling Associates Inc.
Earth Tech Inc.
Ecovation
EMA Inc.
Environ/The ADVENT Group, Inc.
The Eshelman Company Inc.
EWT Holdings Corporation
Fay, Spofford, & Thorndike Inc.
Freese & Nichols Inc.
ftn Associates Inc.
Fuss & O'Neill Inc.
Gannett Fleming Inc.
Geosyntec Consultants
GHD
Golder Associates Ltd.
Greeley and Hansen LLC
Hazen & Sawyer, P.C.
HDR Engineering Inc.
HNTB Corporation
HydroQual Inc.
Infilco Degremont Inc.
Jacobson Helgoth Consultants Inc.
Jason Consultants LLC Inc.
Jordan, Jones, & Goulding Inc.
KCI Technologies Inc.
Kelly & Weaver, P.C.
Kennedy/Jenks Consultants
KMK Consultants
Komline Sanderson Engineering Corporation
Lawler, Matusky & Skelly Engineers, LLP
Limno-Tech Inc.
Lombardo Associates Inc.
Malcolm Pirnie Inc.
Material Matters
McKim & Creed
MEC Analytical Systems Inc.
Metcalf & Eddy Inc.
MPR Engineering Corporation, Inc.
MWH
New England Organics
O'Brien & Gere Engineers Inc.
Odor & Corrosion Technology Consultants Inc.
Oswald Green, LLC
PA Government Services Inc.
Parametrix Inc.
Parsons
Post, Buckley, Schuh & Jernigan
R&D Engineering/Conestoga Rover & Associates
RMC Inc.
R.M. Towill Corporation
Ross & Associates Ltd.
Rothberg, Tamburini & Windsor, Inc.
Royce Technologies
Stantec Consulting Inc.
Stearns & Wheler, LLC
Stone Environmental Inc.
Stormwater360
Stratus Consulting Inc.
Synagro Technologies Inc.
Tetra Tech Inc.
Trojan Technologies Inc.
Trussell Technologies, Inc.
URS Corporation
USfilter
Wade-Trim Inc.
Westin Engineering Inc.
Weston Solutions Inc.
Woodard & Curran
WRc/D&B, LLC
WWETCO, LLC
Zenon Environmental Inc.
Zoeller Pump Company

INDUSTRY

American Electric Power
ChevronTexaco Energy Research & Technology Company
The Coca-Cola Company
Dow Chemical Company
DuPont Company
Eastman Chemical Company
Eastman Kodak Company
Eli Lilly & Company
Merck & Company Inc.
Premier Chemicals LLC
Procter & Gamble Company
RWE Thames Water Plc
Severn Trent Services Inc.
Suez Environnment
United Water Services LLC

Note: List as of 4/1/06

WERF Board of Directors

WERF Research Council

WERF Product Order Form

s a benefit of joining the Water Environment Research Foundation, subscribers are entitled to receive one complimentary copy of all final ports and other products. Additional copies are available at cost (usually $10). To order your complimentary copy of a report, please write ee" in the unit price column. WERF keeps track of all orders. If the charge differs from what is shown here, we will call to confirm the total efore processing.

ame Title

rganization

ddress

ity State Zip Code Country

hone Fax Email

Stock #	Product	Quantity	Unit Price	Total
			Postage & Handling	
			VA Residents Add 4.5% Sales Tax	
			Canadian Residents Add 7% GST	
			TOTAL	

Method of Payment: (All orders must be prepaid.)

- ❑ Check or Money Order Enclosed
- ❑ Visa ❑ Mastercard ❑ American Express

ccount No. Exp. Date

ignature

Shipping & Handling:

Amount of Order	United States	Canada & Mexico	All Others
Up to but not more than:	Add:	Add:	Add:
$20.00	$5.00*	$8.00	50% of amount
30.00	5.50	8.00	40% of amount
40.00	6.00	8.00	
50.00	6.50	14.00	
60.00	7.00	14.00	
80.00	8.00	14.00	
100.00	10.00	21.00	
150.00	12.50	28.00	
200.00	15.00	35.00	
More than $200.00	Add 20% of order	Add 20% of order	

*minimum amount for all orders

lote: Please make checks payable to the Water Environment Research Foundation.

To Order (Subscribers Only):

Log on to www.werf.org and click on the "Product Catalog."

Phone: (703) 684-2470
Fax: (703) 299-0742.

WERF
Attn: Subscriber Services
635 Slaters Lane
Alexandria, VA 22314-1177

To Order (Non-Subscribers):

Non-subscribers may be able to order WERF publications either through WEF (www.wef.org) or IWAP (www.iwapublishing.com). Visit WERF's website at www.werf.org for details.

IWA Publishing's authorised EU representative for General Product Safety Regulations is Diane D'Arras, 15 rue Duret, 75116 Paris, France, e-mail: safety@iwap.co.uk.

Printed and bound by CPI Group (UK) Ltd, Croydon, CR0 4YY
12/05/2026
02108886-0009